RIGID DYNAMICS

AF633633

Written according to the syllabi of BA and BSc [General and Honours] of all Indian universities and all competitive examinations of equivalent standard

RIGID DYNAMICS

Dr M M Rahman, Double Gold Medallist
Reader, PG Department of Mathematics
A N College, Patna 13

New Central Book Agency (P) Ltd
LONDON
HYDERABAD ERNAKULAM BHUBANESWAR
NEW DELHI KOLKATA PUNE GUWAHATI

NCBA

REGD OFFICE
8/1 Chintamoni Das Lane, Kolkata 700 009, India
email: ncbapvtltd@eth.net

OVERSEAS
NCBA (UK) Ltd, 149 Park Avenue North, Northampton, NN3 2HY, UK
email: ncbauk@yahoo.com

EXPORT
NCBA Exports Pvt. Ltd, 212 Shahpur Jat, New Delhi 110 049
email: ncbaexports@ncbapvtltd.com

BRANCHES

New Central Book Agency (P) Ltd
212 Shahpur Jat, **New Delhi** 110 049
email: ncbadel@ncbapvtltd.com

Shop Nos. 3 & 4, Vinayak Towers
681/B Budhwar Peth, Appa Balwant Chowk
Pune 411 002
email: ncbapune@ncbapvtltd.com

Shop No. 15, Bharati Towers, Ground Floor
Forest Park
Bhubaneswar 751 001
email: ncbabub@ncbapvtltd.com

House No. 3-1-315, 1st Floor
Nimboliadda, Kachiguda, **Hyderabad** 500 027
email: ncbahydb@ncbapvtltd.com

GSS Shopping Complex, 1st Floor
Opposite College Post Office, Convent Road
Ernakulam 682 035
0484-4021054 / 53

Radhamayee Complex (Second Floor)
Opp. Axis Bank, Panbazar, Jaswanta Road
Guwahati 781 001
email: ncbaguwahati@ncbapvtltd.com

RIGID DYNAMICS ● Dr M M Rahman

© Copyright reserved by Nelofar Jahan

Publication, Distribution, and Promotion Rights exclusively reserved by the Publisher

All rights reserved. No part of the text in general, and the figures, diagrams, page layout, and cover design in particular, may be reproduced or transmitted in any form or by any means—electronic, mechanical, photocopying, recording, or by any information storage and retrieval system—without the prior written permission of the Publisher

First Published: 1996
Reprinted: 2002, 2004, 2005, 2006
Thoroughly Revised and Updated Second Edition: 2007
Reprinted: 2009, 2011, 2012, 2014, 2015, 2016, 2017, 2018, 2021, **2022**

PUBLISHER AND TYPESETTER
New Central Book Agency (P) Ltd
8/1 Chintamoni Das Lane, Kolkata 700 009

PRINTER
New Central Book Agency (P) Ltd
Web-Offset Division, Dhulagarh, Sankrail, Howrah

TECHNICAL EDITOR
Dipan Roy

PROJECT SUPERVISOR
Abhijit Sen

ISBN: 978 81 7381 224 8

Price ₹ 465.00

Contents

Preface to the Second Edition

This book has been thoroughly revised. Several new examples from various examination papers have been added and a few of them are solved. The book has been revised with the intention to make it more useful to the students. Two new chapters, namely 'Motion of a Rigid Body' and 'Objective Type Questions', have been added in this edition, as they have been included in the syllabus of some universities. Latest university question papers have been included as well.

I am indebted to the teachers and students, without whose generous patronage the book would not have appeared in its second edition. I am also thankful to my wife Mrs Nelofar Jahan and sons Tauseef and Tauqeer for their constant encouragement and inspiration. I am particularly indebted to Dr (Mrs) Sushila Thakur (Retd), Dept of Physiology, Darbhanga Medical College, whose prudent advice had always been a constant source of inspiration to me.

Every effort has been made to avoid the errors and misprints and to improve the text. Even then, there remains always some scope of improvement, and every suggestion for further improvement of the book as well as pointing out of errors by the readers will be thankfully acknowledged.

Last, but not least, I am thankful to Mr Amitabha Sen, the Director of New Central Book Agency, for his cordial cooperation during the preparation of the second edition of this book.

Patna, 05.11.06

Md Motiur Rahman

Preface to the First Edition

The book *Rigid Dynamics* is intended to cover the syllabi of BA and BSc (Hons) of all Indian universities as well as most of the competitive examinations like IAS and PSC. After teaching the Honours students for more than fourteen years, I have experienced that average students find a few of the existing books too difficult, while the others too inadequate to meet their requirements. The present volume is complete as far as the subject matter is concerned and I hope that nothing of much importance has been left out.

I am grateful to my students, whose constant inspiration and help insisted me to complete the present book. I also want to express my sincere thanks to my wife Mrs Nelofar Jahan and sons Tauseef and Tauqeer whose encouragement and inspiration forced me to complete the book on time. I am particularly indebted to Dr (Mrs) Sushila Thakur, Dept of Physiology, Darbhanga Medical College, whose prudent advice has always been a constant source of inspiration to me.

I cannot claim that the book is free from mistakes, though every possible care has been taken to avoid them. I shall appreciate the pointing out of mistakes and constructive suggestions for the improvement of this book in the subsequent editions.

Last, but not least, I am thankful to Mr Amitabha Sen, the Director of New Central Book Agency due to whose cooperation book has come to reality.

01.10.96 Md Motiur Rahman

Some Useful Geometrical Results

Areas

Triangle

- If h be the height and a the length of the base, then its area $= \frac{1}{2}ah$.
- If a and b be the two sides having angle θ between them, then its area $= \frac{1}{2}ab\sin\theta$.

Hexagon. If a be the side of a regular hexagon, then its area $= \frac{3\sqrt{3}}{2}a^2$.

Octagon. If a be the side of a regular octagon, then its area $= 2(1+\sqrt{2})a^2$.

Circle. If r be the radius of the circle, then its area $= \pi r^2$ and circumference $= 2\pi r$.

Sector of a Circle. The area of a sector of a circle of radius r subtending an angle 2θ at the centre $= \frac{1}{2}r^2\theta$.

Ellipse. If $2a$ and $2b$ be the length of the major and minor axes of an ellipse, then its area $= \pi ab$.

Volumes and Surfaces

Rectangular Solid. If a, b and c be the length, breadth and height of a rectangular solid, then

$$\begin{aligned} &\text{its volume} &&= abc \\ &\text{its whole surface} &&= 2(ab+bc+ca) \\ \text{and}\quad &\text{its diagonal} &&= \sqrt{a^2+b^2+c^2}. \end{aligned}$$

Cube. If a be the length of each edge of the cube, then

$$\begin{aligned} &\text{its volume} &&= a^3 \\ &\text{its whole surface} &&= 6a^2 \\ \text{and}\quad &\text{its diagonal} &&= \sqrt{3}a. \end{aligned}$$

Prism. If the base of the prism be a regular polygon, then

$$\begin{aligned} &\text{its volume} = (\text{area of the base}) \times \text{height} \\ \text{and}\quad &\text{its lateral face} = (\text{perimeter of the base}) \times \text{height}. \end{aligned}$$

Cylinder. If r be the radius of the base of the cylinder of height h, then

its curved surface $= 2\pi rh$
its whole surface $= 2\pi r(h+r)$
and its volume $= \pi r^2 h$.

Pyramid.

its volume = 1/3 (area of base) × height

and its slant surface = 1/2 (perimeter of its base) × slant height.

Cone. If h be the height, r the radius, l the slant height, α the semi-vertical angle of the cone, then $r = h\tan\alpha$, $l = h\sec\alpha$.

its volume $= 1/3\ \pi r^2 h = 1/3\ \pi h^3 \tan^2\alpha$

and its curved surface $= \pi r l = \pi h^2 \tan\alpha \sec\alpha$.

Frustum of a Cone. If r_1, r_2 be the radii of its circular ends and d the perpendicular distance between them, then

its volume $= 1/3\ \pi d\left(r_1^2 + r_1 r_2 + r_2^2\right)$.

Sphere. If r be the radius of the sphere, then

its volume $= 4/3\ \pi r^3$
and its curved surface $= 4\pi r^2$.

Zone or Portion of a Sphere. If r be the radius of the sphere of which the zone is given between two parallel planes at distances h_1 and h_2, from the centre of the sphere

its volume $= \pi/3.4\ (h_1 - h_2)[3r^2 - (h_1^2 + h_1 h_2 + h_2^2)]$

and Its curved surface $= 2\pi r d$, where d is the perpendicular distance between the two parallel planes.

Paraboloid of Revolution. Solid formed by the revolution of a paraboloid about its axis is known as *paraboloid of revolution.*

Its volume cut-off by a plane perpendicular to its axis = 1/2 (area of its plane base) × its height.

Centers of Gravity (C.G.)

C.G. of a cylinder is the middle point of its axis.

C.G. of a solid cone of height h is the point on the axis at a distance $3h/4$ from the vertex.

C.G. of curved surface of a circular cone of height h is the point on axis at a distance $3h/4$ from the vertex.

C.G. of a solid half cone base of radius r is a point whose distance from the axis of the whole cone $= r/\pi$.

C.G. of a solid hemisphere of radius r is a point on its axis at a distance $3r/8$ from the centre of the base.

C.G. of the curved surface of a hemisphere of radius r is a point on its axis at a distance $r/2$ from the centre of this base.

C.G. of a semi-circle of radius r is a point on the radius perpendicular to the bounding diameter at a distance $4r/3\pi$ from the centre of the circle.

C.G. of a sector of a circle of radius r subtending an angle 2θ at the centre of the circle is a point on the bisector of the angle 2θ, whose distance from the centre $= \frac{2r}{3}\frac{\sin\theta}{\theta}$.

C.G. of a circular arc of radius r subtending an angle 2θ at the centre is a point on the bisector of the angle 2θ, whose distance from the centre $= \frac{r\sin\theta}{\theta}$.

C.G. of a semi-circular arc of radius r is a point on the radius perpendicular to the bounding diameter at a distance $2r/\pi$ from the centre of the circle.

Radius of Gyration

If AK^2 is the moment of inertia, where K is the radius of gyration, value of K^2 is given below for some cases.

- For a thin rod of length l, about an axis through the middle point perpendicular to the rod, $K^2 = l^2/12$.
- For a rectangular lamina of sides a and b, about a line through its centre parallel to the side a, $K^2 = b^2/12$ and parallel to the side b, $K^2 = a^2/12$.
- For a circular area of radius r about the diameter, $K^2 = r^2/4$.
- For an elliptic area of major and minor axes $2a$ and $2b$

  $K^2 = a^2/4$ about the axis $2b$.

  $K^2 = b^2/4$ about the axis $2a$.
- For a triangular area ABC, about a straight line through A in the plane of the area $K^2 = 1/6\,(\beta^2 + \beta\gamma + \gamma^2)$, where β and γ being the distances of B and C from the axis of rotation through A.

Chapter 1

Moments and Products of Inertia

1.1 Rigid Body

A material body is said to be rigid when the mutual distance of every pair of specified points in it is invariable with respect to time, i.e., a rigid body does not expand or contract or change its shape, though it may change its orientation, i.e., its position with reference to surrounding elements may change.

1.2 Moment of Inertia of a Body about a Line

If m be the mass of an element of a rigid body and r the distance of this element from a given line, then Σmr^2 is called the *moment of inertia of the body* about the given line.

If this sum be denoted by MK^2, where M is the mass of the body, then K is called the *radius of gyration* of the body about the given line.

1.3 Product of Inertia **[LNMU 93H]**

If (x, y) be the coordinates of an element m of the mass referred to two mutually perpendicular lines OX and OY, then Σmxy is called the *product of inertia of the body* with respect to the lines OX and OY.

If three mutually perpendicular axes OX, OY, OZ be taken and (x, y, z) be the coordinates of the element m of the body referred to these axes, then the quantities Σmyz, Σmzx and Σmxy are called the *products of inertia* of the body with respect to the pair of axes, y and z, z and x, x and y, respectively.

1.4 Some Important Symbols

If (x, y, z) be the coordinates of a point P, then its distance from x-axis is

$$PA = \sqrt{y^2 + z^2},$$

since coordinates of A are $(x, 0, 0)$.

Also distance of P from xy-plane is $PM = z$.

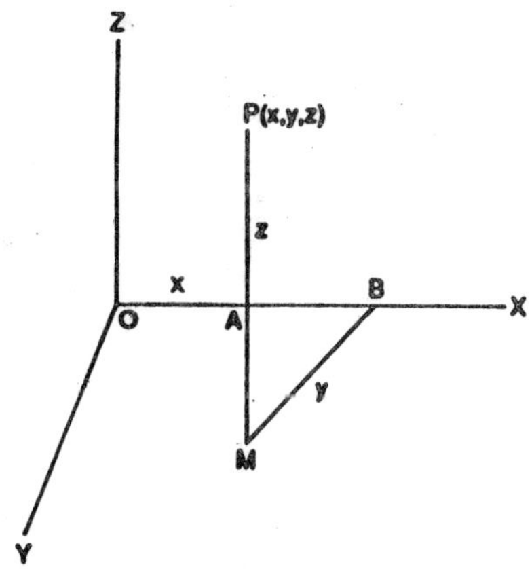

FIG 1.1

- A, B, C denote the moment of inertia of the body about OX, OY and OZ axes.

  $\therefore\ A = \Sigma m(y^2 + z^2),\ B = \Sigma m(z^2 + x^2),\ C = \Sigma m(x^2 + y^2).$

- A', B', C' denote the moment of inertia of the body about yz, zx and xy-planes.

  $A' = \Sigma mx^2,\ B' = \Sigma my^2,\ C' = \Sigma mz^2.$

- D, E, F denote the product of inertia of the body with respect to the pair of axes OY, OZ; OZ, OX and OX, OY.

  $\therefore\ D = \Sigma myz,\ E = \Sigma mzx,\ F = \Sigma mxy.$

- H denotes the moment of inertia of the body with respect to origin O.

  $\therefore\ H = \Sigma m(x^2 + y^2 + z^2).$

1.5 Moment of Inertia in Some Standard Cases

1.5.1 *Moment of inertia of a uniform rod of length* $2a$ *and mass* M *about an axis through an extremity and perpendicular to it.* **[TMBU 95H]**

Let AB be a uniform rod of length $2a$.

Let M be the mass of the rod, then

$$M = 2a\rho, \tag{1.1}$$

where ρ is the density.

We consider an element of breadth δx at a distance x from the end A.

Mass of this element $= \rho\,\delta x$.

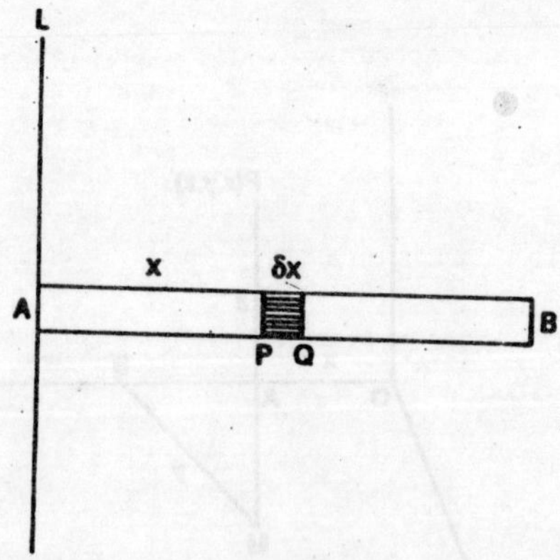

FIG 1.2

Therefore, moment of inertia of the elementary mass about AL $= \rho\,\delta x\, x^2$.

Hence moment of inertia of the rod about AL

$$= \int_0^{2a} \rho x^2\, dx = \frac{\rho}{3}\left[x^3\right]_0^{2a} = \frac{\rho}{3} 8a^3 = \frac{8a^3}{3}\frac{M}{2a} \text{ [From equation (1.1)]}$$

$$= \frac{4Ma^2}{3}.$$

1.5.2 *moment of inertia of a uniform rod of length 2a and mass M about an axis through the middle point and perpendicular to the rod.*

Let AB be a uniform rod of length $2a$ and O its middle point.

Let M be the mass of the rod.

Then

$$M = 2a\rho, \tag{1.2}$$

where ρ is the density.

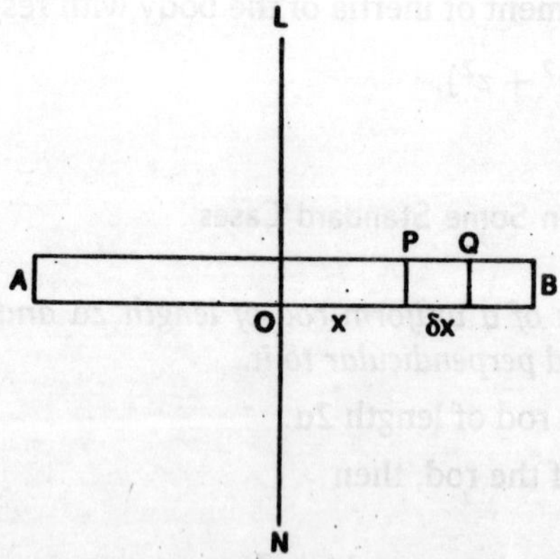

FIG 1.3

We consider an element PQ of breadth δx at a distance x from the axis LN.

Mass of this element $= \rho\,\delta x$.

Therefore, moment of inertia of the elementary mass about LN $= \rho\,\delta x\, x^2$.

Hence moment of inertia of the rod about LN

$$= \int_{-a}^{a} \rho x^2\, dx = \frac{\rho}{3}\left[x^3\right]_{-a}^{a} = \frac{\rho}{3} 2a^3$$

$$= \frac{2}{3}\frac{M}{2a} a^3 \text{ [From equation (1.2)]}$$

$$= \frac{M}{3} a^2.$$

1.5.3 *moment of inertia of a rectangular lamina about a line through its centre and parallel to one of its edges.*

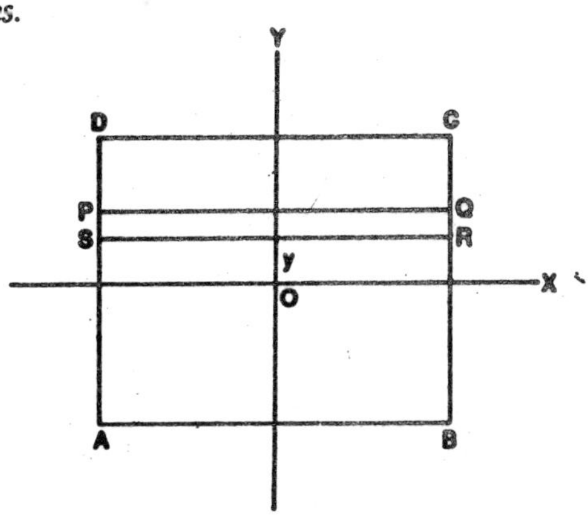

FIG 1.4

Let ABCD be a thin uniform rectangular lamina.

Let $AB = 2a$ and $BC = 2b$.

Let M be the mass of the lamina

$$M = 2a \cdot 2b\rho = 4ab\rho, \tag{1.3}$$

where ρ is the surface density.

Let OX and OY be the axes of symmetry through the centre O, which are parallel to AB and BC, respectively.

We consider an elementary strip PQRS of breadth δx at a distance y from OX.

Mass of the strip $= 2a\, \delta y\, \rho$.

Moment of inertia of the strip about OX $= 2a\, \delta y\, \rho y^2$.

Hence moment of inertia of the whole rectangular lamina about OX

$$= 2a\rho \int_{-b}^{b} y^2\, dy = \frac{2a\rho}{3}\left[y^3\right]_{-b}^{b} = \frac{2a\rho}{3}(b^3 + b^3) = \frac{2a\rho}{3} 2b^3$$

$$= \frac{4ab^3}{3}\rho = \frac{4ab^3}{3}\frac{M}{4ab} \text{ [From equation (1.3)]}$$

$$= \frac{M}{3} b^2.$$

Similarly, the moment of inertia of the lamina about OY is given by $\frac{M}{3}a^2$.

1.5.4 *Moment of inertia of a uniform rectangular lamina about a line through its centre and perpendicular to its plane.* **[MU 98H, BBABU 98H, 02H]**

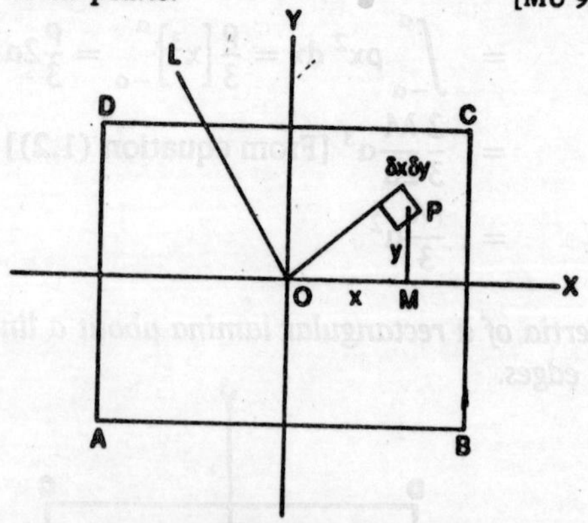

FIG 1.5

Let ABCD be a thin uniform rectangular lamina.

Let AB = 2a and BC = 2b.

Let M be the mass of the lamina

$$M = 2a \cdot 2b\rho = 4ab\rho, \tag{1.4}$$

where ρ is the surface density.

Let OX and OY be the axes of symmetry through the centre O, which are parallel to the sides AB and BC, respectively.

Let OL be a line through the centre O and perpendicular to the lamina.

We consider an elementary area $\delta x\,\delta y$ at $P(x, y)$ at a distance $\sqrt{x^2 + y^2}$ from O.

Mass of this element $= \rho\,\delta x\,\delta y$.

Therefore, moment of inertia of this element about the line OL $= \rho\,\delta x\,\delta y\,(x^2 + y^2)$.

Hence moment of inertia of the rectangular area about OL

$$= \rho \int_{-b}^{b}\int_{-a}^{a} (x^2 + y^2)\,dx\,dy = \rho \int_{-a}^{a} \left\{x^2 y + \frac{y^3}{3}\right\}_{-b}^{b} dx$$

$$= \rho \int_{-a}^{a} \left\{x^2(b+b) + \frac{(b^3 + b^3)}{3}\right\} dx = \rho \left[\frac{2bx^3}{3} + \frac{2b^3}{3}x\right]_{-a}^{a}$$

$$= \rho\left[\frac{2b}{3}2a^3 + \frac{2b^3}{3}2a\right] = \frac{\rho 4ab}{3}(a^2 + b^2)$$

$$= \frac{M}{3}(a^2 + b^2) \text{ [From equation (1.4)]}$$

1.5.5 *Moment of inertia of a rectangular parallelopiped about an axis through its centre and parallel to one of its edges.*

Let 2a, 2b, 2c be the lengths of the sides of the parallelopiped and O be its centre.

Let us take OX, OY, OZ as axes of reference, parallel to the edges of lengths 2a, 2b and 2c, respectively.

Let M be the mass of the lamina, then

$$M = 2a2b2c\rho, \tag{1.5}$$

where ρ is the surface density.

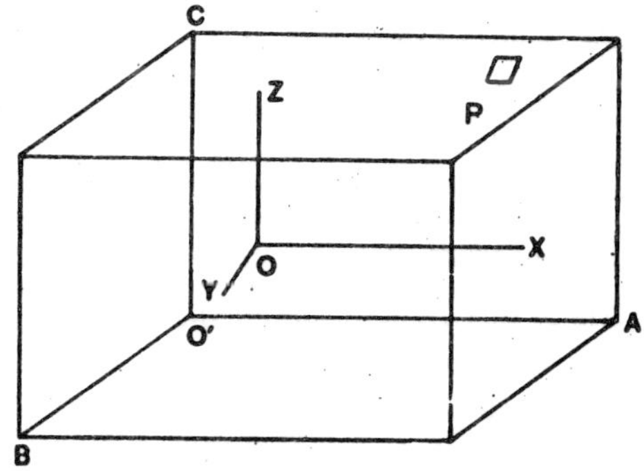

FIG 1.6

We consider an element $\delta x\,\delta y\,\delta z$ at $P(x, y, z)$.
Mass of this element $= \rho\,\delta x\,\delta y\,\delta z$.
Moment of inertia of this element about $OX = \rho(y^2 + z^2)\,\delta x\,\delta y\,\delta z$.
Hence moment of inertia of the parallelopiped about OX

$$= \int_{-a}^{a}\int_{-b}^{b}\int_{-c}^{c} \rho(y^2 + z^2)\,dx\,dy\,dz = \rho\int_{-b}^{b}\int_{-c}^{c}(y^2 + z^2)\,dy\,dz\Big\{x\Big\}_{-a}^{a}$$

$$= 2a\rho\int_{-b}^{b}\int_{-c}^{c}(y^2 + z^2)\,dy\,dz - 2a\rho\int_{-c}^{c}\left\{\frac{y^3}{3} + z^2y\right\}_{-b}^{b} dz$$

$$= 2a\rho\int_{-c}^{c}\left(\frac{2b^3}{3} + 2bz^2\right) dz = 2a\rho\left[\frac{2b^3}{3}z + \frac{2bz^3}{3}\right]_{-c}^{c}$$

$$= 2a\rho\left[\frac{2b^3}{3}2c + \frac{2b}{3}2c^3\right] = \frac{8a\rho bc}{3}(b^2 + c^2)$$

$$= \frac{M}{3}(b^2 + c^2) \text{ [From equation (1.5)]}$$

1.5.6 *Moment of inertia of a circular ring (or hoop) about* (1) *a diameter and* (2) *an axis through the centre and perpendicular to its plane.* **[BBABU 92]**

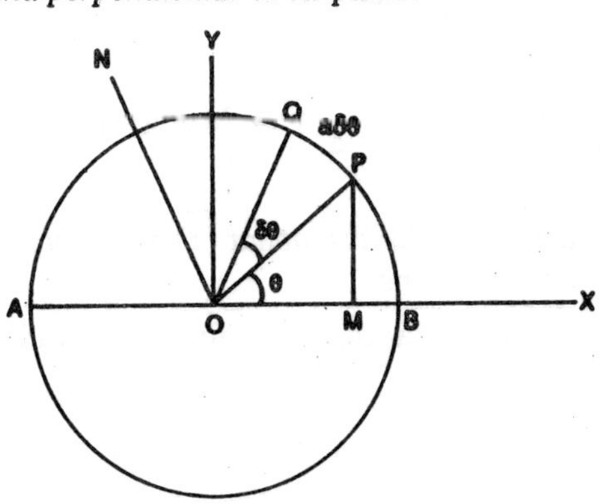

FIG 1.7

Let M be the mass of the circular ring of radius a. Let O be its centre, then

$$M = 2\pi a\rho, \tag{1.6}$$

where ρ is the per unit length of the ring.

Taking O as pole and OX as initial line, we consider an elementary arc $a\delta\theta$ at P, where $\angle XOP = \theta$.

Its mass $= \rho a\delta\theta$.

Distance of this element from the diameter AB $=$ PM $= a\sin\theta$.

∴ moment of inertia of this element about the diameter AB

$$= (a\sin\theta)^2\rho a\delta\theta = \rho a^3\sin^2\theta\delta\theta.$$

Hence moment of inertia of the circular ring about the diameter AB

$$= \int_0^{2\pi}\rho a^3\sin^2\theta d\theta = \rho a^3\int_0^{2\pi}\frac{1-\cos 2\theta}{2}d\theta$$

$$= \frac{\rho a^3}{2}\left[\theta - \frac{\sin 2\theta}{2}\right]_0^{2\pi} = \frac{\rho a^3}{2}2\pi$$

$$= \pi\rho a^3 = \frac{M}{2a}a^3 \text{ [From equation (1.6)]}$$

$$= \frac{M}{2}a^2.$$

1.5.7 *Let ON be the line through the centre O and perpendicular to the plane of the circular ring.*

Moment of inertia of the elementary arc $a\delta\theta$ at P about ON

$$= OP^2 a\delta\theta\rho = a^2\rho a\delta\theta = \rho a^3\delta\theta.$$

Hence moment of inertia of the ring about ON

$$= \int_0^{2\pi}\rho a^3 d\theta = \rho a^3 2\pi = \frac{M}{a}a^3 = Ma^2 \text{ [From equation (1.6)].}$$

1.5.8 *Moment of inertia of a circular disc about* (1) *a diameter and* (2) *an axis through the centre perpendicular to its plane.* **[BBABU 92H, 99H]**

Let M be the mass of a circular disc of centre O and radius a.

Then

$$M = \pi a^2\rho, \tag{1.7}$$

where ρ is the density per unit area.

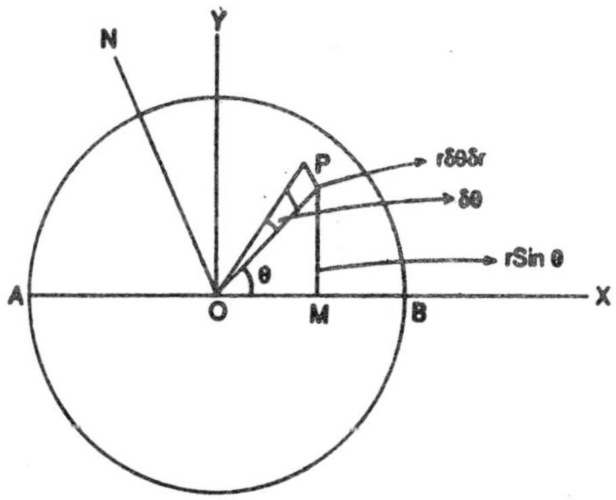

FIG 1.8

Taking the centre O as pole and OX as the initial line, we consider an elementary area $r\,\delta\theta\,\delta r$ at the point $P(r,\theta)$ of the disc.

Mass of the element $= \rho r\,\delta\theta\,\delta r$.

Distance of this element from OX $= PM = r\sin\theta$.

Therefore, moment of inertia of the element about OX $= (r\sin\theta)^2\rho r\delta\theta\,\delta r = \rho r^3\sin^2\theta\,\delta\theta\,\delta r$.

Hence moment of inertia of the circular plate about OX

$$\begin{aligned} &= \int_0^{2\pi}\int_0^a \rho r^3\sin^2\theta\,d\theta\,dr = \rho\int_0^{2\pi}\int_0^a r^3 dr\sin^2\theta d\theta \\ &= \frac{\rho a^4}{4}\int_0^{2\pi}\frac{1-\cos 2\theta}{2}d\theta = \frac{\rho a^4}{8}\left[\theta - \frac{\sin 2\theta}{2}\right]_0^{2\pi} \\ &= \frac{\rho a^4}{8}2\pi = \frac{\pi\rho a^4}{4} = \frac{M}{a^2}\frac{a^4}{4} = \frac{1}{4}Ma^2 \text{ [From equation (1.7)]} \end{aligned}$$

1.5.9 *Let* ON *be the line through the centre* O *and perpendicular to the plane of the disc.*

Moment of inertia of the elementary area about ON

$$= OP^2\rho r\,\delta\theta\,\delta r = r^2\rho r\,\delta\theta\,\delta r = \rho r^3\,\delta\theta\,\delta r.$$

Hence moment of inertia of the circular plate about O

$$\begin{aligned} &= \int_0^{2\pi}\int_0^a \rho r^3\,d\theta\,dr = \rho\int_0^{2\pi} d\theta\int_0^a r^3 dr \\ &= \rho\frac{2\pi a^4}{4} = \frac{M}{2a^2}a^4 = \frac{M}{2}a^2 \text{ [From equation (1.7)]} \end{aligned}$$

1.5.10 *Moment of inertia of an elliptic disc about major axis.* **[LNMU 93H]**

Let the equation of the ellipse be

$$\frac{x^2}{a^2} + \frac{y^2}{b^2} = 1. \tag{1.8}$$

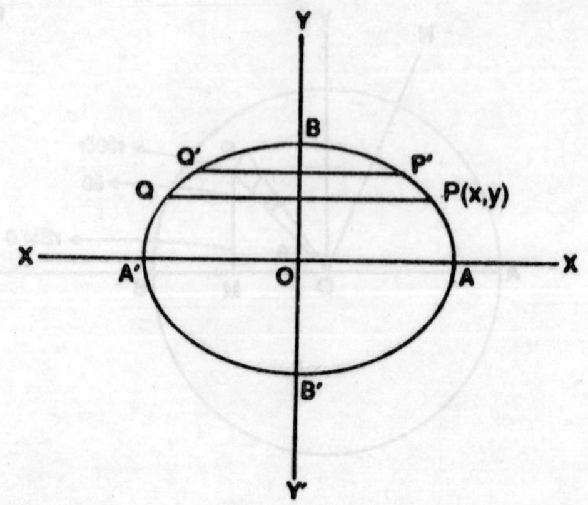

FIG 1.9

Let P(x, y) be a point on the ellipse and let PQQ′P′ be one such strip whose distance from O is y and breadth δy.

Then the area of the elementary strip = 2x δy.

But

$$\frac{x^2}{a^2}+\frac{y^2}{b^2}=1,\text{ or, }\frac{x^2}{a^2}=1-\frac{y^2}{b^2}=\frac{b^2-y^2}{b^2}.\ \therefore\ x=\frac{a}{b}\sqrt{b^2-y^2}.$$

Hence area of the elementary strip = $\frac{2a}{b}\sqrt{b^2-y^2}\,dy$.

Thus mass of the elementary strip = $\frac{2a}{b}\sqrt{b^2-y^2}\,\rho\,dy$.

Therefore, moment of inertia of the elementary strip about the major axis

$$\frac{2a}{b}\sqrt{b^2-y^2}\,\rho\,dy\,y^2.$$

Hence the moment of inertia of the disc about major axis is given by

$$=\int_{-b}^{b}\frac{2a}{b}\sqrt{b^2-y^2}\,\rho\,dy\,y^2=\frac{2a}{b}\rho\int_{-b}^{b}\sqrt{b^2-y^2}\,y^2\,dy.$$

Let $y = b\sin\theta$, ∴ $dy = b\cos\theta\,d\theta$.

When $y = b,\ \theta = \pi/2$,

$$y=-b,\ \theta=-\pi/2.$$

Hence

$$\text{required M.I.}=\frac{2a\rho}{b}\int_{-(\pi/2)}^{\pi/2}\sqrt{b^2-b^2\sin^2\theta}\,b^2\sin^2\theta\,b\cos\theta\,d\theta$$

$$=\frac{2a\rho b^4}{b}\int_{-(\pi/2)}^{\pi/2}\sin^2\theta\cos^2\theta\,d\theta$$

$$=\rho\,ab^3\int_{0}^{\pi/2}4\sin^2\theta\cos^2\theta\,d\theta$$

$$= \rho ab^3 \int_0^{\pi/2} \sin^2 2\theta \, d\theta$$

$$= \rho ab^3 \int_0^{\pi/2} \frac{(1-\cos 4\theta)}{2} d\theta$$

$$= \frac{\rho ab^3}{2}\left[\theta - \frac{\sin 4\theta}{4}\right]_0^{\pi/2}$$

$$= \frac{\rho ab^3}{2}\left[\frac{\pi}{2} - 0\right] = \frac{\pi\rho ab^3}{4}$$

$$= \frac{M}{ab}\frac{ab^3}{4} = \frac{1}{4}Mb^2, \text{ where } M = \pi ab\rho.$$

1.5.11 *Moment of inertia of an elliptic disc about minor axis is* $1/4\ Ma^2$.

1.5.12 *Moment of inertia of a hollow sphere (or spherical shell) about diameter.*

The hollow sphere is generated by the revolution of a semi-circular arc about its bounding diameter.

We consider an elementary arc $PQ = a\,\delta\theta$ at the point P of the semi-circular arc. This arc, when revolved about the diameter AB, generates a circular ring of radius $a\sin\theta$ and width $a\,\delta\theta$.

Mass of this elementary ring $= 2\pi a \sin\theta a\,\delta\theta\rho$, where ρ is the density.

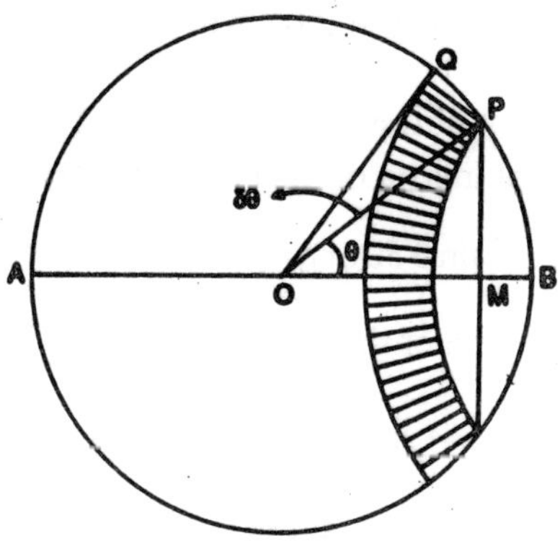

FIG 1.10

Distance of every point of this ring from the line $AB = PM = a\sin\theta$.
Therefore, moment of inertia of the elementary ring about

$$AB = (2\pi a \sin\theta a\delta\theta\rho)a^2\sin^2\theta = 2\pi\rho a^4 \cdot \sin^3\theta\,\delta\theta.$$

Hence moment of inertia of the hollow sphere about the diameter AB

$$= 2\pi\rho a^4 \int_0^{\pi} \sin^3\theta\, d\theta = 2\pi\rho a^4 \int_0^{\pi} \frac{3\sin\theta - \sin 3\theta}{4} d\theta$$

$$= \frac{\pi\rho a^4}{2}\left[-3\cos\theta + \frac{\cos 3\theta}{3}\right]_0^{\pi} = \frac{\pi\rho a^4}{2}\left[-3(-1-1) + \frac{1}{3}(-1-1)\right]$$

$$= \frac{\pi\rho a^4}{2}\left[6 - \frac{2}{3}\right] = \frac{\pi\rho a^4}{2}\frac{16}{3} = \pi\rho a^4\frac{8}{3} = \frac{M}{4a^2}a^4\frac{8}{3}$$

$$= \frac{2}{3}Ma^2 \text{ [since } M = 4\pi^2 a\rho\text{]}.$$

1.5.13 *Moment of inertia of the solid sphere about a diameter.* **[BPSC 88, MU 98H]**

The solid sphere is generated by the revolution of the semi-circular area about its bounding diameter.

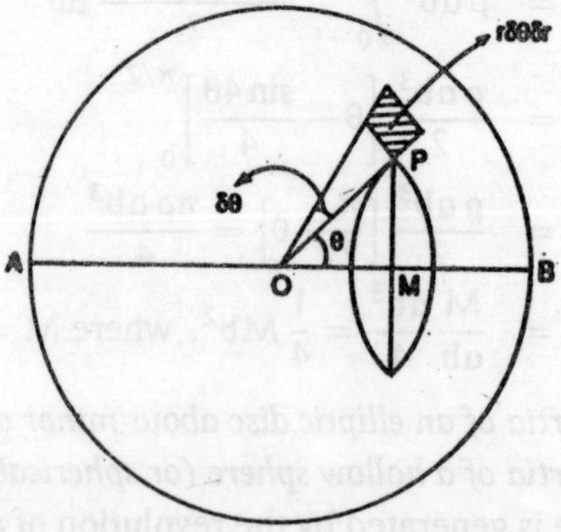

FIG 1.11

We consider an elementary area $r\,\delta\theta\,\delta r$ at the point $P(r, \theta)$ of the semi-circular area.

This element when revolved about the diameter AB, generates a circular ring of radius $r \sin\theta$ and cross-section $r\,\delta\theta\,\delta r$.

Mass of the elementary ring $= 2\pi r \sin\theta r\,\delta\theta\,\delta r\,\rho = 2\pi\rho r^2 \sin\theta\,\delta\theta\,\delta r$, where ρ is the density.

Distance of every point of this ring from $AB = PM = r\sin\theta$.

Therefore, moment of inertia of the elementary ring about AB

$$= (\rho 2\pi r^2 \sin\theta\,\delta\theta\,\delta r)(r\sin\theta)^2 = 2\pi\rho r^4 \sin^3\theta\,\delta\theta\,\delta r.$$

Hence moment of inertia of the sphere about the diameter AB

$$= \int_{\theta=0}^{\pi}\int_{r=0}^{a} 2\pi\rho r^4 \sin^3\theta\,d\theta\,dr = 2\pi\rho\int_{\theta=0}^{\pi}\frac{a^5}{5}\sin^3\theta\,d\theta$$

$$= \frac{2\pi\rho a^5}{5}\int_{\theta=0}^{\pi}\frac{3\sin\theta - \sin 3\theta}{4}\,d\theta = \frac{2\pi\rho a^5}{5\cdot 4}\left[-3\cos\theta + \frac{\cos 3\theta}{3}\right]_0^{\pi}$$

$$= \frac{\pi\rho a^5}{10}\left[-3(-1-1) + \frac{1}{3}(-1-1)\right] = \frac{\pi\rho a^5}{10}\left[6 - \frac{2}{3}\right] = \frac{\pi\rho a^5}{10}\frac{16}{3}$$

$$= \frac{\pi\rho a^5 8}{15} = \frac{3M}{4a^3}a^5\frac{8}{15} = \frac{2}{5}Ma^2 \quad \left[\because M = \frac{4}{3}\pi a^3\rho\right].$$

1.5.14 *Moment of inertia of a solid ellipsoid about principal axis.* **[MU 01H]**

Let the equation of the ellipsoid be

$$\frac{x^2}{a^2} + \frac{y^2}{b^2} + \frac{z^2}{c^2} = 1.$$

We consider an elementary volume $\delta x\,\delta y\,\delta z$ at the point $P(x, y, z)$ of the ellipsoid on the positive octant.

Mass of this element $= \rho\,\delta x\,\delta y\,\delta z$, where ρ is the density.

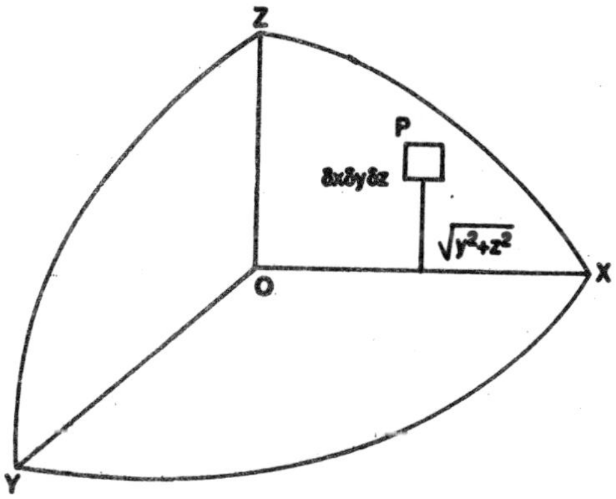

FIG 1.12

Let M be the mass of the ellipsoid.

Hence

$$M = \tfrac{4}{3}\pi abc\rho. \tag{1.9}$$

Distance of the point $P(x, y, z)$ from $OX = \sqrt{y^2 + z^2}$.

Hence moment of inertia of the elementary volume about $OX = (y^2+z^2)\rho\,\delta x\,\delta y\,\delta z$

Therefore, moment of inertia of the ellipsoid about OX

$$= 8 \iiint \rho\, dx\, dy\, dz\, (y^2 + z^2),$$

where $\frac{x^2}{a^2} + \frac{y^2}{b^2} + \frac{z^2}{c^2} \le 1$.

The integration being taken over the positive octant of the ellipsoid and

$$\frac{x^2}{a^2} + \frac{y^2}{b^2} + \frac{z^2}{c^2} \le 1.$$

Putting

$$\frac{x^2}{a^2} = u, \qquad \frac{y^2}{b^2} = v, \qquad \frac{z^2}{c^2} = w,$$

$$x = au^{1/2}, \qquad y = bv^{1/2}, \qquad z = cw^{1/2},$$

$$dx = \tfrac{1}{2}au^{-(1/2)}\,du, \quad dy = \tfrac{1}{2}bv^{-(1/2)}\,dv, \quad dz = \tfrac{1}{2}cw^{-(1/2)}\,dw.$$

Hence

$$\begin{aligned} \text{M.I.} &= \rho 8 \iiint (b^2v + c^2w)\frac{abc}{8}u^{-(1/2)}v^{-(1/2)}w^{-(1/2)}\,du\,dv\,dw \\ &= \rho abc \iiint \left(b^2u^{(1/2)-1}v^{(3/2)-1}w^{(1/2)-1} + c^2u^{(1/2)-1}v^{1/2}w^{(3/2)-1}\right) \\ &\qquad du\,dv\,dw \end{aligned}$$

$$= \quad abc\rho\left[b^2\frac{\Gamma\frac{1}{2}\Gamma\frac{3}{2}\Gamma\frac{1}{2}}{\Gamma\frac{7}{2}}+c^2\frac{\Gamma\frac{1}{2}\Gamma\frac{1}{2}\Gamma\frac{3}{2}}{\Gamma\frac{7}{2}}\right] \text{ [by Dirichlet's Theorem]}$$

$$= \quad \rho abc(b^2+c^2)\left[\frac{\sqrt{\pi}\sqrt{\pi}\Gamma\frac{3}{2}}{\frac{5}{2}\frac{3}{2}\Gamma\frac{3}{2}}\right]$$

$$= \quad \frac{3M}{4\pi}(b^2+c^2)\frac{4}{15}\pi \text{ [From equation (1.9)]}$$

$$= \quad \frac{M}{5}(b^2+c^2).$$

1.5.15 *Moment of inertia of a uniform triangular lamina about one side.*

Let ABC be the given triangular lamina.

We are to find the moment of inertia of ΔABC about its base BC.

Let us divide the lamina ABC into an infinite number of strips parallel to BC. We consider one such strip $PQQ'P'$ of breadth δx.

From A we draw AD perpendicular to BC.

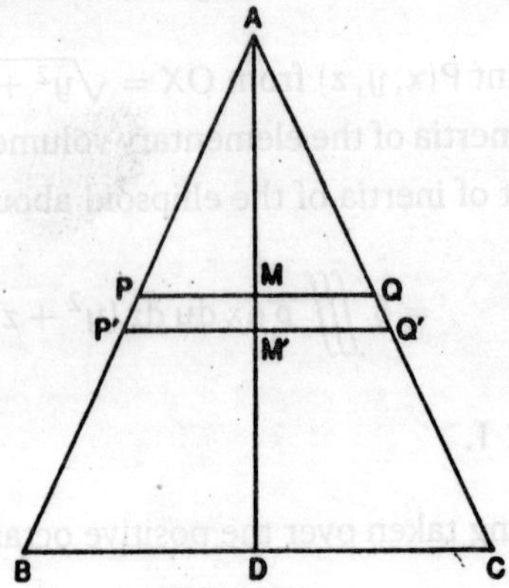

FIG 1.13

Let $AD = p$, $AM = x$.

Now since ΔABC and ΔAPQ are similar.

$$\therefore \frac{PQ}{BC}=\frac{AM}{AD}, \text{ or, } \frac{PQ}{a}=\frac{x}{p}. \quad \therefore PQ=\frac{ax}{p}.$$

Hence mass of the strip

$$= \rho PQ\,\delta x = \frac{ax}{p}\delta x\,\rho,$$

where ρ is the density.

Therefore, moment of inertia of the strip about BC

$$= \frac{ax\,\rho}{p}\delta x(p-x)^2.$$

Hence moment of inertia of the triangle about BC

$$= \frac{a\rho}{p}\int_0^p x(p-x)^2\,dx = \frac{a\rho}{p}\int_0^p x(x^2+p^2-2xp)\,dx$$

$$= \frac{a\rho}{p}\int_0^p (x^3+xp^2-2x^2p)\,dx = \frac{a\rho}{p}\left[\frac{x^4}{4}+\frac{x^2p^2}{2}-\frac{2x^3}{3}p\right]_0^p$$

$$= \frac{a\rho}{p}\left[\frac{p^4}{4}+\frac{p^2p^2}{2}-\frac{2}{3}p^3p\right] = \frac{a\rho p^4}{p}\left[\frac{1}{4}+\frac{1}{2}-\frac{2}{3}\right]$$

$$= a\rho p^3\left[\frac{3+6-8}{12}\right] = \frac{a\rho p^3}{12}. \qquad (1.10)$$

Let M be the mass of the triangle, then

$$M = \tfrac{1}{2}BC\cdot p\rho = \tfrac{1}{2}ap\,\rho.$$

Hence from equation (1.10), the required

$$\text{M.I.} = \frac{2M}{p}\frac{p^3}{12} = \frac{1}{6}Mp^2.$$

1.5.16 *Moment of inertia of a right circular cylinder about its axis.* **[BBABU 01H, 03H, MU 99H, 97H]**

Let h be the height and a be the radius of base of the cylinder, then mass M of the cylinder is given by

$$M = \pi a^2\rho h, \qquad (1.11)$$

where ρ is the density.

We consider an elementary disc of thickness δx at a distance x from O, the centre of its base.

Hence mass of the disc $= \pi a^2\,\delta x\,\rho$.

Therefore, moment of inertia of the disc about the axis $= \pi a^2\,\delta x\,\rho\frac{a^2}{2}$.

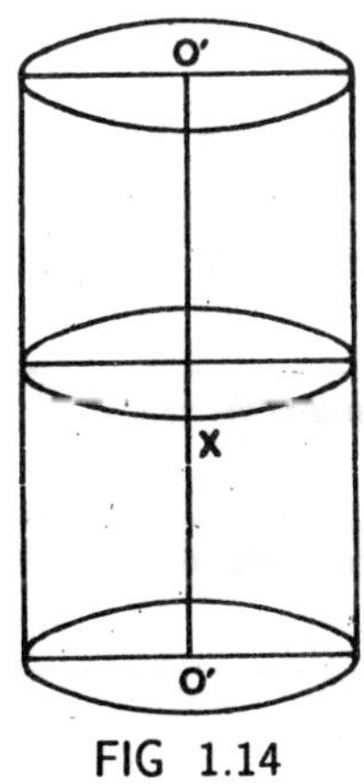

FIG 1.14

Hence moment of inertia of the cylinder about the axis

$$= \frac{a^4\pi\rho}{2}\int_0^h dx = \frac{\pi\rho a^4}{2}h = \frac{M}{a^2}\frac{a^4}{2} = \frac{1}{2}Ma^2.$$

1.6 Reference Table

The following table shows the moments of inertia of various bodies considered above. In all cases, it is assumed that the body has uniform density.

	Rigid Body	Moment of inertia
1.	*Uniform rod of length* 2a *and mass* M	
	– About a line through the middle point and perpendicular to the rod.	$\frac{1}{3}Ma^2$
	– About a line through one end and perpendicular to the rod.	$\frac{4}{3}Ma^2$
2.	*Rectangular plate of sides* 2a, 2b *and mass* M	
	– About a line through the centre and parallel to the side 2a	$\frac{1}{3}Mb^2$
	– About a line through the centre and perpendicular to the plate	$\frac{1}{3}M(a^2+b^2)$
3.	*Rectangular parallelopiped of edges* 2a, 2b, 2c	
	– About a line through its centre and parallel to the edge 2a	$\frac{M}{3}(b^2+c^2)$
4.	*Circular ring of radius* a *and mass* M	
	– About its diameter	$\frac{1}{2}Ma^2$
	– About a line through the centre and perpendicular to its plane	Ma^2
5.	*Circular plate of radius* a *and mass* M	
	– About its diameter	$\frac{1}{4}Ma^2$
	– About a line through the centre and perpendicular to its plane	$\frac{1}{2}Ma^2$
6.	*Elliptic disc of axes* 2a *and* 2b *and mass* M	
	– About the axis 2a	$\frac{1}{4}Mb^2$
	– About the axis 2b	$\frac{1}{4}Ma^2$
	– About a line through the centre and perpendicular to its plane	$\frac{1}{4}M(a^2+b^2)$
7.	*Hollow sphere of radius* a *and mass* M	
	– About a diameter	$\frac{2}{3}Ma^2$
8.	*Solid sphere of radius* a *and mass* M	
	– About a diameter	$\frac{2}{5}Ma^2$
9.	*Ellipsoid of axes* 2a, 2b, 2c *and mass* M	
	– About the axis 2a	$\frac{1}{5}M(b^2+c^2)$

1.7 Routh's Rule

Results of all the above moment of inertia's may be remembered with the help of Routh's rule which is as follows. Moment of inertia about an axis of symmetry is

$$\frac{\text{mass} \times \text{sum of squares of the perpendicular semi-axes}}{3,\ 4 \text{ or } 5},$$

the denominator to be 3, 4 or 5 according as the body is rectangular, elliptical or ellipsoidal.

1.8 Theorem of Parallel Axes

[TMBU 91H, 93H, 95H, 99H; MU 94H, 96H, 99H, 02H;
BBABU 92H, 99H, 01H, 03H]

Statement. *The moment of inertia of a body about any axis is equal to its moment of inertia about a parallel axis through its centre of gravity, together with the moment of inertia of the whole mass at its centre of gravity about the axis.*

Proof. Let OX, OY, OZ be a set of coordinate axes through any point O. Let G be the C.G. of the body. Through G we draw three coordinate axes GX′, GY′, GZ′ parallel to the axes OX, OY, OZ, respectively.

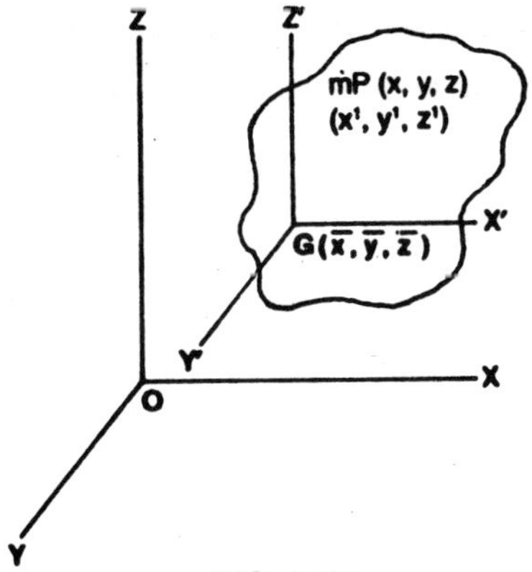

FIG 1.15

Let $(\bar{x}, \bar{y}, \bar{z})$ be the coordinates of the centre of gravity G of the body with respect to the rectangular axes OX, OY, OZ, respectively.

Let (x, y, z) and (x', y', z') be the coordinates of a particle of mass m at P referred to the coordinate axes OX, OY, OZ and parallel axes GX′, GY′, GZ′, respectively, then

$$x = \bar{x} + x', \quad y = \bar{y} + y', \quad z = \bar{z} + z'.$$

Now the coordinates of G referred to G as origin are $(0, 0, 0)$, which are also given by

$$\left(\frac{\Sigma m x'}{\Sigma m}, \frac{\Sigma m y'}{\Sigma m}, \frac{\Sigma m z'}{\Sigma m}\right).$$

It follows that

$$\frac{\Sigma mx'}{\Sigma m} = 0, \text{ or, } \Sigma mx' = 0.$$

Similarly,

$$\Sigma my' = 0, \Sigma mz' = 0.$$

Now, moment of inertia of the body OX

$$\begin{aligned}
&= \Sigma m\,(y^2 + z^2) = \Sigma m\left[(\bar{y} + y')^2 + (\bar{z} + z')^2\right] \\
&= \Sigma m\left[\bar{y}^2 + y'^2 + 2y'\bar{y} + \bar{z}^2 + z'^2 + 2\bar{z}z'\right] \\
&= \Sigma m\left(y'^2 + z'^2\right) + \Sigma m\,(\bar{y}^2 + \bar{z}^2) + 2\bar{y}\Sigma my' + 2\bar{z}\Sigma mz' \\
&= \Sigma m\left(y'^2 + z'^2\right) + \Sigma m\,(\bar{y}^2 + \bar{z}^2) \quad (\because\ \Sigma my' = 0 = \Sigma mz')
\end{aligned}$$

= M.I. of the body about GX′ + M.I. of the total mass M at G about OX.

Also, the product of inertia of the body about OX and OY

$$\begin{aligned}
&= \Sigma mxy = \Sigma m\,(\bar{x} + x')\,(\bar{y} + y') \\
&= \Sigma m\,(\bar{x}\bar{y} + \bar{x}y' + \bar{y}x' + x'y') \\
&= \bar{x}\bar{y}\Sigma m + \bar{x}\Sigma my' + \bar{y}\Sigma mx' + \Sigma mx'y' \\
&= \bar{x}\bar{y}\Sigma m + \Sigma mx'y' \quad (\because\ \Sigma my' = \Sigma mx' = 0)
\end{aligned}$$

= P.I. of the body about GX′ and GY′+ P.I. of the total mass M at G about OX and OY.

1.9 Theorem of Perpendicular Axes **[MU 97H, TMBU 93H, BBABU 2000H, 02H, 04H]**

Statement. *The moment of inertia of a plane lamina about an axis perpendicular to its plane is equal to the sum of the moments of inertia about any two perpendicular axes in the plane that intersect on the first axis.*

Proof. Let OX, OY, OZ be a system of rectangular axes.

Let a particle of mass m be at $P(x, y)$.

From P we draw PM and PN perpendicular to OX and OY.

So PM $= y$, PN $= x$.

Then sum of the moment of inertia of the particle of mass m about OX and OY $= mx^2 + my^2$.

Hence moment of inertia of the plane lamina about OX and OY

$$\begin{aligned}
&= \Sigma mx^2 + \Sigma my^2 = \Sigma m(x^2 + y^2) = \Sigma m \cdot OP^2 \\
&= \text{M.I. of the plane lamina about OZ.}
\end{aligned}$$

Hence the theorem.

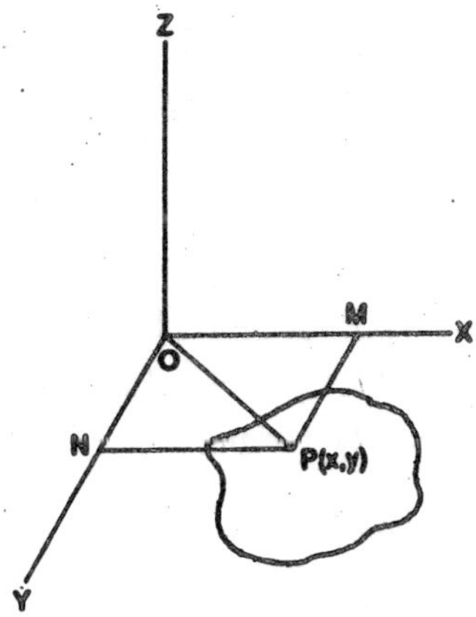

FIG 1.16

1.10 Moment of Inertia of the Body about a Line

If the moments and products of inertia of a body about three mutually perpendicular axes are known, to find the moment of inertia about any line through their meeting point. [MU 91H, 05H; CU 96H]

Let OX, OY, OZ be three mutually perpendicular axes.

We consider an element of mass m' of the body at P whose coordinates are (x, y, z), then

$$A = \Sigma m'(y^2 + z^2) \quad B = \Sigma m'(z^2 + x^2) \quad C = \Sigma m'(x^2 + y^2)$$
$$D = \Sigma m'yz \quad E = \Sigma m'zx \quad F = \Sigma m'xy.$$

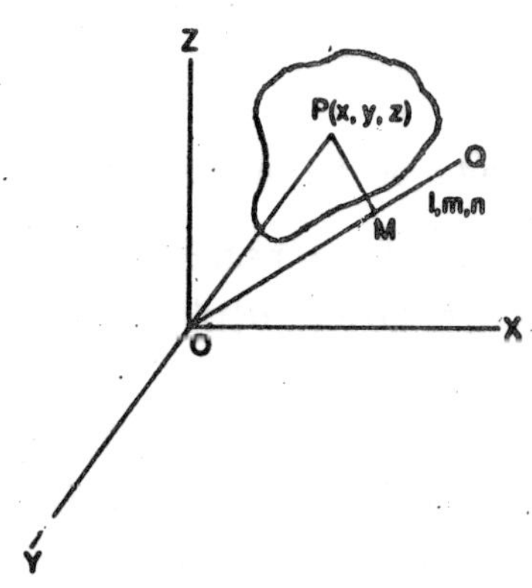

FIG 1.17

Let OQ be the line with direction cosines l, m, n and passing through O, so that

$$l^2 + m^2 + n^2 = 1.$$

From P we draw PM perpendicular to OQ.

$$\therefore OP^2 = OM^2 + PM^2$$
$$PM^2 = OP^2 - OM^2. \tag{1.12}$$

Now,

$$\begin{aligned} OP^2 &= x^2 + y^2 + z^2 \\ OM &= \text{Projection of OP on OQ} \\ &= l(x-0) + m(y-0) + n(z-0) \\ &= lx + my + nz. \end{aligned}$$

Hence from equation (1.12)

$$\begin{aligned} PM^2 &= x^2 + y^2 + z^2 - (lx + my + nz)^2 \\ &= x^2 + y^2 + z^2 - (l^2x^2 + m^2y^2 + n^2z^2 + 2lmny + 2mnyz + 2lnzx) \\ &= x^2(1-l^2) + y^2(1-m^2) + z^2(1-n^2) \\ &\quad -2lmxy - 2mnyz - 2lnzx \\ &= x^2(m^2+n^2) + y^2(n^2+l^2) + z^2(l^2+m^2) \\ &\quad -2lmxy - 2mnyz - 2lnzx \\ &= l^2(y^2+z^2) + m^2(z^2+x^2) + n^2(x^2+y^2) \\ &\quad -2lmxy - 2mnyz - 2lnzx. \end{aligned} \tag{1.13}$$

Hence the required moment of inertia about $OQ = \Sigma m'PM^2$.
From equation (1.13)

$$\begin{aligned} &= l^2\Sigma m'(y^2+z^2) + m^2\Sigma m'(z^2+x^2) + n^2\Sigma m'(x^2+y^2) \\ &\quad -2lm\Sigma m'xy - 2mn\Sigma m'yz - 2nl\Sigma m'zx \\ &= Al^2 + Bm^2 + Cn^2 - 2Flm - 2Dmn - 2Enl. \end{aligned}$$

Vector Method

Let $\angle POM = \theta$, $\overrightarrow{OP} = \vec{r}$, and $\vec{a}$ be the unit vector along OQ. Then we have

$$\begin{aligned} p &= OP\sin\theta = r\sin\theta \\ &= |\vec{r}|\sin\theta = |\vec{a}||\vec{r}|\sin\theta \ (\because\ |\vec{a}| = 1) \\ &= |\vec{a}\times\vec{r}|. \end{aligned}$$

Now,

$$\vec{a}\times\vec{r} = \begin{vmatrix} \vec{i} & \vec{j} & \vec{k} \\ l & m & n \\ x & y & z \end{vmatrix},$$

where $\vec{a} = l\vec{i} + m\vec{j} + n\vec{k} = \vec{i}(mz - ny) - \vec{j}(lz - nx) + \vec{k}(ly - mx)$.
Hence

$$|\vec{a}\times\vec{r}| = \sqrt{(mz-ny)^2 + (nx-lz)^2 + (ly-mx)^2}.$$

Hence moment of inertia of the body about OQ

$$\begin{aligned}
&= \Sigma m' p^2 \\
&= \Sigma m'[m^2z^2 + n^2y^2 - 2mnyz + n^2x^2 + l^2z^2 - 2nlxz \\
&\qquad + l^2y^2 + m^2x^2 - 2lmxy] \\
&= \Sigma m'[l^2(y^2+z^2) + m^2(z^2+x^2) + n^2(x^2+y^2) \\
&\qquad -2mnyz - 2nlzx - 2lmxy] \\
&= l^2\Sigma m'(y^2+z^2) + m^2\Sigma m'(z^2+x^2) + n^2\Sigma m'(x^2+y^2) \\
&\qquad -2mn\Sigma m'yz - 2nl\Sigma m'zx - 2lm\Sigma m'xy \\
&= Al^2 + Bm^2 + Cn^2 - 2Dmn - 2Enl - 2Flm.
\end{aligned}$$

1.11 Moments and Product of Inertia of a Plane Lamina [CU 98H, 04H]

If the moments and products of inertia of a plane lamina about two perpendicular axes in its plane are given, to find the moment and product of inertia about any perpendicular lines through their point of intersection.

We consider an elementary mass m at the point P whose coordinates with reference to the axes OX, OY be (x, y). Let A and B be the moments of inertia and F the product of inertia of the plane lamina about these axes in the plane, then we have $A = \Sigma my^2$, $B = \Sigma mx^2$, $F = \Sigma mxy$.

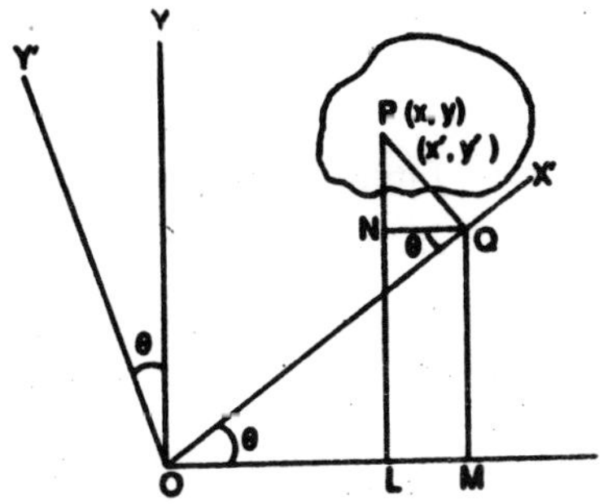

FIG 1.18

Let OX′ and OY′ be the perpendicular axes in the plane of the lamina and inclined at an angle θ to OX and OY, respectively. If (x', y') be the coordinates of the point P with respect to these axes, then

$$\begin{aligned}
x' &= x\cos\theta + y\sin\theta, \\
y' &= y\cos\theta - x\sin\theta.
\end{aligned}$$

Hence moment of inertia of the lamina about OX′

$$\begin{aligned}
&= \Sigma m y'^2 \\
&= \Sigma m(y\cos\theta - x\sin\theta)^2 \\
&= \Sigma m(y^2\cos^2\theta + x^2\sin^2\theta - 2xy\cos\theta\sin\theta) \\
&= \cos^2\theta\Sigma my^2 + \sin^2\theta\Sigma mx^2 - 2\cos\theta\sin\theta\Sigma mxy \\
&= A\cos^2\theta + B\sin^2\theta - F\sin 2\theta.
\end{aligned}$$

Similarly, moment of inertia of the lamina about OY′

$$
\begin{aligned}
&= \Sigma m x'^2 \\
&= \Sigma m(x\cos\theta + y\sin\theta)^2 \\
&= \Sigma m(x^2\cos^2\theta + y^2\sin^2\theta + 2xy\sin\theta\cos\theta) \\
&= \cos^2\theta\Sigma m x^2 + \sin^2\theta\Sigma m y^2 + 2\sin\theta\cos\theta\Sigma mxy \\
&= B\cos^2\theta + A\sin^2\theta + F\sin 2\theta.
\end{aligned}
$$

The P.I. of the lamina about OX′ and OY′

$$
\begin{aligned}
&= \Sigma m x'y' = \Sigma m(x\cos\theta + y\sin\theta)(y\cos\theta - x\sin\theta) \\
&= \Sigma m[xy\cos^2\theta - x^2\sin\theta\cos\theta + y^2\sin\theta\cos\theta - xy\sin^2\theta] \\
&= \Sigma m[xy(\cos^2\theta - \sin^2\theta) + \sin\theta\cos\theta(y^2 - x^2)] \\
&= \cos 2\theta\Sigma mxy + \sin\theta\cos\theta(\Sigma my^2 - \Sigma mx^2) \\
&= F\cos 2\theta + (A - B)\sin\theta\cos\theta.
\end{aligned}
$$

Note 1. $A + B > C$.

We have

$$A = \Sigma m(y^2 + z^2),\ B = \Sigma m(z^2 + x^2),\ C = \Sigma m(x^2 + y^2).$$

Now,

$$
\begin{aligned}
A + B - C &= \Sigma m[y^2 + z^2 + z^2 + x^2 - x^2 - y^2] \\
&= 2\Sigma m z^2 = \text{a positive quantity.}
\end{aligned}
$$

Hence $A + B > C$.

Note 2. *The sum of the moment of inertia's about any three rectangular axes meeting at a given point is always constant and is equal to twice the moment of inertia about that point.*

[BBABU 2000H, 04H]

We have

$$
\begin{aligned}
A + B + C &= \Sigma m(y^2 + z^2 + z^2 + x^2 + x^2 + y^2) \\
&= 2\Sigma m(x^2 + y^2 + z^2) = 2\Sigma m r^2 \\
&= 2\ (\text{M.I. of the body about the given point}).
\end{aligned}
$$

$\because r = \sqrt{x^2 + y^2 + z^2}$ = distance of the mass m at (x, y, z) from the given point O as origin, hence $A + B + C$ is constant.

Note 3. *The sum of moment of inertia's of a body with respect to any plane through a given point and its normal at that point is constant and is equal to moment of inertia of the body about that point.*

Let the given point O be taken as the origin and the plane as xy-plane.

It C' be the moment of inertia of the body about the xy-plane and C the moment of inertia of the body about its normal at O which is z-axis, then $C' = \Sigma m z^2$ and $C = \Sigma m(x^2 + y^2)$.

$$\begin{aligned} \therefore\ C' + C &= \Sigma m(x^2 + y^2 + z^2) = \Sigma m r^2 \\ &= \text{M.I. of the body about O} \\ &= \text{constant} \end{aligned}$$

as $C + C'$ is independent of the plane through O.

Since $A + B + C = 2\Sigma m r^2$ and $C + C' = \Sigma m r^2$, hence

$$\begin{aligned} A + B + C &= 2C + 2C' \\ A + B - C &= 2C' \\ C' &= \tfrac{1}{2}(A + B - C). \end{aligned}$$

Thus of A', B', C' denote the moment of inertia's of the body with respect to the planes yz, zx and xy, respectively, then

$$A' = \tfrac{1}{2}(B + C - A),\ B' = \tfrac{1}{2}(C + A - B),\ C' = \tfrac{1}{2}(A + B - C).$$

Note 4. $A > 2D, B > 2E$ *and* $C > 2F$.

Since A.M. > G.M.

$$\therefore\ \frac{y^2 + z^2}{2} > \sqrt{y^2 z^2} = yz,\ \text{or, } y^2 + z^2 > 2yz,\ \text{or, } \Sigma m(y^2 + z^2) > 2\Sigma m yz.$$

Hence $A > 2D$.

Similarly $B > 2E$ and $C > 2F$.

1.12 Momental Ellipsoid

[MU 93H, BBABU 99H, 2000H, 01H, 02H, 03H, TMBU 97H, 01H, 2000 H, 02H, CU 96H]

If A, B, C, D, E, F are moments and products of inertia of the body about axes, then the moment of inertia of the body about a line OQ whose direction cosines are l, m, n is given by

$$Al^2 + Bm^2 + Cn^2 - 2Dmn - 2Enl - 2Flm.$$

Let P be a point on OQ such that the moment of inertia of the body about OQ may be inversely proportional to OP^2,

$$\text{i.e., } Al^2 + Bm^2 + Cn^2 - 2Dmn - 2Enl - 2Flm \propto \frac{1}{OP^2},$$

$$\text{or, } Al^2 + Bm^2 + Cn^2 - 2Dmn - 2Enl - 2Flm = \frac{Mk^4}{r^2},$$

where OP = r, M is the mass of the body and k is some constant having the dimension of length.

$$\text{or, } Ar^2l^2 + Bm^2r^2 + Cn^2r^2 - 2Dmnr^2 - 2Enlr^2 - 2Flmr^2 = Mk^4.$$

If (x, y, z) be the coordinates of P on OQ, then equation of OP is

$$\frac{x}{l} = \frac{y}{m} = \frac{z}{n} = r.$$

Hence locus of P is

$$Ax^2 + By^2 + Cz^2 - 2Dyz - 2Ezx - 2Fxy = Mk^4. \tag{1.14}$$

Since A, B, C are essentially positive, therefore, equation (1.14) represents an ellipsoid.

This ellipsoid is called the *momental ellipsoid of the body* at the point O.

Note 1. *We know from solid Geometry that for every ellipsoid there exists three mutually perpendicular diameters such that, if they be taken as axes of reference, the transformed equation has no terms involving yz, zx and xy.*

These new axes of coordinates are called the *Principal Axes* of the ellipsoid.

Let referred to principal axes, the equation of the momental ellipsoid equation (1.14) becomes

$$A_1x^2 + B_1y^2 + C_1z^2 = Mk^4. \tag{1.15}$$

Evidently the products of inertia about these new axes are zero, for otherwise the equation would contain the terms of xy, yz and zx.

Hence we conclude that, *for every body there exists at every point O a set of three mutually perpendicular axes, which are three principal diameters of the momental ellipsoid at O, such that the products of inertia of the body about them, taken two at a time, vanish.*

Note 2. *If the three principal moments at any point O are equal to each other, ellipsoid becomes a sphere. In this case, every diameter is a principal diameter and all radii vectors are equal.*

Also every straight line through O becomes principal axis at O and moment of inertia about them all are equal.

For example, the perpendiculars from the centre of gravity of a cube on three conterminous faces are principal axes, since referred to them as axes $\Sigma mxy = 0$, $\Sigma myz = 0$, $\Sigma mzx = 0$.

Also there moment of inertia's about them are equal, each being $\frac{2Ma^2}{3}$, where $2a$ is the side of the cube.

Hence moment of inertia about any line through the C.G. of the cube and whose d.c.'s are l, m, n is given by

$$\frac{2Ma^2}{3}l^2 + \frac{2Ma^2}{3}m^2 + \frac{2Ma^2}{3}n^2 = \frac{2Ma^2}{3},$$

which is always the same.

1.13 Momental Ellipse

Let OX and OY be two mutually perpendicular axes and OQ be a line through O in the plane of the lamina. Then moment of inertia of the plane lamina about OQ is given by

$$A\cos^2\theta - 2F\sin\theta\cos\theta + B\sin^2\theta,$$

where A, B denote the moment of inertia of the lamina about OX, OY and F the P.I. of the lamina about OX and OY.

Let P be a point on OQ such that the moment of inertia of the lamina about OQ may be inversely proportional to OP^2,

$$\text{i.e.,}\quad A\cos^2\theta - 2F\sin\theta\cos\theta + B\sin^2\theta \propto \frac{1}{OP^2},$$

$$\text{or,}\quad A\cos^2\theta - 2F\sin\theta\cos\theta + B\sin^2\theta = \frac{Mk^4}{r^2},\ \text{where } OP = r$$

$$\text{or,}\quad Ar^2\cos^2\theta - 2Fr^2\sin\theta\cos\theta + Br^2\sin^2\theta = Mk^4$$

$$\text{or,}\quad Ax^2 - 2Fxy + By^2 = Mk^4. \tag{1.16}$$

Since A and B are essentially positive, therefore, equation (1.16) represents an ellipse. This is called a *momental ellipse* of the lamina at O.

1.14 Equimomental Systems or Bodies

[TMBU 92H, 95H, 03H; BBABU 97H, 99H, 02H; CU 92H, 94H, 96H, 2000H, 02H]

Two systems or bodies are said to be equi-momental or kinetically (or dynamically) equivalent when moments and products of inertia of one system or body about all axes are each equal to the moments and products of inertia of the other system or body about the same axes.

The Conditions are Sufficient

Hence we assume that the given conditions hold. Let M be the total mass of each system, G the centroid, A_1, B_1, C_1 the principal moments of inertia at G.

In both cases, the moment of inertia about the line through G having d.c.'s λ, μ, V with respect to the principal axes is $A_1\lambda^2 + B_1\mu^2 + C_1V^2$. Hence by the theorem of parallel axes, the moment of inertia about a parallel line distant h from this line is $A_1\lambda^2 . B_1\mu^2 + C_1V^2 + Mh^2$ in both cases. Thus the two systems have equal moments of inertia about any line of space and the systems are equi-momental.

1.14.1 *The necessary and sufficient conditions for two systems to be equi-momental are that:*

- the centre of gravity of the two systems is the same point.
- both the systems have the same mass.

the two systems have the same principal axes and some principal moments about centre of gravity.

Proof. The conditions are necessary.

Here we assume that the two systems are equi-momental.

Let G_1, G_2 be the centroids of the two systems, and M_1, M_2 their total masses. Since the systems are equi-momental, they have equal moments of inertia I, say, about G_1G_2. The moment of inertia's about any parallel line at a distance d from G_1G_2 are respectively $I + M_1d^2$, $I + M_2d^2$ by the theorem of parallel axes. Since the systems are equi-momental.

$\therefore\ I + M_1d^2 = I + M_2d^2$ and so $M_1 = M_2 = M$ (say).

Let J be the moment of inertia of either system about the line through G_1 perpendicular to G_1G_2. Then the moment of inertia of the first system about the parallel line through G_2 is $J + M \cdot G_1G_2^2$ and that of the second system is $J - M \cdot G_1G_2^2$.

$\therefore\ J + M \cdot G_1G_2^2 = J - M \cdot G_1G_2^2 \Rightarrow G_1 = G_2$.

The systems have equal moments of inertia about any line through their common centroid G. Thus their momental ellipsoids of G are the same and so are the principal axes and principal moments of inertia at G.

1.14.2 *A triangular lamina is equi-momental with three particles each of mass one third of the mass of the lamina placed at the middle points of its sides.*

[MU 92H, BBABU 95H, 99H, MU 03H, CU 92H, 02H; TMBU 91H]

Let ABC be the given triangular lamina of mass M.

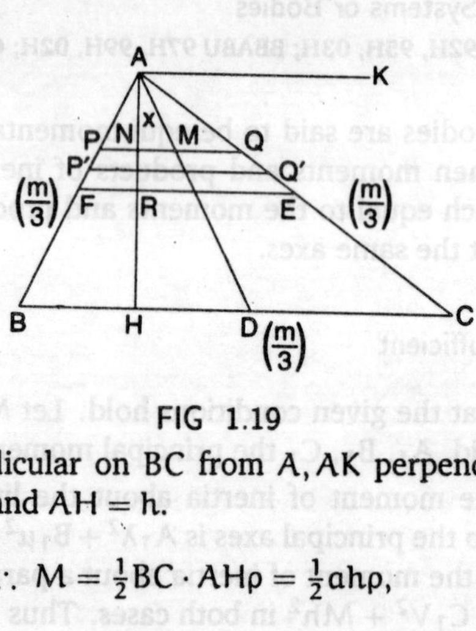

FIG 1.19

Let AH be the perpendicular on BC from A, AK perpendicular to AN in the plane of the triangle ABC and AH = h.

$$\therefore\ M = \tfrac{1}{2}BC \cdot AH\rho = \tfrac{1}{2}ah\rho, \tag{1.17}$$

where ρ is the density.

We divide the triangle ABC into a large number of thin strips parallel to BC.

Let PQQ′P′ be one of such elementary strip of thickness δx at a distance x = AN from A.

Since ΔAPQ and ΔABC are similar

$$\therefore\ \frac{PQ}{BC} = \frac{AN}{AH};\ \frac{PQ}{a} = \frac{x}{h};\ PQ = \frac{ax}{h}. \tag{1.18}$$

Now mass of the strip

$$PQQ'P' = \rho PQ\,\delta x = \rho\frac{ax}{h}\,\delta x.$$

$\therefore$ moment of inertia of the strip about AK

$$= \text{M.I. of the strip about PQ + M.I. of its mass at its C.G. about AK}$$
$$= 0 + x^2\frac{\rho a x}{h}\,\delta x = \frac{\rho a}{h}x^3\,\delta x.$$

Also moment of inertia of the strip about AH

$$= \text{M.I. of the strip about parallel line through its C.G. D}$$
$$+ \text{ M.I. of its mass at its C.G. M about AH}$$
$$= \frac{1}{3}\left(\frac{1}{2}PQ\right)^2\frac{\rho a x}{h}\,\delta x + \frac{\rho a x}{h}\,\delta x\,(NM)^2$$
$$= \left[\frac{1}{3}\left(\frac{1}{2}\frac{ax}{h}\right)^2 + NM^2\right]\frac{\rho a x}{h}\,\delta x \text{ [From equation (1.18)]}$$

Since ΔANM and ΔAHD are similar

$$\frac{NM}{HD} = \frac{AN}{AH};\ \frac{NM}{HD} = \frac{x}{h},\ \therefore\ NM = \frac{x}{h}HD.$$

Hence moment of inertia of the strip PQQ′P′ about AH

$$= \left[\frac{1}{3}\left(\frac{1}{2}\frac{ax}{h}\right)^2 + \frac{x^2}{h^2}(HD)^2\right]\frac{\rho a x}{h}\,\delta x = \left[\frac{a^2}{12} + HD^2\right]\frac{\rho a x}{h}\,\delta x\,\frac{x^2}{h^2}$$
$$= \left[a^2 + 12HD^2\right]\frac{\rho a x^3}{12h^3}\,\delta x = \frac{\rho a}{12h^3}\left(a^2 + 12HD^2\right)x^3\,\delta x.$$

Hence moment of inertia of the ΔABC about AK

$$= \int_0^h \frac{\rho a}{h}x^3\,dx = \frac{\rho a}{h}\frac{h^4}{4} = \frac{\rho a}{4}h^3 = \frac{2M}{4h}h^3 \text{ [From equation (1.17)]}$$
$$= \frac{M}{2}h^2. \tag{1.19}$$

Also moment of inertia of the ΔABC about AH

$$= \frac{\rho a}{12h^3}\int_0^h \left(a^2 + 12HD^2\right)x^3\,dx = \frac{\rho a}{12h^3}\left[\frac{a^2h^4}{4} + 12HD^2\frac{h^4}{4}\right]$$
$$= \frac{\rho a}{12h^3}\frac{h^4}{4}\left[a^2 + 12HD^2\right] = \frac{\rho a h}{48}\left(a^2 + 12HD^2\right).$$

Now

$$HD = HC - DC = HC - \tfrac{1}{2}BC = HC - \tfrac{1}{2}a.$$

In $\triangle AHC$

$$\cos C = \frac{HC}{AC} = \frac{HC}{b}$$
$$HC = b\cos C$$
$$HD = b\cos C - \tfrac{1}{2}(b\cos C + c\cos B)$$
$$= \tfrac{1}{2}(b\cos C - c\cos B).$$

Hence moment of inertia of the $\triangle ABC$ about AH

$$= \frac{\rho a h}{48}\left[a^2 + 3(b\cos C - c\cos B)^2\right]$$
$$= \frac{\rho a h}{48}\left[(b\cos C + c\cos B)^2 + 3(b\cos C - c\cos B)^2\right]$$
$$= \frac{2M}{48}\left[4b^2\cos^2 C + 4c^2\cos^2 B - 4bc\cos B\cos C\right]$$

[From equation (1.17)]

$$= \frac{M}{6}\left(b^2\cos^2 C + c^2\cos^2 B - bc\cos B\cos C\right). \tag{1.20}$$

Now P.I. of the $\triangle ABC$ about AK and AH

$$= \int_0^h (AN\cdot NM)\frac{\rho a}{h}x\,dx = \int_0^h \left(x\cdot\frac{x}{h}HD\right)\frac{\rho a}{h}x\,dx$$
$$= \frac{\rho a}{h^2}\int_0^h x^3\,dx\,HD = \frac{\rho a}{h^2}\frac{h^4}{4}HD = \frac{\rho a h^2}{4}HD$$
$$= \frac{2Mh}{4}\frac{1}{2}(b\cos C - c\cos B) \text{ [From equation (1.17)]}$$
$$= \frac{Mh}{4}(b\cos C - c\cos B). \tag{1.21}$$

Now we consider three particlés each of mass $M/3$ placed at the mid-points D, E, F of the sides.

Moment of inertia of the three particles about

$$AK = \frac{M}{3}\left[h^2 + \left(\frac{h}{2}\right)^2 + \left(\frac{h}{2}\right)^2\right] = \frac{M}{3}\left[h^2 + \frac{h^2}{2}\right] = \frac{Mh^2}{2}. \tag{1.22}$$

Moment of inertia of the three particles about AH

$$= \frac{M}{3}\left[DH^2 + ER^2 + RF^2\right]$$
$$= \frac{M}{3}\left[DH^2 + \left(\frac{1}{2}CH\right)^2 + \left(\frac{1}{2}BH\right)^2\right]$$
$$= \frac{M}{3}\left[\frac{1}{4}(b\cos C - c\cos B)^2 + \frac{1}{4}(b\cos C)^2 + \frac{1}{4}(c\cos B)^2\right]$$
$$= \frac{M}{12}\left[2b^2\cos^2 C + 2c^2\cos^2 B - 2bc\cos B\cos C\right]$$
$$= \frac{M}{6}\left[b^2\cos^2 C + c^2\cos^2 B - bc\cos B\cos C\right] \tag{1.23}$$

Again product of inertia of the three particles about AK and AH

$$= \frac{M}{3}\left[AH \cdot HD + \tfrac{1}{2}AH \cdot \tfrac{1}{2}HC - \tfrac{1}{2}AH \cdot \tfrac{1}{2}BH\right]$$

$$= \frac{M}{3} \cdot \frac{h}{4}\left[2(b\cos C - c\cos B) + b\cos C - c\cos B\right]$$

$$= \frac{Mh}{12} 3(b\cos C - c\cos B)$$

$$= \frac{Mh}{4}(b\cos C - c\cos B). \tag{1.24}$$

Comparing results of equations (1.19), (1.20), (1.21) to the results of equations (1.22), (1.23), (1.24) we see that these results are respectively identical, i.e., moments and products of inertia of the triangle about AK, AH are the same as those of three particles, each of mass M/3 placed at the middle points of the sides.

Also the C.G.'s of the two systems coincide and their masses are also the same. Hence the two systems are equi-momental.

Hence the triangle of mass M is equi-momental to three particles each of mass M/3 placed at the middle points of the sides.

1.15 Principal Axes

A set of rectangular axes through any point O are said to be principal axes if the products of inertia of a given body about them, taken in pairs vanish.

[CU 95H, 99H, 03H, TMBU 94H, 98H, 01H]

To find whether a given straight line is, at any point of its length, a principal axis of a given material system and, if so, to find the other two principal axes.

or

Necessary and sufficient condition for a given straight line to be a principal at some point of its length. **[MU 96H, 04H, BBABU 98H, 03H, CU 91H, 93H, 95H, 97H, 99H, 01H, 03H]**

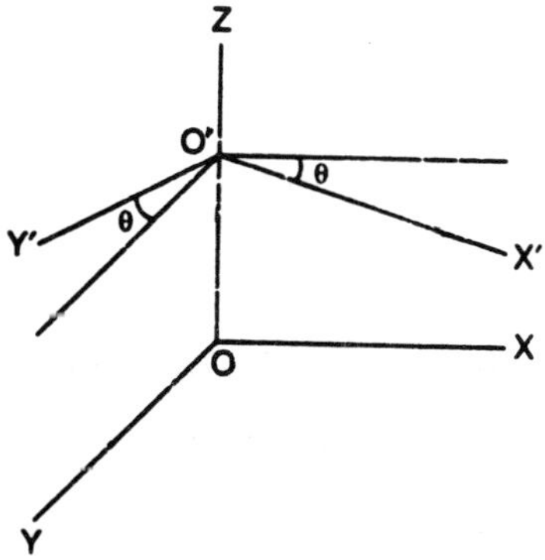

FIG 1.20

Let the given straight line be taken as z-axis and a point O on it as origin. Let the two perpendicular lines OX and OY, perpendicular to OZ be taken as the axes of x and y, respectively.

Now let the line OZ be the principal axis of the system at O′, where OO′ = h.

Also let O′X′, O′Y′ be the other two principal axes, O′X′ being inclined at an angle θ to a line parallel to OX.

Let (x, y, z) and (x', y', z') be the coordinates of the point P with reference to the two set of axes OX, OY, OZ and OX′, OY′, OZ′, respectively. Then we have

$$\begin{aligned} x' &= x\cos\theta + y\sin\theta, \\ y' &= y\cos\theta - x\sin\theta, \\ z' &= z - h. \end{aligned}$$

Now the necessary and sufficient conditions; that O′X′, O′Y′, O′Z may be the principal axes of the system, the products of inertia $\Sigma my'z' = 0$, $\Sigma mz'x' = 0$, $\Sigma mx'y' = 0$.

Now

$$\begin{aligned} \Sigma my'z' &= \Sigma m(y\cos\theta - x\sin\theta)(z - h) \\ &= \Sigma m[zy\cos\theta - hy\cos\theta - zx\sin\theta + hx\sin\theta] \\ &= -\sin\theta\Sigma mzx + \cos\theta\Sigma myz + h\sin\theta\Sigma mx - h\cos\theta\Sigma my \\ &= (D\cos\theta - E\sin\theta) + h(\Sigma mx)\sin\theta - h(\Sigma my)\cos\theta. \end{aligned}$$

If $(\bar{x}, \bar{y}, \bar{z})$ be the coordinates of C.G., then

$$\bar{x} = \frac{\Sigma mx}{\Sigma m} = \frac{\Sigma mx}{M},\ \bar{y} = \frac{\Sigma my}{M},\ \bar{z} = \frac{\Sigma mz}{M}. \tag{1.25}$$

Hence,

$$\Sigma my'z' = (D\cos\theta - E\sin\theta) + hM(\bar{x}\sin\theta - \bar{y}\cos\theta).$$

Also

$$\begin{aligned} \Sigma mz'x' &= \Sigma m(z - h)(x\cos\theta + y\sin\theta) \\ &= \Sigma m[xz\cos\theta + zy\sin\theta - hx\cos\theta - hy\sin\theta] \\ &= (\Sigma mxz)\cos\theta + (\Sigma myz)\sin\theta - h\cos\theta\Sigma mx - h\sin\theta\Sigma my \\ &= (D\sin\theta + E\cos\theta) - Mh(\bar{x}\cos\theta + \bar{y}\sin\theta)\ \text{[From equation (1.25)]} \end{aligned}$$

Also

$$\begin{aligned} \Sigma mx'y' &= \Sigma m(x\cos\theta + y\sin\theta)(y\cos\theta - x\sin\theta) \\ &= \Sigma m\left[xy\cos^2\theta - x^2\sin\theta\cos\theta + y^2\sin\theta\cos\theta - xy\sin^2\theta\right] \\ &= \Sigma m[xy(\cos^2\theta - \sin^2\theta) - x^2\sin\theta\cos\theta + y^2\sin\theta\cos\theta] \\ &= \Sigma mxy(\cos^2\theta - \sin^2\theta) + \sin\theta\cos\theta(\Sigma my^2 - \Sigma mx^2) \\ &= F\cos 2\theta + \tfrac{1}{2}\sin 2\theta\left[\Sigma m(y^2 + z^2) - \Sigma m(x^2 + y^2)\right] \\ &= F\cos 2\theta + \tfrac{1}{2}\sin 2\theta(A - B). \end{aligned}$$

Now $\Sigma mx'y' = 0$.

Hence

$$F\cos 2\theta + \tfrac{1}{2}\sin 2\theta(A - B) = 0 \Rightarrow 2F\cos 2\theta = \sin 2\theta(B - A)$$

$$\Rightarrow \tan 2\theta = \frac{2F}{B - A} \Rightarrow 2\theta = \tan^{-1}\frac{2F}{B - A}. \quad (1.26)$$

$$\Rightarrow \theta = \frac{1}{2}\tan^{-1}\frac{2F}{B - A}.$$

Also $\Sigma my'z' = 0$.
Thus $(D\cos\theta - E\sin\theta) + hM(\bar{x}\sin\theta - \bar{y}\cos\theta) = 0$.

$$\Rightarrow Mh = \frac{E\sin\theta - D\cos\theta}{\bar{x}\sin\theta - \bar{y}\cos\theta}. \quad (1.27)$$

Again $\Sigma mz'x' = 0$.
Hence $(D\sin\theta + E\cos\theta) - Mh(\bar{x}\cos\theta + \bar{y}\sin\theta) = 0$.

$$\Rightarrow Mh = \frac{D\sin\theta + E\cos\theta}{\bar{x}\cos\theta + \bar{y}\sin\theta} = 0. \quad (1.28)$$

Hence from equations (1.26) and (1.27), we get

$$\frac{E\sin\theta - D\cos\theta}{\bar{x}\sin\theta - \bar{y}\cos\theta} = \frac{D\sin\theta + E\cos\theta}{\bar{x}\cos\theta + \bar{y}\sin\theta},$$

or, $E\bar{x}\sin\theta\cos\theta + E\bar{y}\sin^2\theta - \bar{x}D\cos^2\theta - D\bar{y}\sin\theta\cos\theta$
$= D\bar{x}\sin^2\theta + E\bar{x}\sin\theta\cos\theta - D\bar{y}\sin\theta\cos\theta - E\bar{y}\cos^2\theta,$

or, $E\bar{y} = D\bar{x}$,

or, $$\frac{E}{\bar{x}} = \frac{D}{\bar{y}}. \quad (1.29)$$

Now

$$\frac{E}{\bar{x}} = \frac{D}{\bar{y}} = \frac{E\sin\theta - D\cos\theta}{\bar{x}\sin\theta - \bar{y}\cos\theta} = Mh \text{ [From equation (1.26)]}$$

or, $$h = \frac{E}{M\bar{x}} = \frac{D}{M\bar{y}}. \quad (1.30)$$

The equation (1.29) gives the condition so that line OZ may be principal axis of the system at some point of its length and if this condition is satisfied, then equation (1.30) gives the point at which the line OZ is a principal axis. Also equation (1.26) gives the direction of the other principal axes.

Conversely, if we assume the given conditions, then we will show that the axes are principal axes.

Now

$$\begin{aligned} \Sigma my'z' &= (D\cos\theta - E\sin\theta) + hM(\bar{x}\sin\theta - \bar{y}\cos\theta) \\ &= (M\bar{y}h\cos\theta - M\bar{x}h\sin\theta) + Mh(\bar{x}\sin\theta - \bar{y}\cos\theta) \\ &= 0 \text{ [From equation (1.30)]} \end{aligned}$$

Also

$$\begin{aligned}\Sigma mz'x' &= (D\sin\theta + E\cos\theta) - Mh(\bar{x}\cos\theta + \bar{y}\sin\theta) \\ &= (Mh\bar{y}\sin\theta + hM\bar{x}\cos\theta) - Mh(\bar{x}\cos\theta + \bar{y}\sin\theta) \\ &= 0 \text{ [From equation (1.30)]}\end{aligned}$$

Again

$$\begin{aligned}\Sigma mx'y' &= \frac{A-B}{2}\sin 2\theta + F\cos 2\theta \\ &= -F\cot 2\theta \sin 2\theta + F\cos 2\theta \\ &= 0 \text{ [From equation (1.26)]}\end{aligned}$$

Thus we have shown that

$$\Sigma my'z' = 0,\ \Sigma mz'x' = 0,\ \Sigma mx'y' = 0.$$

Thus the given system of axes are principal axes.

Corollary 1.15.1 *Through each point of a plane lamina there exists a pair of principal axes of the lamina.*

Let OZ be a normal to the plane of the lamina through any point O of it.

Then z coordinate of the C.G. vanishes and hence $D = E = 0$.

Therefore, the conditions $E/\bar{x} = D/\bar{y}$ are satisfied and $h = E/\bar{x} = D/\bar{y}$ gives $h = 0$.

Hence z-axis is a principal axis of the lamina at O and the other two principal axes which will be in the plane of the lamina be OX′ and OY′. If θ be the inclination of OX′ to OX, then

$$\theta = \tfrac{1}{2}\tan^{-1}\frac{2F}{B-A}.$$

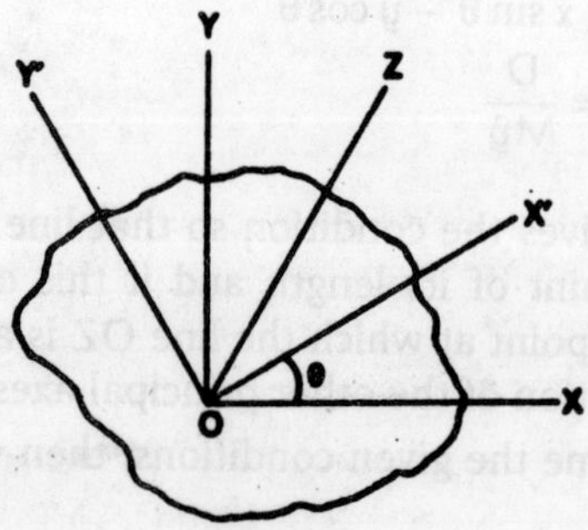

FIG 1.21

Corollary 1.15.2 *If an axis passes through the centre of gravity of a body and is a principal axis at any point of its length, then it is a principal axis at all points of its length.* [BU 94H]

If axis of z is a principal axis at O, then $D = E = 0$.

Hence the equation

$$h = \frac{E}{M\bar{x}} = \frac{D}{M\bar{y}} \text{ gives } h = 0,$$

which shows that there is no such other point as O'.

But if z-axis is a principal axis at O and passes through the C.G. of the body. Then $\bar{x} = 0, \bar{y} = 0$ and $D = E = 0$. Hence the value of h becomes indeterminate.

Hence if an axis passes through the C.G. of the body and is a principal axis at any point of its length, then it is a principal axis at all points of its length. [CU 97H]

1.16 Principal Moments

Moments of inertia of a body about its principal axes at any point are called its principal moments at that point.

The equation of the ellipsoid at any point is given by

$$Ax^2 + By^2 + Cz^2 - 2Dyz - 2Ezx - 2Fxy = MK^4. \tag{1.31}$$

Taking the principal axes as the coordinate axes equation (1.31) reduces to the form

$$A'x^2 + B'y^2 + C'z^2 = Mk^4,$$

where A', B', C' are the principal moments and are the values of λ in the cubic equation

$$\begin{vmatrix} A-\lambda & H & G \\ H & B-\lambda & F \\ G & F & C-\lambda \end{vmatrix} = 0.$$

1.17 Worked out Examples

Type I: Problems based on Definition

Problem 1. *Find the moment of inertia of a truncated cone about its axes, the radii of its ends being a and b.*

Solution: Let ABCD be the truncated cone with the vertex at V and of semi-vertical angle α.

We consider a circular disc of breadth δx at a distance x from the vertex V.

Then radius of the disc $= EN = x\tan\alpha$.

Its mass $= \pi x^2 \tan^2\alpha\, \delta x \rho$.

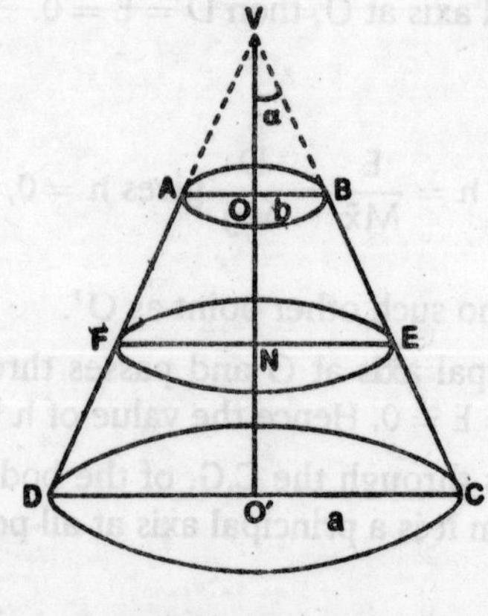

FIG 1.22

If M be the total mass of the truncated cone, then

$$M = \int_{b\cot\alpha}^{a\cot\alpha} \pi x^2 \tan^2 \alpha \rho \, dx$$

$$VO = b\cot\alpha$$

$$VO' = a\cot\alpha$$

$$M = \pi\rho\tan^2\alpha\left[\frac{x^3}{3}\right]_{b\cot\alpha}^{a\cot\alpha} = \frac{\pi\rho}{3}\tan^2\alpha\,[a^3 - b^3]\cot^3\alpha$$

$$M = \frac{\pi\rho}{3}(a^3 - b^3)\cot\alpha. \quad (1.32)$$

Now moment of inertia of the elementary disc about the axis OO′

$$= (\pi x^2 \tan^2\alpha\,\delta x\rho)\frac{x^2\tan^2\alpha}{2}$$

$$= \tfrac{1}{2}\pi\rho\tan^4\alpha x^4\,\delta x.$$

∴ moment of inertia of the truncated cone about OO′

$$= \frac{\pi\rho}{2}\tan^4\alpha\int_{b\cot\alpha}^{a\cot\alpha} x^4\,dx = \frac{\pi\rho}{2}\tan^4\alpha\left[\frac{x^5}{5}\right]_{b\cot\alpha}^{a\cot\alpha}$$

$$= \frac{\pi\rho}{2}\frac{\tan^4\alpha}{5}\cot^5\alpha(a^5 - b^5) = \frac{\pi\rho}{10}\cot\alpha(a^5 - b^5)$$

$$= \frac{3M(a^5 - b^5)}{10(a^3 - b^3)} \quad \text{[From equation (1.32)]}$$

Problem 2. *Find the moment of inertia of a solid right circular cone about its axis.*

Solution: Let OC be the axis of the cone, O being the vertex. Let h be the height of the cone and a be the radius of the base.

Let M be the mass of the cone, then

$$M = \tfrac{1}{3}\pi a^2 h\rho, \quad (1.33)$$

where ρ is the density.

We consider an elementary disc PQ of thickness δx, parallel to the base AB and at a distance x from the vertex O.

Therefore, mass of the disc $= \rho\pi r^2\,\delta x$.

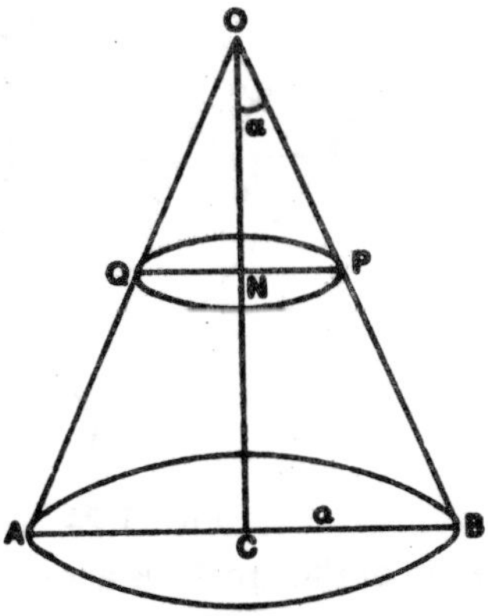

FIG 1.23

Hence moment of inertia of the elementary disc about axis OC

$$= \pi\rho r^2\,\delta x\frac{r^2}{2} = \frac{\pi\rho}{2}r^4\,\delta x = \frac{\pi\rho}{2}x^4\tan^4\alpha\,\delta x,$$

where α is the semi-vertical angle of the cone.

Hence moment of inertia of the cone about axis OC

$$= \int_0^h \frac{\pi\rho}{2}\tan^4\alpha x^4\,dx = \frac{\pi\rho}{2}\tan^4\alpha\frac{h^5}{5} = \frac{\pi\rho}{10}h^5\tan^4\alpha$$

$$= \frac{3M}{10a^2h}h^5\tan^4\alpha = \frac{3M}{10a^2}h^4\tan^4\alpha = \frac{3M}{10a^2}a^4 = \frac{3Ma^2}{10}.$$

Problem 3. *Find the moment of inertia of the arc of a circle about:*

(a) *the diameter bisecting the arc;*

(b) *an axis through the centre, perpendicular to its plane;*

(c) *an axis through its middle point perpendicular to its plane.*

Solution: Let OC be the diameter bisecting the circular arc ACB, subtending an angle 2α at the centre O.

We consider an elementary arc $a\delta\theta$ at the point P of the arc.

Its mass $= \rho a\delta\theta$.

If M be the mass of the arc ACB, then

$$M = \rho a 2\alpha. \tag{1.34}$$

(a) Distance of P from the diameter OC $=$ PM $= a\sin\theta$.

Hence moment of inertia of the elementary arc about OC

$$= PM^2\rho a\,\delta\theta = a^2\sin^2\theta\rho a\,\delta\theta = \rho a^3\sin^2\theta\,\delta\theta.$$

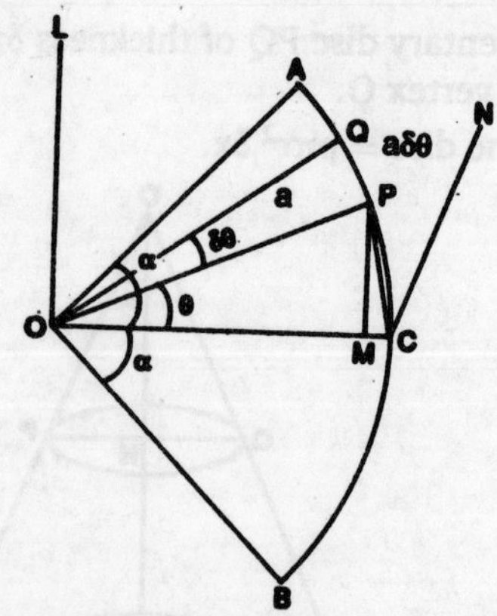

FIG 1.24

Hence moment of inertia of the arc ACB about the diameter OB

$$= \int_{-\alpha}^{\alpha} \rho a^3 \sin^2\theta\, d\theta = \rho a^3 \int_{-\alpha}^{\alpha} \frac{1-\cos 2\theta}{2}\, d\theta = \frac{\rho a^3}{2}\left[\theta - \frac{\sin 2\theta}{2}\right]_{-\alpha}^{\alpha}$$

$$= \frac{\rho a^3}{2}[2\alpha - \sin 2\alpha] = \frac{\rho a^3}{2}[2\alpha - 2\sin\alpha\cos\alpha] = \rho a^3[\alpha - \sin\alpha\cos\alpha]$$

$$= \frac{Ma^2}{2\alpha}(\alpha - \sin\alpha\cos\alpha) \text{ [From equation (1.34)]}$$

(*b*) Let OL be a line through centre O perpendicular to the plane of the arc.
Distance of the point P from OL = OP = a.
Moment of inertia of the elementary arc about OL $= (a + \delta\theta)a^2 = \rho a^3\,\delta\theta$.
∴ moment of inertia of the arc ACB about OL

$$= \int_{-\alpha}^{\alpha} \rho a^3\, d\theta = \rho a^3 2\alpha = a^2 M \text{ [From equation (1.34)]}$$

(*c*) Let CN be the line through the middle point C of the arc and perpendicular to the plane of the arc. Then distance of P from CN

$$= \text{PC} = \sqrt{\text{OC}^2 + \text{OP}^2 - 2\text{OCOP}\cos\theta} = \sqrt{a^2 + a^2 - 2a^2\cos\theta}$$

$$= \sqrt{2a^2(1-\cos\theta)} = \sqrt{2a^2 2\sin^2\frac{\theta}{2}} = 2a\sin\frac{\theta}{2}.$$

Thus moment of inertia of the elementary arc about CN

$$= \rho a\,\delta\theta\left(2a\sin\frac{\theta}{2}\right)^2 = 4a^3\rho\sin^2\frac{\theta}{2}\,\delta\theta.$$

Hence moment of inertia of the arc ACB about CN

$$= \int_{-\alpha}^{\alpha} 4a^3\rho\sin^2\frac{\theta}{2}\, d\theta = 4a^3\rho\int_0^{\alpha} 2\sin^2\frac{\theta}{2}\, d\theta = 4a^3\rho\int_0^{\alpha}(1-\cos\theta)\, d\theta$$

$$= 4a^3\rho\left[\theta - \sin\theta\right]_0^{\alpha} = 4a^3\rho(\alpha - \sin\alpha)$$

$$= \frac{2a^2M}{\alpha}(\alpha - \sin\alpha) \text{ [From equation (1.34)]}$$

Problem 4. *Show that the moment of inertia of parabolic area (of latus rectum* 4a*) cut off by an ordinate at distance* h *from the vertex is* 3/7 Mh^2 *about the tangent at the vertex and* 4/5 Mah *about the axis.*

Solution: Let the equation of the parabola of latus rectum 4a be $y^2 = 4ax$.

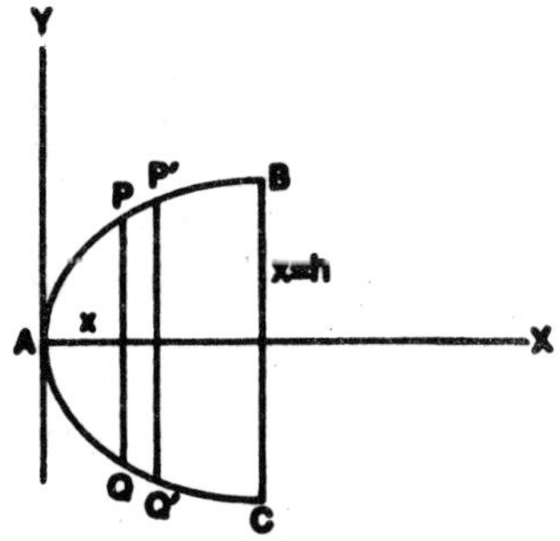

FIG 1.25

Let ABC be the portion of the parabola cut off by an ordinate at a distance h from the vertex.

We consider an elementary strip PQQ′P′ of breadth δx at a distance x from the vertex.

Mass of the strip $= \rho 2y\, \delta x$.

Hence

$$\begin{aligned} M &= \text{Mass of the portion ABCA} \\ &= \int_0^h 2\rho y\, dx = 2\rho \int_0^h 2\sqrt{ax}\, dx = 4\rho\sqrt{a}\frac{2}{3}\left[x^{3/2}\right]_0^h \\ &= 4\rho\sqrt{a}\frac{2}{3}h^{3/2} = \frac{8}{3}\rho a^{1/2}h^{3/2}. \qquad (1.35) \end{aligned}$$

Now every point of this strip is at a distance x from AY, the tangent at the vertex.
Hence moment of inertia of the strip about AY $= (2y\, \delta x \rho)x^2$.
Hence moment of inertia of the whole area about AY

$$\begin{aligned} &= \int_0^h 2y\rho x^2\, dx = 2\rho \int_0^h \sqrt{4ax}x^2\, dx \\ &= 4\rho\sqrt{a}\int_0^h x^{5/2}\, dx = 4\rho\sqrt{a}\frac{2}{7}\left[x^{7/2}\right]_0^h \\ &= \frac{8}{7}\rho\sqrt{a}h^{7/2} = \frac{3}{7}Mh^2 \text{ [From equation (1.35)]} \end{aligned}$$

Again moment of inertia of the strip PQRS about AX $= (2y\, \delta x \rho)\, y^2/3$.
Thus moment of inertia of the whole area about the axis

$$\begin{aligned} &= \frac{2}{3}\rho\int_0^h y^3\, dx = \frac{2}{3}\rho\int_0^h (4ax)^{3/2}\, dx = \frac{2}{3}\rho 8a^{3/2}\frac{2}{5}\left[x^{5/2}\right]_0^h \\ &= \frac{4}{15}\rho 8a^{3/2}h^{5/2} = \frac{32}{15}\rho a^{3/2}h^{5/2} \\ &= \frac{4}{5}Mah \text{ [From equation (1.35)]} \end{aligned}$$

Problem 5. *Show that the moment of inertia of the a paraboloid of revolution about its axis is $M/3 \times$ the square of the radius of its base.* **[MU 2000H]**

Solution: The paraboloid of revolution is generated by the revolution of arc APB of parabola $y^2 = 4ax$ about x-axis.

If $BM = b$, then $AM = b^2/4a$.

Now consider a strip $PQQ'P'$ of breadth δx at a distance x from the vertex. This strip when revolved about x-axis generates a circular disc of radius y.

Mass of the circular disc $= \pi y^2\,\delta x\rho$.

If M be the mass of the whole paraboloid of revolution, then

$$M = \rho\int_0^{AM}\pi y^2\,dx = \rho\pi\int_0^{AM} 4ax\,dx = \rho 4a\pi\left[\frac{x^2}{2}\right]_0^{b^2/4a}$$

$$= \rho 2a\pi\frac{b^4}{4\cdot 4a^2} = \frac{\pi\rho b^4}{8a}. \tag{1.36}$$

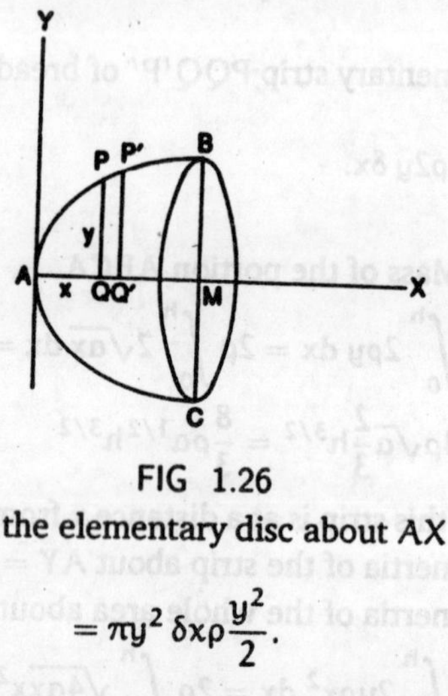

FIG 1.26

Also moment of inertia of the elementary disc about AX

$$= \pi y^2\,\delta x\rho\frac{y^2}{2}.$$

Hence moment of inertia of the paraboloid of revolution about AX

$$= \frac{\pi\rho}{2}\int_0^{AM} y^4\,dx = \frac{\pi\rho}{2}\int_0^{b^2/4a} y^4\,dx = \frac{\pi\rho}{2}\int_0^{b^2/4a} 16a^2x^2\,dx$$

$$= \pi\rho 8a^2\int_0^{b^2/4a} x^2\,dx = \pi\rho 8a^2\left[\frac{x^3}{3}\right]_0^{b^2/4a} = \frac{\pi\rho 8a^2}{3}\frac{b^6}{64a^3}$$

$$= \frac{\pi\rho b^6}{24a} = \frac{8aM}{b^4}\frac{b^6}{24a} \quad \text{[From equation (1.36)]}$$

$$= \frac{Mb^2}{3} = \frac{M}{3} \times \text{square on the radius of the base.}$$

Problem 6. *Find the moment of inertia of the area bounded by $r^2 = a^2\cos 2\theta$ about its axes.*

Solution: The curve is as shown in the figure.

The loop of the curve is formed between $\theta = -\pi/4$ and $\theta = \pi/4$.

We consider an elementary area $r\,\delta\theta\,\delta r$ at the point $P(r, \theta)$ of the curve.

Its mass $= \rho r\,\delta\theta\,\delta r$.

Its distance from $OX = PM = r\sin\theta$.

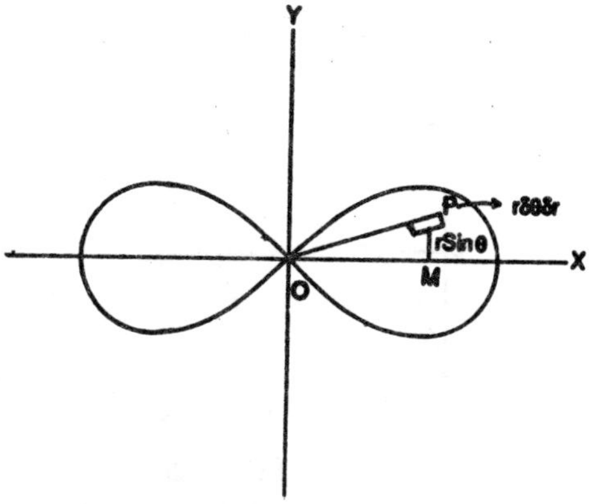

FIG 1.27

Area of a loop

$$= 2\int_0^{\pi/4} \frac{1}{2} r^2\, d\theta = \int_0^{\pi/4} a^2 \cos 2\theta\, d\theta = \frac{a^2}{2}\Big[\sin 2\theta\Big]_0^{\pi/4} = \frac{a^2}{2}[1-0] = \frac{a^2}{2}.$$

$\therefore$ area of both the loops $= a^2$.

Hence the mass M of the whole area is given by

$$M = a^2\rho. \qquad (1.37)$$

Now moment of inertia of the elementary mass about axis OX

$$= \rho r\,\delta\theta\,\delta r (r\sin\theta)^2 = \rho r^3 \sin^2\theta\,\delta\theta\,\delta r.$$

Hence moment of inertia of the whole area about OX

$$= 4\int_{\theta=0}^{\pi/4}\int_{r=0}^{a\sqrt{\cos 2\theta}} \rho r^3 \sin^2\theta\, d\theta\, dr = 4\rho\int_0^{\pi/4}\left\{\frac{r^4}{4}\right\}_0^{a\sqrt{\cos 2\theta}} \sin^2\theta\, d\theta$$

$$= \frac{4\rho}{4}\int_0^{\pi/4} a^4\cos^2 2\theta \sin^2\theta\, d\theta - \rho a^4\int_0^{\pi/4}\cos^2 2\theta \frac{(1-\cos 2\theta)}{2}\, d\theta$$

$$= \frac{\rho a^4}{2}\int_0^{\pi/4}(\cos^2 2\theta - \cos^3 2\theta)\, d\theta.$$

Let $\alpha = 2\theta$, $d\alpha = 2d\theta$.

When $\theta = \frac{\pi}{4}$, $\alpha = \frac{\theta}{2}$, $\theta = 0$, $\alpha = 0$, hence

$$\text{M.I.} = \frac{\rho a^4}{4}\int_0^{\pi/2}(\cos^2\alpha - \cos^3\alpha)\, d\alpha = \frac{\rho a^4}{4}\left[\frac{1}{2}\frac{\pi}{2} - \frac{2}{3}\right]$$

$$= \frac{\rho a^4}{4}\left(\frac{\pi}{4} - \frac{2}{3}\right) = \frac{Ma^2}{4}\left(\frac{\pi}{4} - \frac{2}{3}\right) = \frac{Ma^2}{16}\left(\pi - \frac{8}{3}\right).$$

Problem 7. *Find the product of inertia of a semi-circular wire about diameter and tangent at its extremity.*

Solution: Let M be the mass, a the radius and OA the diameter of a semi-circular arc. Let OB be the tangent at extremity.

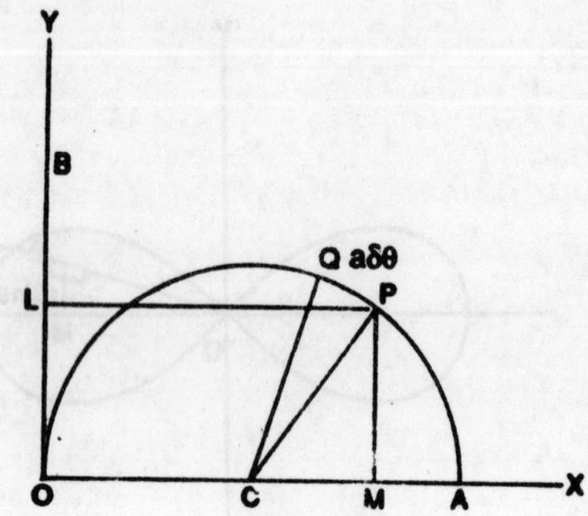

FIG 1.28

We consider an elementary arc $a\delta\theta$ at the point P of the wire.
Its mass $= \rho a\ \delta\theta$, where ρ is the density.
If M be the total mass of the semi-circular wire, then

$$M = \pi a\rho. \tag{1.38}$$

Now the distance of the elementary mass from OX and OY are PM and PL.
Now

$$PM = a\sin\theta.$$

$$PL = OM = OC + CM = a + a\cos\theta.$$

Its P.I. about OX and OY

$$= a\,\delta\theta\rho(a\sin\theta)(a + a\cos\theta) = \rho a^3 \sin\theta(1 + \cos\theta)\,\delta\theta.$$

Hence P.I. of the wire about OX and OY

$$= \rho a^3 \int_0^\pi \sin\theta(1+\cos\theta)\,d\theta = \rho a^3 \int_0^\pi \left(\sin\theta + \frac{1}{2}\sin 2\theta\right) d\theta$$

$$= \rho a^3 \left[-\cos\theta - \frac{\cos 2\theta}{4}\right]_0^\pi = -\rho a^3\left[-1 + \frac{1}{4} - \left(1 + \frac{1}{4}\right)\right]$$

$$= 2\rho a^3 = \rho a 2a^2 = \frac{M}{\pi}2a^2 = \frac{2Ma^2}{\pi} \quad \text{[From equation (1.38)]}$$

Type I

Problem 8. *Find the moment of inertia of a uniform rod of length $2a$ with respect to an axis, perpendicular to the rod and at a distance d from its middle point.*

[BBABU 2000, 04]

Solution: Let AB be a uniform rod of length $2a$ whose mid-point is C. If ρ be the linear density then mass of the rod

$$M = 2a\rho \tag{1.39}$$

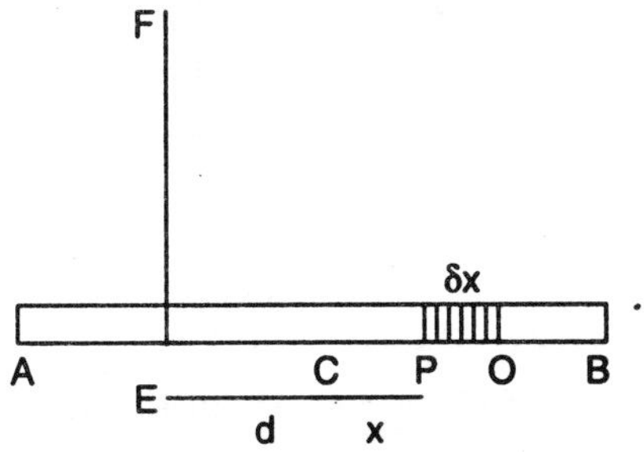

FIG 1.29

Let EF be the line about which moment of inertia is to be found. We consider an element of breadth δx at a distance x from C.

Mass of the element $= \rho\delta x$.

Therefore, moment of inertia of the element about EF

$$= \rho\delta x EP^2 = \rho\delta x(EC + CP)^2 = \rho\delta x(d + x)^2 = \rho\delta x(d^2 + x^2 + 2dx).$$

Hence moment of inertia of the rod about EF

$$\begin{aligned} &= \rho\int_{-a}^{a}(d^2 + x^2 + 2dx)dx = \rho\left[d^2x + \frac{x^3}{3} + 2d\frac{x^2}{2}\right]_{-a}^{a} \\ &= \rho\left[d^2 2a + \frac{2a^3}{3} + d(a^2 - a^2)\right] = \rho\left(2ad^2 + \frac{2a^3}{3}\right) \\ &= 2a\rho\left(d^2 + \frac{a^2}{3}\right) = M\left(d^2 + \frac{a^2}{3}\right) \quad \text{[From equation (1)].} \end{aligned}$$

Type II: Problems based on the theorem of Parallel Axes

Problem 9. *Prove that the moment of inertia of a uniform solid right circular cone of mass M, height h and vertical angle 2α, about a diameter of its base is*

$$\frac{Mh^2}{2}(3\tan^2\alpha + 20). \qquad \textbf{[IAS 88]}$$

Solution: Let O be the vertex of a right circular cone of mass M, height h and base radius a.

If α is the semi-vertical angle and ρ be the density of the cone, then

$$M = \tfrac{1}{3}\pi h^3 \tan^2\alpha\rho. \tag{1.40}$$

We consider an elementary disc of radius CQ whose thickness is δx and parallel to the base at a distance x from the vertex O.

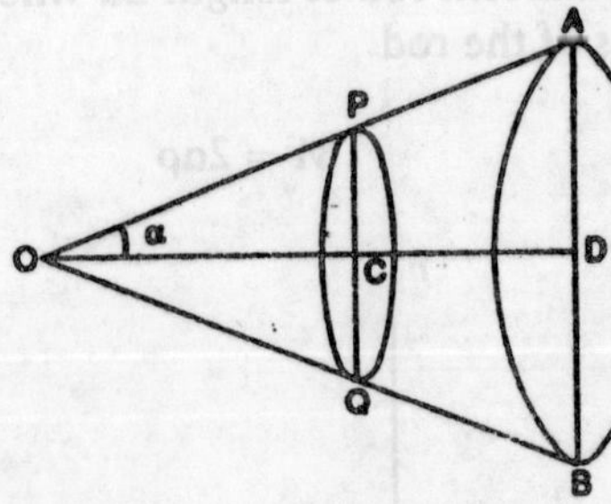

FIG 1.30

Mass of the disc $= \pi\rho x^2 \tan^2 \alpha\, \delta x$.

Hence moment of inertia of the disc about the diameter AB of the base of the cone

$$= \text{M.I. of the disc about PQ} + \text{M.I. of the total mass at centre C about AB}$$

$$= \frac{1}{4}\pi\rho x^2 \tan^2 \alpha\, \delta x \cdot CP^2 + \pi\rho x^2 \tan^2 \alpha\, \delta x \cdot CD^2$$

$$= \frac{\pi\rho x^2}{4} \tan^2 \alpha \left[CP^2 + 4CD^2\right] \delta x$$

$$= \frac{\pi\rho x^2}{4} \tan^2 \alpha \left[x^2 \tan^2 \alpha + 4(h-x)^2\right] \delta x.$$

Hence moment of inertia of the cone about the diameter of the base

$$= \int_0^h \frac{\pi\rho x^2 \tan^2 x}{4} \left[x^2 \tan^2 \alpha + 4(h-x)^2\right] dx$$

$$= \frac{\pi\rho}{4} \tan^2 \alpha \int_0^h \left[x^4 \tan^2 \alpha + 4x^2(h^2 + x^2 - 2hx)\right] dx$$

$$= \frac{\pi\rho}{4} \tan^2 \alpha \int_0^h \left[x^4 \tan^2 \alpha + 4(x^2h^2 + x^4 - 2hx^3)\right] dx$$

$$= \frac{\pi\rho}{4} \tan^2 \alpha \left[\frac{h^5}{5} \tan^2 \alpha + 4\left(\frac{h^3}{3}h^2 + \frac{h^5}{5} - \frac{2hh^4}{4}\right)\right]$$

$$= \frac{\pi\rho \tan^2 \alpha h^5}{4} \left[\frac{\tan^2 \alpha}{5} + 4\left(\frac{1}{3} + \frac{1}{5} - \frac{1}{2}\right)\right]$$

$$= \frac{\pi\rho h^5 \tan^2 \alpha}{4} \left[\frac{\tan^2 \alpha}{5} + 4\frac{(10+6-15)}{30}\right]$$

$$= \frac{\pi\rho h^5 \tan^2 \alpha}{4} \left[\frac{\tan^2 \alpha}{5} + \frac{2}{15}\right] = \frac{\pi\rho h^5 \tan^2 \alpha}{60}\left[3\tan^2 \alpha + 2\right]$$

$$= \frac{\pi\rho h^5 \tan^2 \alpha\, (3\tan^2 \alpha + 2)}{60}$$

$$= \frac{3Mh^2\,(3\tan^2 \alpha + 2)}{60} \quad \text{[From equation (1.39)]}$$

$$= \frac{Mh^2}{20}\,(3\tan^2 \alpha + 2).$$

Problem 10. *Find the moment of inertia of a right circular cylinder about a straight line through its C.G. perpendicular to its axis.* **[BBABU 01H, 03H]**

Solution: Let h be the height and a the radius of the base of the cylinder. Then mass M of the cylinder is given by

$$M = \pi a^2 h\rho. \tag{1.41}$$

We consider an elementary disc of breadth δx at a distance x from the centre of gravity G.

Moment of inertia of the disc about OX $= (\pi a^2 \delta x \rho)\ a^2/4$.

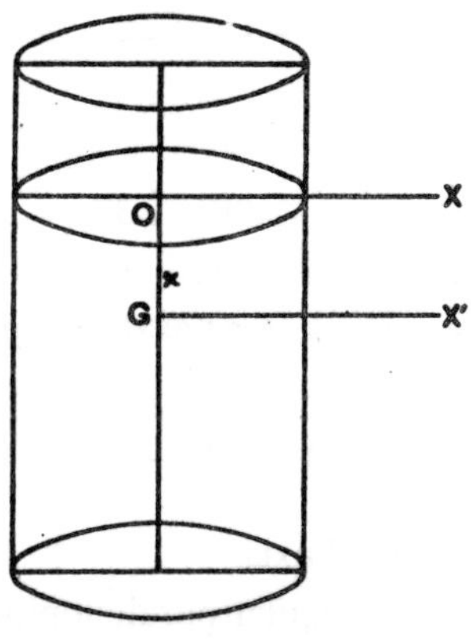

FIG 1.31

Now by the theorem of parallel axis,

Moment of inertia of the disc about GX′

$$= \text{M.I. of the disc about OX} + \text{M.I. of the total mass of disc placed at its C.G. about GX}'$$

$$= (\pi a^2 \delta x \rho)\frac{a^2}{4} + (\pi a^2 \delta x \rho)x^2 = \pi a^2 \rho\left(\frac{a^2}{4} + x^2\right)\delta x.$$

Hence moment of inertia of the whole cylinder about GX′

$$= \pi a^2 \rho \int_{-h/2}^{h/2}\left(\frac{a^2}{4} + x^2\right)dx = \pi a^2 \rho\left[\frac{a^2}{4}x + \frac{x^3}{3}\right]_{-h/2}^{h/2}$$

$$= \pi a^2 \rho\left[\frac{a^2}{4}\cdot h + \frac{h^3}{12}\right] = \frac{\pi a^2 \rho h}{12}(3a^2 + h^2)$$

$$= \frac{Mh}{12}(3a^2 + h^2) \quad \text{[From equation (1.40)]}$$

Problem 11. *Find the moment of inertia of a rectangular parallelopiped about an edge.*

Solution: 1st Method. Let $2a$, $2b$, $2c$ be the lengths of the edges of a rectangular parallelopiped of mass M.

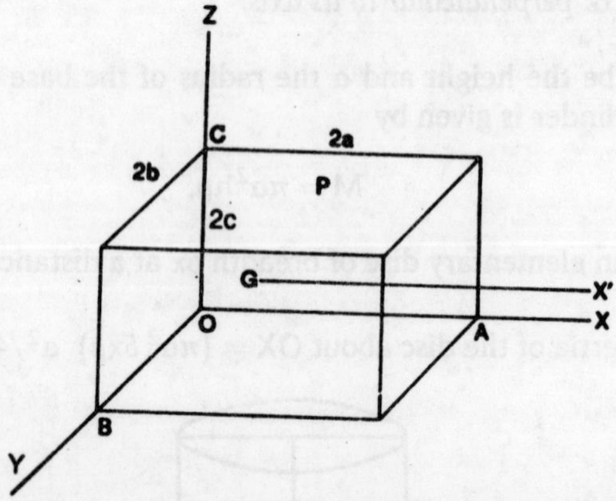

FIG 1.32

Then we have to determine the moment of inertia about an edge, say OA.

Moment of inertia of the rectangular parallelopiped about an axis through C.G. and parallel to

$$OA = M\frac{b^2 + c^2}{3}.$$

Now by the theorem of parallel axis,

Moment of inertia of the rectangular parallelopiped about OA

$$= \text{M.I. of the rectangular parallelopiped about } GX' + \text{M.I. of total mass M at C.G. about OA}$$

$$= \frac{M}{3}(b^2 + c^2) + M(b^2 + c^2) = \frac{4}{3}M(b^2 + c^2).$$

2nd Method. Referred to O as origin.

We consider an element $\delta x\, \delta y\, \delta z$ at. The point P whose coordinates referred to the rectangular axes along OA, OB, OC are (x, y, z).

$\therefore$ moment of inertia of this element about OA $= (\rho\, \delta x\, \delta y \delta z)(y^2 + z^2)$.

Hence moment of inertia of the rectangular parallelopiped about OA

$$= \int_{x=0}^{2a}\int_{y=0}^{2b}\int_{z=0}^{2c} \rho(y^2 + z^2)\, dx\, dy\, dz = \rho\left[\frac{y^3}{3}xz + \frac{z^3}{3}xy\right]_{x=0,\ y=0,\ z=0}^{2a\ 2b\ 2c}$$

$$= \frac{\rho}{3}[8b^3 2a2c + 8c^3 2a2b] = \frac{32abc\rho}{3}(b^2 + c^2)$$

$$= \frac{4}{3}M(b^2 + c^2)\ [\because\ M = 8abc\rho].$$

Problem 12. *Find the moment of inertia of the triangle* ABC *about a perpendicular to its plane through* A. [MU 92H]

Solution: Let AK be a line through A and perpendicular to the plane of the triangle ABC of mass M and density ρ.

Let AE $= h$ be the height of the triangle.

Hence

$$M = \tfrac{1}{2}BC \cdot AD\rho = \tfrac{1}{2}ah\rho. \tag{1.42}$$

We consider an elementary strip $PQQ'P'$ of breadth δx at a distance $x = AN$ from A and parallel to BC.

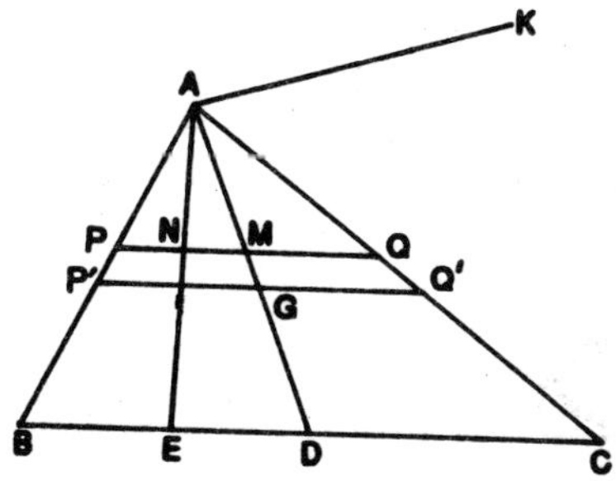

FIG 1.33

From A we draw AE perpendicular to BC.

Let AD and AE meet the strip at M and N.

Now from similar triangles, ΔAPQ and ΔABC, we have

$$\frac{PQ}{BC} = \frac{AN}{AE} = \frac{AM}{AD}, \text{ or, } \frac{PQ}{a} = \frac{x}{h} = \frac{AM}{AD}; \ \therefore\ PQ = \frac{ax}{h};\ AM = \frac{ADx}{h}.$$

In ΔADE

$$\begin{aligned} AD^2 &= AE^2 + DE^2 = AE^2 + (BE - BD)^2 \\ &= AE^2 + BE^2 + BD^2 - 2BE \cdot BD \\ &= AB^2 + BD^2 - 2BE \cdot BD \\ &= c^2 + \left(\frac{1}{2}BC\right)^2 - 2BE \cdot BD = c^2 + \frac{a^2}{4} - 2AB\cos B\frac{a}{2} \\ &= c^2 + \frac{a^2}{4} - \frac{c(a^2 + c^2 - b^2)a}{2ac} = c^2 + \frac{a^2}{4} - \frac{a^2 + c^2 - b^2}{2} \\ &= \frac{4c^2 + a^2 - 2a^2 - 2c^2 + 2b^2}{4}. \end{aligned}$$

$$AD^2 = \frac{2b^2 + 2c^2 - a^2}{4}. \tag{1.43}$$

Now mass of the elementary strip PQ

$$= \rho PQ\,\delta x = \rho\frac{ax}{h}\,\delta x.$$

Hence by the theorem of parallel axes,
Moment of inertia of the strip PQQ′P′ about the line AK

$$= \text{M.I. of the strip about the line parallel to AK through its C.G. M}$$
$$+ \text{M.I. of total mass of the strip placed at its C.G. M about AK.}$$

$$= \frac{1}{3}\frac{\rho a x}{h}\delta x \frac{1}{4}(PQ)^2 + \frac{\rho a x}{h}\delta x AM^2$$

$$= \frac{\rho a x}{h}\delta x\left[\frac{1}{12}\frac{a^2x^2}{h^2} + \frac{AD^2x^2}{h^2}\right] = \frac{\rho a x^3}{h^3}\delta x\left[\frac{a^2}{12} + AD^2\right]$$

$$= \frac{\rho a x^3}{h^3}\delta x\left[\frac{a^2}{12} + \frac{2b^2 + 2c^2 - a^2}{4}\right] \quad \text{[From equation (1.42)]}$$

$$= \frac{\rho a x^3}{h^3}\delta x\left[\frac{a^2 + 6b^2 + 6c^2 - 3a^2}{12}\right]$$

$$= \frac{\rho a x^3}{h^3}\delta x\frac{2(3b^2 + 3c^2 - a^2)}{12} = \frac{\rho a x^3(3b^2 + 3c^2 - a^2)}{6h^3}.$$

Hence moment of inertia of the ΔABC about AK

$$= \int_0^h \frac{\rho a x^3}{h^3}\frac{(3b^2 + 3c^2 - a^2)}{6}\,dx = \frac{\rho a(3b^2 + 3c^2 - a^2)}{6h^3}\frac{h^4}{4}$$

$$= \frac{\rho a h(3b^2 + 3c^2 - a^2)}{24} = \frac{2M(3b^2 + 3c^2 - a^2)}{24} \quad \text{[From equation (1.41)]}$$

$$= \frac{M(3b^2 + 3c^2 - a^2)}{12}.$$

Problem 13. *A solid body of density ρ is in the shape of the solid formed by the revolution of the cardioid $r = a(1 + \cos\theta)$ about the initial line; show that its moment of inertia about a straight line through the pole perpendicular to the initial line is $\frac{353}{105}\pi\rho a^5$.*

[VU 93H]

Solution: Let OX be the initial line and OY be the line perpendicular to it through the pole O.

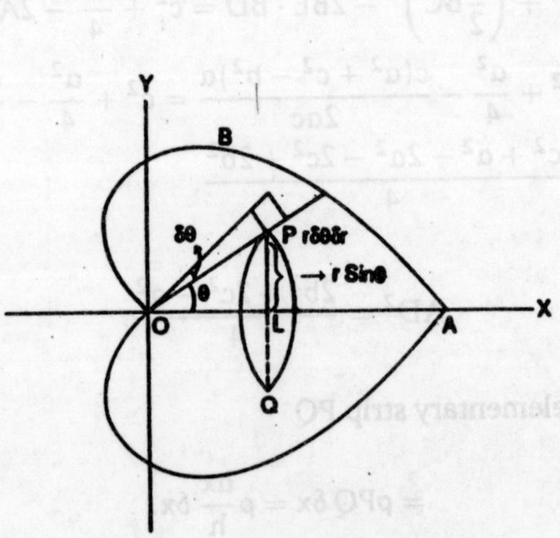

FIG 1.34

We have to find the moment of inertia of the body formed by the revolution of area OBA about OY.

We consider an elementary area $r\,\delta\theta\,\delta r$ at the point $P(r,\theta)$. This area when revolved about OA forms a circular ring of radius $r\sin\theta$.

Mass of the ring $= 2\pi r\sin\theta r\,\delta\theta\,\delta r\rho = 2\pi\rho r^2\sin\theta\,\delta\theta\,\delta r$, where ρ is the density of the solid body.

Moment of inertia of this elementary ring about a line parallel to OY

$$= (2\pi\rho r^2\sin\theta\,\delta\theta\,\delta r)\,\frac{r^2\sin^2\theta}{2}.$$

Hence by the theorem of parallel axis,

Moment of inertia of the elementary ring about OY

$$\begin{aligned}
&= \text{M.I. of the ring about a line parallel to OY}\\
&\quad + \text{M.I. of total mass of the ring at centre L about OY}\\
&= \pi\rho r^4\sin^3\theta\,\delta\theta\,\delta r + 2\pi\rho r^2\sin\theta\,\delta\theta\,\delta r(OL)^2\\
&= \pi\rho r^4\sin^3\theta\,\delta\theta\,\delta r + 2\pi\rho r^2\sin\theta\,\delta\theta\,\delta r r^2\cos^2\theta\\
&= \pi\rho r^4\sin\theta\,\delta\theta\,\delta r(\sin^2\theta + 2\cos^2\theta)\\
&= \pi\rho r^4\sin\theta\,\delta\theta\,\delta r(1+\cos^2\theta).
\end{aligned}$$

Thus M.I. of the solid of revolution about OY

$$\begin{aligned}
&= \pi\rho\int_{\theta=0}^{\pi}\int_{r=0}^{a(1+\cos\theta)} r^4(1+\cos^2\theta)\sin\theta\,d\theta\,dr\\
&= \pi\rho\int_{\theta=0}^{\pi}\frac{a^5}{5}(1+\cos\theta)^5(1+\cos^2\theta)\sin\theta\,d\theta\\
&= \frac{\pi\rho a^5}{5}\int_0^{\pi}(1+\cos\theta)^5(1+\cos^2\theta)\sin\theta\,d\theta.
\end{aligned}$$

Let $y = 1+\cos\theta$, $\therefore\ dy = -\sin\theta\,d\theta$.

When $\theta = \pi,\quad y = 0;$

$\qquad\theta = 0,\quad y = 2.$

Hence

$$\begin{aligned}
\text{M.I.} &= \frac{\pi\rho a^5}{5}\int_2^0 y^5\left[1+(y-1)^2\right](\ dy) - \frac{\pi\rho a^5}{5}\int_0^2 y^5\left[2-2y+y^2\right]dy\\
&= \frac{\pi\rho a^5}{5}\int_0^2\left(2y^5-2y^6+y^7\right)dy = \frac{\pi\rho a^5}{5}\left[\frac{2y^6}{6}-\frac{2y^7}{7}+\frac{y^8}{8}\right]_0^2\\
&= \frac{\pi\rho a^5}{5}\left[\frac{2^6}{3}-\frac{2\cdot 2^7}{7}+\frac{2^8}{8}\right] = \frac{\pi\rho a^5 2^6}{5}\left(\frac{1}{3}-\frac{2^2}{7}+\frac{2^2}{8}\right)\\
&= \frac{\pi\rho a^5 64}{5}\left[\frac{1}{3}-\frac{4}{7}+\frac{1}{2}\right] = \frac{\pi\rho a^5 64}{5}\left[\frac{14-24+21}{42}\right] = \frac{\pi\rho a^5 64}{5}\left[\frac{11}{42}\right]\\
&= \frac{\pi\rho a^5 11\cdot 32}{5\cdot 21} = \frac{\pi\rho a^5 352}{105}.
\end{aligned}$$

Problem 14. *Find the moment of inertia of a uniform solid sphere about any tangent line.* [TMBU 99H, 02H]

Solution: Let O be the centre of the solid sphere whose mass is M. Let a be the radius of the sphere. Then $M = \frac{4}{3}\pi a^2 \rho$, where ρ is the density.

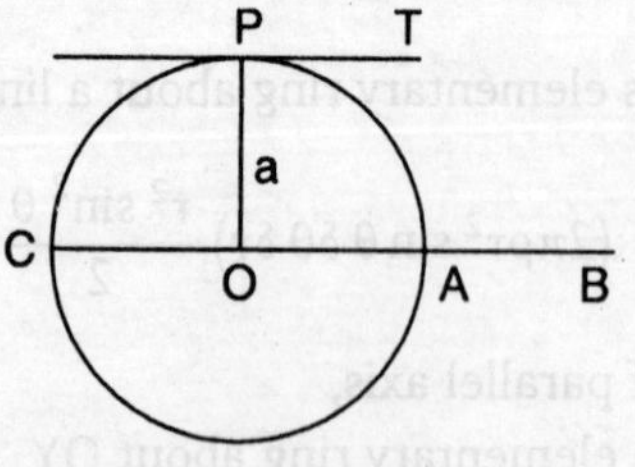

FIG 1.35

Let PT be the tangent at P to the sphere. Then by the theorem of parallel axes, moment of inertia of the sphere about PT

$$= \text{M.I. of the sphere about OB} + \text{M.I. of the whole mass M at O about PT}$$

$$= \frac{2}{5}Ma^2 + Ma^2 = \frac{7}{5}Ma^2.$$

Type III: Problems based on moment of inertia of a body about a line

Problem 15. *Show that the moment of inertia of the elliptic area of mass M and semi-axes a and b about a diameter of length r is $\frac{1}{4}M\frac{a^2b^2}{r^2}$.* [IAS 53, 73]

Solution: Let PP′ be the diameter of length r of an elliptic area of mass M and semi-axes a and b.

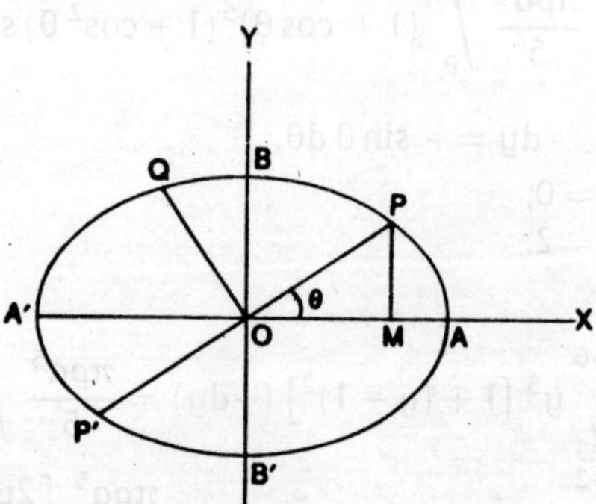

FIG 1.36

Equation of the ellipse is

$$\frac{x^2}{a^2} + \frac{y^2}{b^2} = 1. \tag{1.44}$$

If PP′ makes angle θ with OX, then coordinates of P are $(r\cos\theta, r\sin\theta)$.

Since P lies on equation (1.43)

$$\frac{r^2\cos^2\theta}{a^2}+\frac{r^2\sin^2\theta}{b^2}=1,\text{ or, } b^2\cos^2\theta+a^2\sin^2\theta=\frac{a^2b^2}{r^2}. \tag{1.45}$$

Now moment of inertia of the ellipse about $OX = A = \frac{1}{4}Mb^2$,
and moment of inertia of the ellipse about $OY = B = \frac{1}{4}Ma^2$.
F = P.I. of the ellipse about OX and OY = 0, since C.G. O is at (0, 0).
Hence moment of inertia of the ellipse about the diameter PP′

$$= A\cos^2\theta + B\sin^2\theta - F\sin 2\theta = \tfrac{1}{4}Mb^2\cos^2\theta + \tfrac{1}{4}Ma^2\sin^2\theta - 0$$

$$= \frac{M}{4}\frac{a^2b^2}{r^2} \quad \text{[From equation (1.45)]}$$

Problem 16. *Show that moment of inertia of a rectangle of mass M and sides 2a, 2b about a diagonal is* $\frac{2M}{3}\frac{a^2b^2}{a^2+b^2}$.

Solution: Let ABCD be a rectangle of mass M and AB = 2a, BC – 2b.

Then moment of inertia of rectangle about $OX = A = \frac{1}{3}Mb^2$ and moment of inertia of rectangle about $OY = B = \frac{1}{3}Ma^2$.

Also F = P.I. of rectangle about OX, OY = 0.

Let the diagonal AC make an angle θ with AB, then

$$\sin\theta = \frac{CM}{OC} = \frac{b}{\sqrt{a^2+b^2}} \text{ and } \cos\theta = \frac{OM}{OC} = \frac{a}{\sqrt{a^2+b^2}}.$$

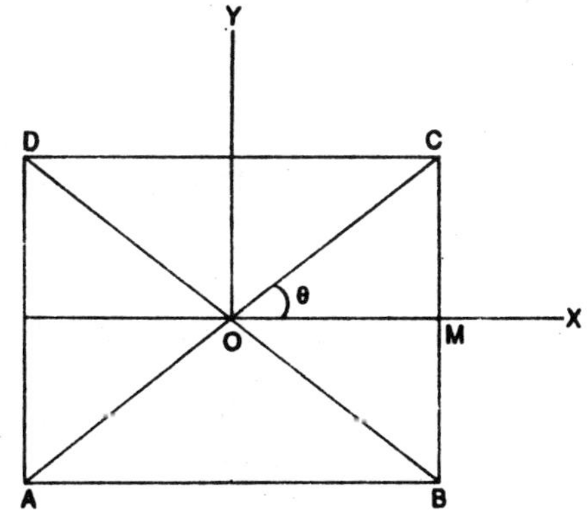

FIG 1.37

Now moment of inertia of the rectangle about AC

$$= A\cos^2\theta + B\sin^2\theta - F\sin 2\theta = \frac{Mb^2}{3}\frac{a^2}{a^2+b^2} + \frac{Ma^2}{3}\frac{b^2}{a^2+b^2}$$

$$= \frac{2Ma^2b^2}{3(a^2+b^2)}. \tag{1.46}$$

Problem 17. *If k_1, k_2 be the radii of gyration of an elliptic lamina about two conjugate diameters, then* $\frac{1}{k_1^2} + \frac{1}{k_2^2} = 4\left(\frac{1}{a^2} + \frac{1}{b^2}\right)$. [BPSC 82]

Solution: See Fig. 1.36.

Let OP and OP′ conjugate semi-diameters of lengths r_1 and r_2, then

$$Mk_1^2 = \frac{M}{4}\frac{a^2b^2}{r_1^2} \text{ and } Mk_2^2 = \frac{M}{4}\frac{a^2b^2}{r_2^2}.$$

$$\therefore \frac{1}{k_1^2} + \frac{1}{k_2^2} = \frac{4r_1^2}{a^2b^2} + \frac{4r_2^2}{a^2b^2} = \frac{4(r_1^2 + r_2^2)}{a^2b^2} = \frac{4(a^2 + b^2)}{a^2b^2} \quad \text{[by property of ellipse } r_1^2 + r_2^2 = a^2 + b^2]$$

$$= 4\left(\frac{1}{a^2} + \frac{1}{b^2}\right).$$

Type IV: Problems based on Momental Ellipsoid

Problem 18. *Show that the momental ellipsoid at the centre of an elliptic plate is* $\frac{x^2}{a^2} + \frac{y^2}{b^2} + z^2\left(\frac{1}{a^2} + \frac{1}{b^2}\right) =$ *constant.*

Solution: Let M be the mass of an elliptic plate of semi-axes a and b. Let the axes OX and OY be taken along the major and minor axes of the elliptic plate in its plane and the axis OZ perpendicular to its plane.

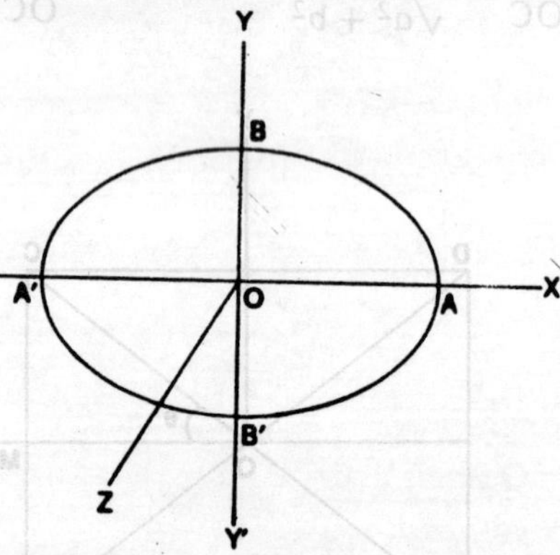

FIG 1.38

Then

$$A = \text{M.I. of the plate about OX} = \tfrac{1}{4}Mb^2$$
$$B = \text{M.I. of the plate about OY} = \tfrac{1}{4}Ma^2$$
$$C = \text{M.I. of the plate about OZ} = \tfrac{1}{4}M(b^2 + a^2).$$

If D, E and F are P.I.'s of the plate about the axes y, z; z, x and x, y, then $D = 0 = E = F$ as the plate is symmetrical about OX and OY.

Hence the equation of the momental ellipsoid at O is

$$Ax^2 + By^2 + Cz^2 - 2Dyz - 2Ezx - 2Fxy = \text{constant},$$

$$\text{or,} \quad \frac{M}{4}b^2x^2 + \frac{M}{4}a^2y^2 + \frac{M}{4}(a^2+b^2)z^2 = \text{constant},$$

$$\text{or,} \quad \frac{x^2}{a^2} + \frac{y^2}{b^2} + \left(\frac{1}{a^2} + \frac{1}{b^2}\right)z^2 = \text{constant}.$$

Problem 19. *Show that the momental ellipsoid at the centre of an ellipsoid is*

$$(b^2+c^2)x^2 + (c^2+a^2)y^2 + (a^2+b^2)z^2 = \textit{constant}.$$

[BBABU 94H, 96H, 01H, 03H; TMBU 92H, 96H; MU 95H]

Solution: Referred to principal axes as the reference, the equation of the ellipsoid is $\frac{x^2}{a^2} + \frac{y^2}{b^2} + \frac{z^2}{c^2} = 1$.

$$A = \text{M.I. about } OX = \tfrac{1}{5}M(b^2+c^2)$$
$$B = \text{M.I. about } OY = \tfrac{1}{5}M(c^2+a^2)$$
$$A = \text{M.I. about } OZ = \tfrac{1}{5}M(a^2+b^2).$$

Products of inertia $D = E = F = 0$.

Hence equation of the momental ellipsoid at the centre O is

$$Ax^2 + By^2 + Cz^2 - 2Dyz - 2Ezx - 2Fxy = \text{constant},$$

$$\text{or,} \quad \frac{M}{5}(b^2+c^2)x^2 + \frac{M}{5}(c^2+a^2)y^2 + \frac{M}{5}(a^2+b^2)z^2 = \text{constant},$$

$$\text{or,} \quad x^2(b^2+c^2) + y^2(c^2+a^2) + z^2(a^2+b^2) = \text{constant}$$

Problem 20. *Show that the equation of the momental ellipsoid at the corner of a cube of side $2a$ referred to its principal axes is $2x^2 + 11(y^2+z^2) = c$, where c is a constant.* **[VU 88H, MU 04H]**

Solution: Let the cube be as shown in figure. Let G be its C.G. Let O be a corner of the cube at which we have to determine the equation of the momental ellipsoid.

We take the line OX through G as the x-axis and two mutually perpendicular lines OY and OZ through O as the axes of y and z.

The coordinates of G referred to OX, OY, OZ as axes are $(a\sqrt{3}, 0, 0)$ and the P.I.'s of the cube about any two mutually perpendicular lines through G is zero.

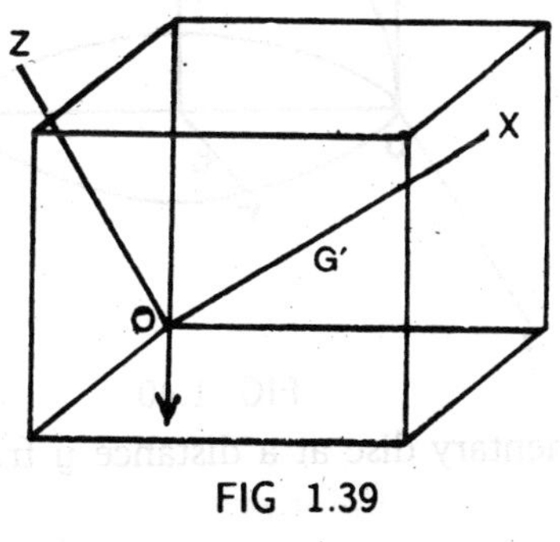

FIG 1.39

Hence D = E = F = 0.

Also OX, OY, OZ are the principal axes of the momental ellipsoid at O.

Since the moment of inertia of the cube about any axis through G $= \frac{2}{3}Ma^2$, hence

$$\begin{aligned} A &= \text{M.I. about OX} = \tfrac{2}{3}Ma^2. \\ B &= \text{M.I. about OY} \\ &= \text{M.I. about parallel axis through G} \\ &\quad + \text{M.I. of total mass M at G about OY} \\ &= \tfrac{2}{3}Ma^2 + M\cdot OG^2 = \tfrac{2}{3}Ma^2 + M3a^2 = \tfrac{11}{3}Ma^2. \end{aligned}$$

Similarly, C = moment of inertia about OZ $= \frac{11}{3}Ma^2$.

Hence equation of the momental ellipsoid at O is

$$Ax^2 + By^2 + Cz^2 - 2Dyz - 2Ezx - 2Fxy = \text{constant},$$

$$\text{or,}\quad \tfrac{2}{3}Ma^2x^2 + \tfrac{11}{3}Ma^2y^2 + \tfrac{11}{3}Ma^2z^2 = \text{constant},$$

$$\text{or,}\quad 2x^2 + 11((y^2 + z^2) = c,$$

where c is a constant.

Problem 21. *Show that the momental ellipsoid at a point on the rim of a semi-sphere is* $2x^2 + 7(y^2 + z^2) - \frac{15}{4}xz =$ *constant* **[VU 93H; MU 01H, 03H; BBABU 97H, 02H]**

Solution: Let O be the point on the rim of the hemisphere, at which equation of the momental ellipsoid is to be determined.

Let the diameter through O be taken as x-axis; a line in the plane of the base and perpendicular to OX be taken as y-axis, a line perpendicular to the plane of the base as z-axis.

∴ coordinates of the C.G. at G are $\left(a, 0, \frac{3a}{8}\right)$.

Let A, B, C be moments and D, E, F products of inertia about these axes.

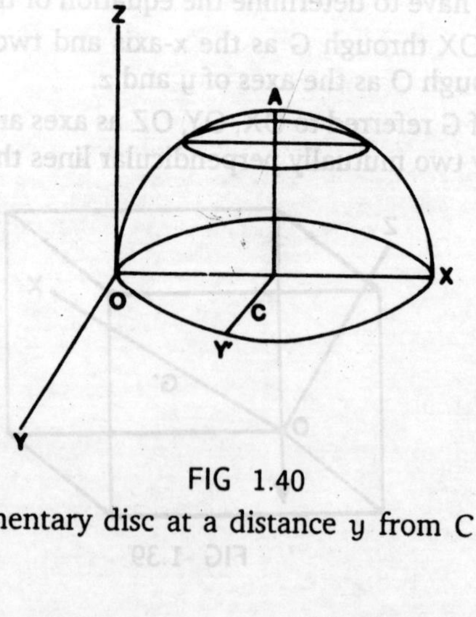

FIG 1.40

We consider an elementary disc at a distance y from C, radius of the disc $= \sqrt{a^2 - y^2}$.

Its moment of inertia about a radius parallel to OX is $\pi(a^2 - y^2)\delta y\rho\frac{(a^2-y^2)}{4}$.
Hence by the theorem of parallel axes,
Moment of inertia of the disc about OX

$$= \text{M.I. of the disc about parallel axis through its centre} + \text{M.I. of its mass at the centre about OX}$$

$$= \pi(a^2 - y^2)\,\delta y\rho\left[\frac{a^2-y^2}{4} + y^2\right]$$

$$= \frac{\pi}{4}(a^2 - y^2)\,\delta y\rho(a^2 + 3y^2).$$

Hence moment of inertia of the hemisphere about OX,

$$A = \int_0^a \frac{\pi\rho}{4}(a^4 + 2a^2y^2 - 3y^4)\,dy = \frac{\pi\rho}{4}\left[a^4y + \frac{2a^2y^3}{3} - \frac{3y^5}{5}\right]_0^a$$

$$= \frac{\pi\rho}{4}\left[a^4a + \frac{2a^2a^3}{3} - \frac{3a^5}{5}\right] = \frac{\pi\rho}{4}a^5\left[1 + \frac{2}{3} - \frac{3}{5}\right]$$

$$= \frac{\pi\rho}{4}a^5\left[\frac{15+10-9}{15}\right] = \frac{\pi\rho}{4}a^5\frac{16}{15} = \frac{4}{15}\pi\rho a^5.$$

Also

$$M = \frac{2}{3}\pi\rho a^3.$$

Hence

$$A = \frac{4}{15}a^5\frac{3M}{2a^3} = \frac{2}{5}Ma^2.$$

Also moment of inertia of the disc about an axis GA perpendicular to its plane is

$$= \pi(a^2 - y^2)\rho\frac{\delta y(a^2 - y^2)}{2}.$$

Hence moment of inertia of the hemisphere about GA

$$= \frac{\pi\rho}{2}\int_0^a (a^4 - 2a^2y^2 + y^4)\,dy = \frac{\pi\rho}{2}\left[a^4y - \frac{2a^2y^3}{3} + \frac{y^5}{5}\right]_0^a$$

$$= \frac{\pi\rho}{2}\left[a^5 - \frac{2}{3}a^5 + \frac{a^5}{5}\right] = \frac{\pi\rho}{2}a^5\left[1 - \frac{2}{3} + \frac{1}{5}\right]$$

$$= \frac{\pi\rho a^5}{2}\frac{15-10+3}{15} = \frac{\pi\rho a^5}{2}\frac{8}{15} = \frac{4\pi\rho a^5}{15}$$

$$= \frac{2}{5}Ma^2 \quad \left[\because M = \frac{2}{3}\pi a^3\rho\right]$$

$$\begin{aligned} C &= \text{M.I. about OZ} \\ &= \text{M.I. about GA} + \text{M.I. of M at G about OZ} \\ &= \tfrac{2}{5}Ma^2 + Ma^2 = \tfrac{7}{5}Ma^2. \end{aligned}$$

CY′ and OY are parallel axes at distance **CG** and **OG** from C.G. G.

$\therefore$ (moment of inertia about axis through O) $- M \cdot OG^2$ = moment of inertia about axis through C $- M \cdot CG^2$.

Hence

$$B = \text{M.I. about } OY = \tfrac{2}{5}Ma^2 + M(OG^2 - CG^2) = \tfrac{2}{5}Ma^2 + Ma^2 = \tfrac{7}{5}Ma^2.$$

Also y coordinate of G is zero.

$\therefore$ $D = 0$, $F = 0$ and $E = M\left(a \cdot \frac{3a}{8}\right) = \frac{3Ma^2}{8}$.

Hence the equation of momental ellipsoid at O is

$$Ax^2 + By^2 + Cz^2 - 2Dyz - 2Ezx - 2Fxy = \text{constant},$$

$$\text{or,} \quad Ma^2\left[\frac{2}{5}x^2 + \frac{7}{5}y^2 + \frac{7}{5}z^2 - 2 \cdot \frac{3}{8}zx\right] = \text{constant},$$

$$\text{or,} \quad 2x^2 + 7(y^2 + z^2) - \frac{15}{4}zx = \text{constant}.$$

Type V: Problems based on equi-momental bodies

Problem 22. *Obtain the moment of inertia for a triangular lamina ABC about a straight line through A (or any vertex) in the plane of the triangle.* [BBABU 93H]

Solution: Let m be the mass of the triangle. Then the triangle is equi-momental to three particles, each of mass $m/3$ placed at the mid-points D, E, F of the sides.

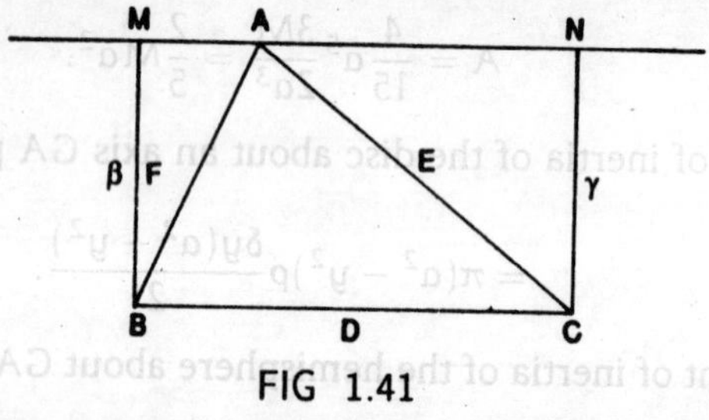

FIG 1.41

Let MN be any line through the vertex A and in the plane of the triangle ABC. Let β and γ be the distances of the vertices B and C from the straight line MN.

Distance of D from MN $= \frac{1}{2}(\beta + \gamma)$.

Distance of E from MN $= \frac{1}{2}\gamma$.

Distance of F from MN $= \frac{1}{2}\beta$.

Hence moment of inertia of the triangle ABC about MN

$$= \text{sum of M.I. of masses } \tfrac{m}{3} \text{ each at D, E, F about LM}$$

$$= \frac{m}{3}\left[\frac{(\beta+\gamma)^2}{4} + \frac{\gamma^2}{4} + \frac{\beta^2}{4}\right] = \frac{m}{12}\left[2(\beta^2 + \gamma^2 + \beta\gamma)\right]$$

$$= \frac{m}{6}\left(\beta^2 + \gamma^2 + \beta\gamma\right).$$

Problem 23. *If α, β, γ be the distances of the vertices of a triangle of mass m from any straight line in its plane, show that the moment of inertia of the triangle about this line is*

$$\frac{m}{6}\left[\alpha^2+\beta^2+\gamma^2+\beta\gamma+\gamma\alpha+\alpha\beta\right].$$

Hence deduce that if h be the distance of the centre of inertia of the triangle from the line, then moment of inertia about this line is

$$\frac{m}{12}(\alpha^2+\beta^2+\gamma^2+9h^2).$$

Solution: Let ABC be the triangular lamina of mass m. From A, B, C we draw AL, BM and CN perpendicular to a line GH in its plane, then $AL=\alpha$, $BM=\beta$, $CN=\gamma$.

If DP, EQ, FR are the perpendiculars from the middle points D, E, F of the sides on GH, then

$$\begin{aligned} DP &= \tfrac{1}{2}(BM+CN)=\tfrac{1}{2}(\beta+\gamma) \\ EQ &= \tfrac{1}{2}(AL+CN)=\tfrac{1}{2}(\alpha+\gamma) \\ FR &= \tfrac{1}{2}(AL+BM)=\tfrac{1}{2}(\alpha+\beta). \end{aligned}$$

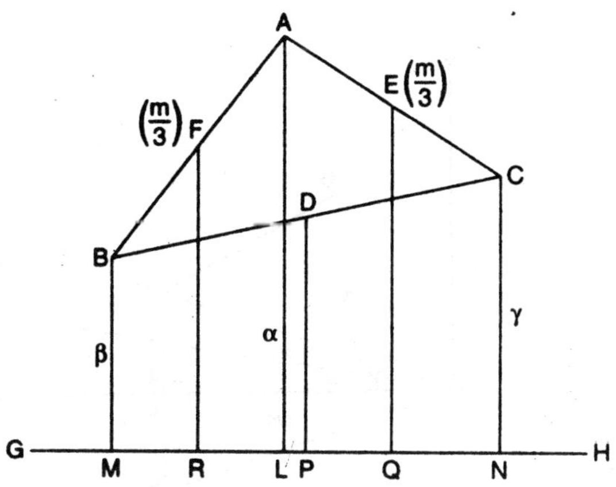

FIG 1.42

Since the triangle is equi-momental to the three particles, each of mass $m/3$ placed at the middle points D, E, F of the triangle, hence moment of inertia of the $\triangle ABC$ about GH

$$\begin{aligned} &= \text{sum of moment of inertia of masses } \frac{m}{3} \text{ each at } D, E, F \text{ about } GH \\ &= \frac{m}{3}(DP)^2+\frac{m}{3}(EQ)^2+\frac{m}{3}(FR)^2 \\ &= \frac{m}{12}\left[(\beta+\gamma)^2+(\gamma+\alpha)^2+(\alpha+\beta)^2\right] \\ &= \frac{m}{6}\left(\alpha^2+\beta^2+\gamma^2+\beta\gamma+\gamma\alpha+\alpha\beta\right). \qquad (1.47) \end{aligned}$$

If h is the distance of the centre of inertia of the $\triangle ABC$ from GH, then

$$h=\tfrac{1}{3}(\alpha+\beta+\gamma).$$

Hence from equation (1.47), moment of inertia of the $\triangle ABC$ about GH

$$= \frac{m}{12}[2\alpha^2 + 2\beta^2 + 2\gamma^2 + 2\beta\gamma + 2\gamma\alpha + 2\alpha\beta]$$

$$= \frac{m}{12}[\alpha^2 + \beta^2 + \gamma^2 + (\alpha + \beta + \gamma)^2]$$

$$= \frac{m}{12}[\alpha^2 + \beta^2 + \gamma^2 + 9h^2].$$

Problem 24. *Show that the moment of inertia of a regular polygon of n sides about any straight line through its centre is* $\frac{Mc^2}{24}\frac{2+\cos(2\pi/n)}{1-\cos(2\pi/n)}$, *where n is the number of sides and c is the length of each side.* **[IAS 90]**

Solution: Let ABCD be a regular polygon of n sides, each of length c, O being the centre of the polygon and lines OX (bisecting BC) and OY (perpendicular to OX) be taken in its plane as the axes of X and Y, respectively.

Let M be the mass of the polygon. So the polygon can be divided into n equal isosceles triangles each of mass M/n.

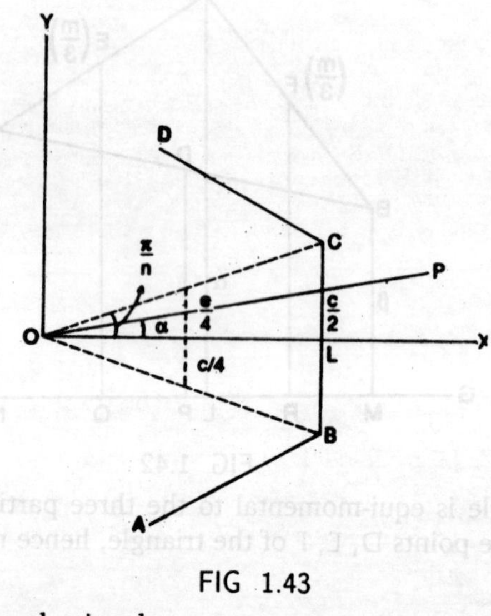

FIG 1.43

Let OBC be one of such triangle.
Also

$$\angle BOC = \frac{2\pi}{n}, \ \angle BOX = \angle COX = \frac{1}{2}\angle BOC = \frac{\pi}{n}.$$

Now the triangle OBC is equi-momental to three particles each of mass $\frac{1}{3}\frac{M}{n}$ at the middle points of its sides.

$\therefore$ moment of inertia of the triangle OBC about OX

$$= \frac{M}{3n}\cdot 0 + \frac{M}{3n}\left(\frac{c}{4}\right)^2 + \frac{M}{3n}\left(\frac{c}{4}\right)^2 = \frac{M}{3n}\cdot\frac{C^2}{8} = \frac{Mc^2}{24n} = A_1 \text{ (say)}.$$

Also moment of inertia of the triangle OBC about OY

$$= \frac{M}{3n}\left(\frac{1}{2}c\cot\frac{\pi}{n}\right)^2 + \frac{M}{3n}\left(\frac{1}{4}c\cot\frac{\pi}{n}\right)^2 + \frac{M}{3n}\left(\frac{c}{4}\cot\frac{\pi}{n}\right)^2 \qquad \left[\because OL = \frac{c}{2}\cot\frac{\pi}{n}\right]$$

$$= \frac{M}{3n}\left(c\cot\frac{\pi}{n}\right)^2\left[\frac{1}{4}+\frac{1}{16}+\frac{1}{16}\right] = \frac{M}{3n}\left(c\cot\frac{\pi}{n}\right)^2\left[\frac{1}{4}+\frac{1}{8}\right]$$

$$= \frac{M}{3n}\left(c\cot\frac{\pi}{n}\right)^2\frac{3}{8} = \frac{Mc^2}{8n}\cot^2\frac{\pi}{n} = B_1 \text{ (say)}$$

and P.I. of the triangle OBC about OX and OY $= 0 = F_1$ (say).

Let OP be a line inclined at an angle α to OX, then moment of inertia of $\triangle OBC$ about OP

$$= A_1\cos^2\alpha + B_1\sin^2\alpha - 2F_1\sin 2\alpha$$

$$= \left(\frac{M}{24n}c^2\right)\cos^2\alpha + \left(\frac{Mc^2}{8n}\cot^2\frac{\pi}{n}\right)\sin^2\alpha. \tag{1.48}$$

The moment of inertia of the other two triangles about OP are obtained by replacing α by $\alpha+\frac{2\pi}{n}$, $\alpha+\frac{4\pi}{n}$, in equation (1.48) successively, then,

Moment of inertia of the polygon about OP

$$= \frac{Mc^2}{24n}\left[\cos^2\alpha + \cos^2\left(\alpha+\frac{2\pi}{n}\right) + \cos^2\left(\alpha+\frac{4\pi}{n}\right) + \cdots + \text{to } n \text{ terms}\right]$$

$$+\frac{Mc^2}{8n}\cot^2\frac{\pi}{n}\left[\sin^2\alpha + \sin^2\left(\alpha+\frac{2\pi}{n}\right) + \sin^2\left(\alpha+\frac{4\pi}{n}\right) + \cdots + \text{to } n \text{ terms}\right]$$

$$= \frac{Mc^2}{24n}\frac{1}{2}\left[\{1+\cos 2\alpha\} + \left\{1+\cos\left(2\alpha+\frac{4\pi}{n}\right)\right\} + \cdots + \text{to } n \text{ terms}\right]$$

$$+\frac{Mc^2}{8n}\cot^2\frac{\pi}{n}\frac{1}{2}\left[\{1-\cos 2\alpha\} + \left\{1-\cos\left(2\alpha+\frac{4\pi}{n}\right)\right\} + \cdots + \text{to } n \text{ terms}\right]$$

$$= \frac{Mc^2}{24n}\frac{1}{2}\left[n + \left\{\cos 2\alpha + \cos\left(2\alpha+\frac{4\pi}{n}\right) + \cdots + \text{to } n \text{ terms}\right\}\right]$$

$$+\frac{Mc^2}{8n}\cot^2\frac{\pi}{n}\frac{1}{2}\left\{n - \left\{\cos 2\alpha + \cos\left(2\alpha+\frac{4\pi}{n}\right) + \cdots + \text{to } n \text{ terms}\right\}\right\}$$

$$= \frac{Mc^2}{24n}\frac{1}{2}\left[n + \frac{\sin\frac{n\cdot 4\pi}{2n}}{\sin\frac{4\pi}{2n}}\cos\left\{2\alpha + (n-1)\frac{4\pi}{2n}\right\}\right]$$

$$+\frac{Mc^2}{8n}\cot^2\frac{\pi}{n}\frac{1}{2}\left[n - \frac{\sin\frac{n\cdot 4\pi}{2n}}{\sin\frac{4\pi}{2n}}\cos\left\{2\alpha + (n-1)\frac{4\pi}{2n}\right\}\right]$$

$$= \frac{Mc^2}{24n}\frac{n}{2} + \frac{Mc^2}{8n}\cot^2\frac{\pi}{n}\frac{n}{2} = \frac{Mc^2}{48}\left[1+3\cot^2\frac{\pi}{n}\right]$$

$$= \frac{Mc^2}{48}\left[1+\frac{3\left(1+\cos\frac{2\pi}{n}\right)}{1-\cos\frac{2\pi}{n}}\right] = \frac{Mc^2}{48}\left[\frac{4+2\cos\frac{2\pi}{n}}{1-\cos\frac{2\pi}{n}}\right] = \frac{Mc^2}{24}\frac{\left(2+\cos\frac{2\pi}{n}\right)}{\left(1-\cos\frac{2\pi}{n}\right)}$$

Problem 25. *Show that there is a momental ellipse at the centre of inertia of a uniform triangle which touches the sides of the triangle at the middle points.* **[LNMU 93H]**

Solution: Let ABC be a triangle of mass m and G be its C.G. Let D, E and F be the middle points of the sides BC, CA and AB.

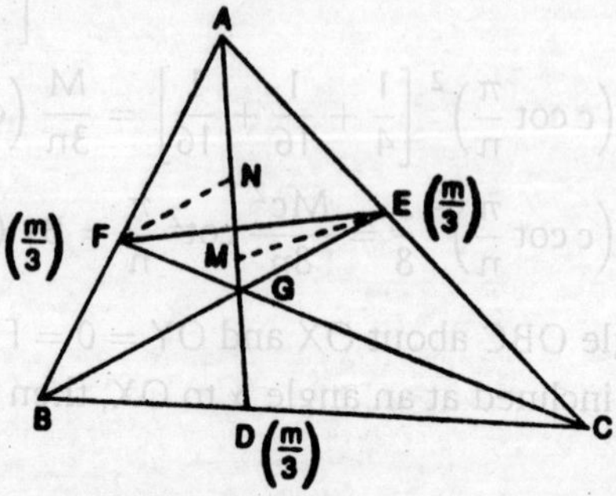

FIG 1.44

Now the momental ellipse at the centre of inertia G will pass through D, E and F of the moments of inertia of the triangle ABC about GD, GE and GF are equal to

$$\frac{Mk^4}{GD^2}, \frac{Mk^4}{GE^2} \text{ and } \frac{Mk^4}{GF^2}, \text{ respectively.}$$

Let the $\triangle ABC$ be replaced by three particles, each of mass $m/3$ placed at the middle points D, E, F of the sides of the triangle.

Then moment of inertia of the triangle ABC about AD

$$= \frac{m}{3}EM^2 + \frac{m}{3}FN^2 = \frac{m}{3}\left[\left(\frac{c}{2}\sin BAD\right)^2 + \left(\frac{b}{2}\sin CAD\right)^2\right]$$

$$= \frac{m}{12}\left[c^2\sin^2 BAD + b^2\sin^2 CAD\right]. \tag{1.49}$$

In $\triangle ABD$ and $\triangle ACD$,

$$\frac{\sin BAD}{\frac{a}{2}} = \frac{\sin B}{AD} \text{ and } \frac{\sin CAD}{\frac{a}{2}} = \frac{\sin C}{AD}$$

$$\therefore \sin BAD = \frac{a}{2}\frac{\sin B}{AD}, \quad \sin CAD = \frac{a}{2}\frac{\sin C}{AD}.$$

Therefore, from equation (1.48) moment of inertia of the $\triangle ABC$ about AD

$$= \frac{m}{12}\left[\frac{c^2a^2}{4}\frac{\sin^2 B}{AD^2} + \frac{b^2a^2}{4}\frac{\sin^2 C}{AD^2}\right] = \frac{a^2m}{AD^2 48}\left[c^2\sin^2 B + b^2\sin^2 C\right]$$

$$= \frac{m}{12AD^2}\left[\left(\frac{ac\sin B}{2}\right)^2 + \left(\frac{ab\sin C}{2}\right)^2\right] = \frac{m}{12AD^2}(\Delta^2 + \Delta^2)$$

$$= \frac{m\Delta^2}{6AD^2}, \text{ where } \Delta = \text{Area of } \triangle ABC$$

$$= \frac{m\Delta^2}{6 \cdot 9GD^2} = \left(\frac{m\Delta^2}{54}\right)\frac{1}{GD^2}.$$

Similarly, moment of inertia of the triangle about GE and GF are

$$\left(\frac{m\Delta^2}{54}\right)\frac{1}{GE^2} \text{ and } \left(\frac{m\Delta^2}{54}\right)\frac{1}{GF^2}.$$

Thus the momental ellipse at G will pass through D, E and F. Also GD is the diameter of the ellipse and bisects EF, so the tangent at P will be parallel to EF which is parallel to BC. Hence BC is tangent to the momental ellipse at D. Similarly, the sides CA and AB are tangents to the momental ellipse at E and F, respectively.

Type VI: Problems based on Principal Axes

Problem 26. *The lengths* AB *and* AD *of the sides of a rectangle* ABCD *are* $2a$ *and* $2b$. *Show that the inclination to* AB *of one of the principal axes at* A *is* $\frac{1}{2}\tan^{-1}\frac{3ab}{2(a^2-b^2)}$.

[MU 97H]

Solution: Let AB and AD be taken as x-axis and y-axis, respectively.

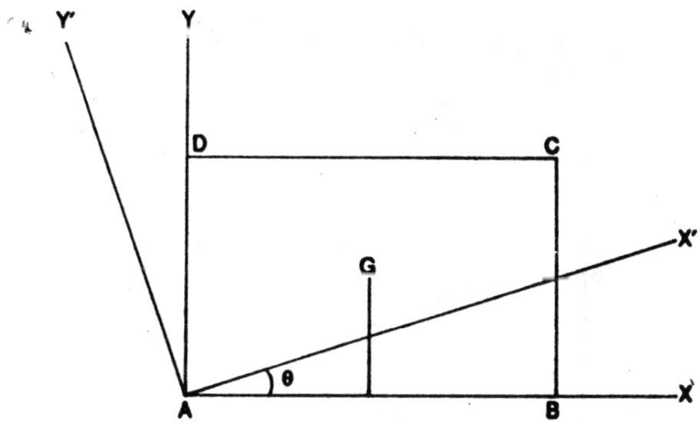

FIG 1.45

Then

$$\begin{aligned} A &= \text{M.I. of the rectangle about AB} \\ &= \text{M.I. of the rectangle about the axis parallel to AB through C.G. at G} \\ &\qquad + \text{M.I. of the whole mass M at G about AB} \\ &= \tfrac{1}{3}Mb^2 + Mb^2 = \tfrac{4}{3}Mb^2. \end{aligned}$$

Similarly,

$$\begin{aligned} B &= \text{M.I. of the rectangle about AD} = \tfrac{4}{3}Ma^2 \\ F &= \text{P.I. of the rectangle about AB and AD} \\ &= \text{P.I. of the rectangle about axes parallel to AX, AY through C.G. at G} \\ &\qquad + \text{P.I. of whole mass M at G about AB and AD} \\ &= 0 + Mab = Mab. \end{aligned}$$

If θ is the inclination of one principal axis to AB, then

$$\tan 2\theta = \frac{2F}{B-A} = \frac{2Mab}{\frac{4}{3}Ma^2 - \frac{4}{3}Mb^2} = \frac{3ab}{2(a^2-b^2)}.$$

Hence

$$\theta = \frac{1}{2}\tan^{-1}\frac{3ab}{2(a^2-b^2)}.$$

Problem 27. *Show that at the centre of a quadrant of an ellipse, the principal axis in its plane are inclined at an angle* $\frac{1}{2}\tan^{-1}\left(\frac{4}{\pi}\frac{ab}{(a^2-b^2)}\right)$ *to the axes.*

Solution: Let OAB be the quadrant of an ellipse

$$\frac{x^2}{a^2}+\frac{y^2}{b^2}=1.$$

We consider an element $\delta x\,\delta y$ at the point $P(x, y)$ of the quadrant.

Now A = M.I. of the quadrant about OX

$$= \int_0^a\int_0^y \rho y^2\,dx\,dy = \rho\int_0^a \frac{y^3}{3}\,dx.$$

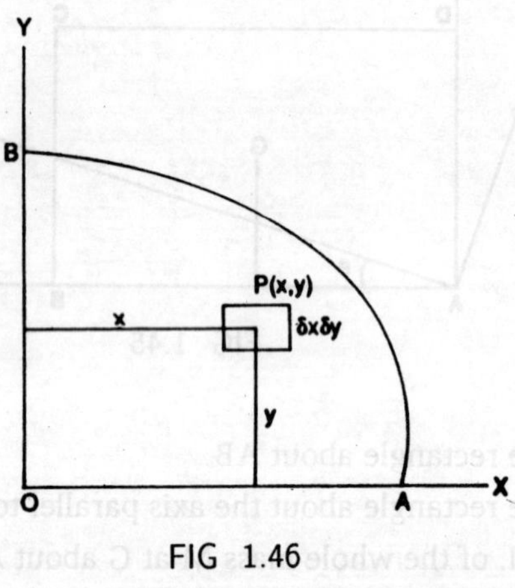

FIG 1.46

Putting

$$\begin{aligned} x &= a\cos\theta, & y &= b\sin\theta, & dx &= -a\sin\theta\,d\theta \\ x &= a, & \theta &= 0 \\ x &= 0, & \theta &= \tfrac{\pi}{2}. \end{aligned}$$

Hence

$$\begin{aligned} A &= \frac{\rho}{3}\int_{\pi/2}^{0} b^3\sin^3\theta(-a\sin\theta\,d\theta) = \frac{\rho}{3}ab^3\int_0^{\pi/2}\sin^4\theta\,d\theta \\ &= \frac{\rho}{3}ab^3\frac{4-1}{4}\frac{4-3}{4-2}\frac{\pi}{2} = \frac{\rho}{3}ab^3\frac{3}{4}\frac{1}{2}\frac{\pi}{2} \\ &= \frac{\pi\rho ab^3}{16} = \frac{4Mb^3}{16b} = \frac{Mb^2}{4} \quad (\because\ M = \pi ab\rho). \end{aligned}$$

Similarly,

$$\begin{aligned} B &= \tfrac{1}{4}Ma^2 \\ F &= \text{P.I. of the quadrant about OX and OY} \\ &= \int_0^a \int_0^y \rho xy\,dx\,dy = \rho \int_0^a x\frac{y^2}{2}\,dx \\ &= \frac{\rho}{2}\int_0^{\pi/2} a\cos\theta b^2\sin^2\theta a\sin\theta\,d\theta \\ &= \frac{\rho a^2 b^2}{2}\int_0^{\pi/2} \sin^3\theta\cos\theta\,d\theta. \end{aligned}$$

Let $z = \sin\theta$, $\therefore\ dz = \cos\theta\,d\theta$.
When $\theta = \pi/2,\ z = 1,$
$\theta = 0,\ z = 0.$

Hence

$$F = \frac{\rho a^2 b^2}{2}\int_0^1 z^3\,dz = \frac{\rho a^2 b^2}{2\cdot 4} = \frac{4M}{\pi ab}\frac{a^2b^2}{2\cdot 4} = \frac{M}{2\pi}ab \quad (\because\ M = \pi ab\rho).$$

If θ be the inclination of the principal axis with OX, then

$$\tan 2\theta = \frac{2F}{B-A} = \frac{\frac{2M}{2\pi}ab}{\frac{M}{4}(a^2-b^2)} = \frac{4ab}{\pi(a^2-b^2)}.$$

Hence

$$\theta = \frac{1}{2}\tan^{-1}\left[\frac{4ab}{\pi(a^2-b^2)}\right].$$

Problem 28. *A wire is in the form of a semi-circle of radius a. Show that at an end of its diameter the principal axes in its plane are inclined to the diameter at angles* $\frac{1}{2}\tan^{-1}\frac{4}{\pi}$ *and* $\frac{\pi}{2} + \frac{1}{2}\tan^{-1}\frac{4}{\pi}$. [IAS 73]

Solution: Let O be the centre and OA be the diameter of a semi-circular wire of radius a. Let the axes OX and OY be taken along and perpendicular to the diameter OA.

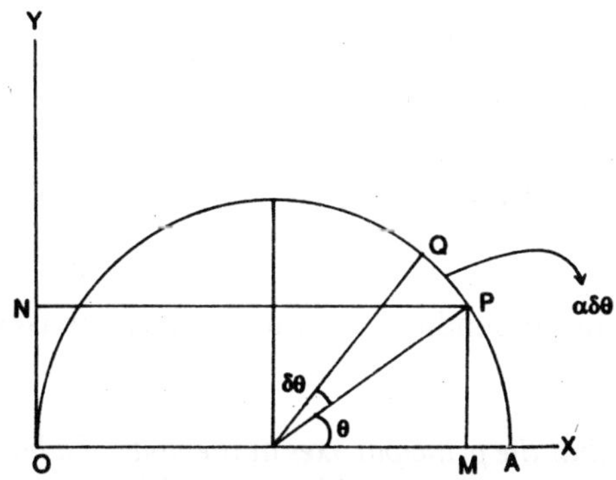

FIG 1.47

We consider an elementary area $a\,\delta\theta$ at P.
Mass of the elementary area $= \rho a\,\delta\theta$, where ρ is the density.
If M be the mass of semi-circular wire, then $M = \frac{1}{2}(2\pi a\rho) = \pi a\rho$.
Now

$$\begin{aligned} A &= \text{M.I. about OX} \\ &= \int_0^{\pi} a\rho\, d\theta PM^2 = a\rho \int_0^{\pi} (a\sin\theta)^2 d\theta \\ &= a^3\rho \int_0^{\pi} \frac{1-\cos 2\theta}{2} d\theta = \frac{a^3\rho}{2}\left[\theta - \frac{\sin 2\theta}{2}\right]_0^{\pi} \\ &= \frac{a^3\rho}{2}[\pi - 0] = \frac{\pi a^3\rho}{2} = \frac{M}{2a}a^3 = \frac{Ma^2}{2} (\because M = \pi a\rho). \end{aligned}$$

Now

$$\begin{aligned} B &= \text{M.I. about OY} = \int_0^{\pi} \rho a\, d\theta PN^2 = \rho a \int_0^{\pi} (a + a\cos\theta)^2 d\theta \\ &= \rho a^2 \int_0^{\pi} [1 + 2\cos\theta + \cos^2\theta]\, d\theta = \rho a^3 \int_0^{\pi} \left[1 + 2\cos\theta + \frac{1+\cos 2\theta}{2}\right] d\theta \\ &= \rho a^3 \int_0^{\pi} \left(\frac{3}{2} + 2\cos\theta + \frac{\cos 2\theta}{2}\right) d\theta = \rho a^3 \left[\frac{3}{2}\pi + \left(2\sin\theta + \frac{\sin 2\theta}{4}\right)_0^{\pi}\right] \\ &= \rho a^3\left(\frac{3\pi}{2} + 0\right) = \frac{3\pi\rho a^3}{2} = \frac{3}{2}a^3\frac{M}{a} = \frac{3}{2}Ma^2. \end{aligned}$$

F = P.I. of the wire about OX and OY

$$\begin{aligned} &= \int_0^{\pi} PM \cdot PN a\rho\, d\theta = a\rho \int_0^{\pi} a\sin\theta(a + a\cos\theta)\, d\theta \\ &= a^3\rho \int_0^{\pi} \left(\sin\theta + \frac{1}{2}\sin 2\theta\right) d\theta = a^3\rho\left[-\cos\theta - \frac{\cos 2\theta}{4}\right]_0^{\pi} \\ &= -a^3\rho\left[-1 + \frac{1}{4} - \left(1 + \frac{1}{4}\right)\right] = -a^3\rho[-2] \\ &= 2a^3\rho = 2a\rho a^2 = \frac{2a^2 M}{\pi} \quad (\because M = \pi a\rho). \end{aligned}$$

Thus if θ be the inclination of a principal axis to the diameter OX, then

$$\theta = \frac{1}{2}\tan^{-1}\frac{2F}{B-A} = \frac{1}{2}\tan^{-1}\frac{\frac{2\cdot 2Ma^2}{\pi}}{\frac{3}{2}Ma^2 - \frac{Ma^2}{2}} = \frac{1}{2}\tan^{-1}\frac{4Ma^2}{\pi Ma^2} = \frac{1}{2}\tan^{-1}\frac{4}{\pi}.$$

The other axis being perpendicular to the above axis makes an angle $\frac{\pi}{2} + \frac{1}{2}\tan^{-1}\frac{4}{\pi}$ with OX.

Problem 29. *Show that the principal axes at the node of a half loop of the lemniscate $r^2 = a^2\cos 2\theta$ are inclined to the initial line at angles $\frac{1}{2}\tan^{-1}\frac{1}{2}$ and $\frac{\pi}{2} + \frac{1}{2}\tan^{-1}\frac{1}{2}$.*

[IAS 59, 75]

Solution: We consider an elementary area $r\,\delta\theta\,\delta r$ at $P(r, \theta)$.

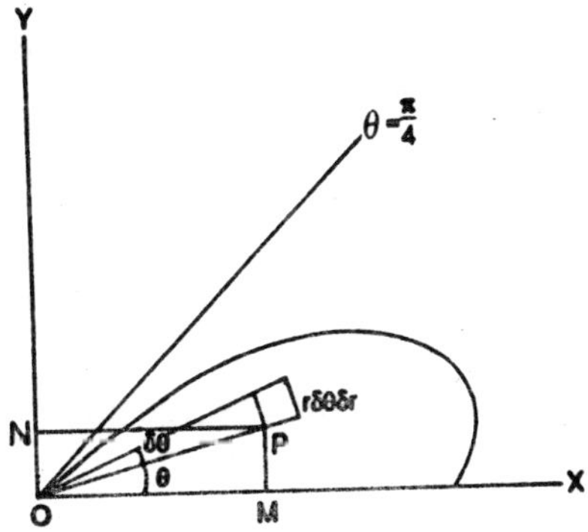

FIG 1.48

Mass of the elementary area $= \rho r\,\delta\theta\,\delta r$.
Its moment of inertia about OX

$$= \rho r\,\delta\theta\,\delta r (r\sin\theta)^2 = \rho r^3 \sin^2\theta\,\delta\theta\,\delta r.$$

Its moment of inertia about OY

$$= \rho r\,\delta\theta\,\delta r (r\cos\theta)^2 = \rho r^3 \cos^2\theta\,\delta r\,\delta\theta.$$

Hence A = moment of inertia of half loop of the lemniscate about OX

$$= \int_{\theta=0}^{\pi/4}\int_{r=0}^{a\sqrt{\cos 2\theta}} \rho r^3 \sin^2\theta\,d\theta\,dr = \rho\int_{\theta=0}^{\pi/4} \frac{a^4}{4}\cos^2 2\theta \sin^2\theta\,d\theta$$

$$= \frac{\rho a^4}{4}\int_{\theta=0}^{\pi/4} \frac{\cos^2 2\theta(1-\cos 2\theta)}{2}\,d\theta = \frac{\rho a^4}{8}\int_{\theta=0}^{\pi/4} (\cos^2 2\theta - \cos^3 2\theta)\,d\theta.$$

Let $2\theta = \alpha$; $2d\theta = d\alpha$.
When $\theta = \frac{\pi}{4}$, $\alpha = \frac{\pi}{2}$,
$\theta = 0$, $\alpha = 0$

Thus

$$A = \frac{\rho a^4}{8}\int_0^{\pi/2} (\cos^2\alpha - \cos^3\alpha)\frac{d\alpha}{2} = \frac{\rho a^4}{16}\left[\frac{2-1}{2}\frac{\pi}{2} - \frac{3-1}{3}\right]$$

$$= \frac{\rho a^4}{16}\left(\frac{\pi}{4} - \frac{2}{3}\right) = \frac{\rho a^4}{192}(3\pi - 8).$$

B = moment of inertia of half loop of the lemniscate about OY

$$= \int_{\theta=0}^{\pi/4}\int_{r=0}^{a\sqrt{\cos 2\theta}} \rho r^3 \cos^2\theta\,d\theta\,dr = \rho\int_{\theta=0}^{\pi/4} \frac{a^4}{4}\cos^2 2\theta \cos^2\theta\,d\theta$$

$$= \frac{\rho a^4}{4}\int_0^{\pi/4} \frac{\cos^2 2\theta(1+\cos 2\theta)}{2}\,d\theta = \frac{\rho a^4}{8}\int_0^{\pi/4} (\cos^2 2\theta + \cos^3 2\theta)\,d\theta$$

$$= \frac{\rho a^4}{8.2}\int_0^{\pi/2} (\cos^2\alpha + \cos^3\alpha)\,d\theta = \frac{\rho a^4}{8.2}\left[\frac{2-1}{2}\frac{\pi}{2} + \frac{3-1}{3}\right]$$

$$= \frac{\rho a^4}{8.2}\left(\frac{\pi}{4} + \frac{2}{3}\right) = \frac{\rho a^4(3\pi+8)}{192}.$$

F = P.I. of half loop of the lemniscate about OX and OY

$$= \int_0^{\pi/4}\int_0^{a\sqrt{\cos 2\theta}} \rho r\, d\theta\, dr (r\sin\theta)(r\cos\theta)$$

$$= \rho\int_0^{\pi/4}\int_0^{a\sqrt{\cos 2\theta}} r^3 \sin\theta\cos\theta\, d\theta\, dr$$

$$= \rho\int_0^{\pi/4} \frac{a^4}{4}\cos^2 2\theta\sin\theta\cos\theta\, d\theta$$

$$= \frac{\rho a^4}{4.2}\int_0^{\pi/4} \cos^2 2\theta \sin 2\theta\, d\theta.$$

Let $y = \cos 2\theta$, $dy = -2\sin 2\theta\, d\theta$.

$\therefore$ when $\theta = \frac{\pi}{4}$, $y = 0$;
$\theta = 0$, $y = 1$.

Hence

$$F = \frac{\rho a^4}{8}\int_1^0 \frac{y^2(-dy)}{2} = \frac{\rho a^4}{16}\int_0^1 y^2\, dy = \frac{\rho a^4}{16}\cdot\frac{1}{3} = \frac{\rho a^4}{48}.$$

If the principal axis at O makes an angle ϕ with OX, then

$$\tan 2\phi = \frac{2F}{B-A} = \frac{\frac{2\rho a^2}{48}}{\frac{\rho a^4}{192}[3\pi + 8 - (3\pi - 8)]} = \frac{192}{24\times 16} = \frac{1}{2}.$$

Hence $\phi = \frac{1}{2}\tan^{-1}\frac{1}{2}$.

The other principal axis being at right angle to the former makes an angle $\frac{\pi}{2} + \frac{1}{2}\tan^{-1}\frac{1}{2}$ with OX.

Problem 30. *If the vertical angle of the cone is 90°, the point at which a generator is a principal axis divides the generator in the ratio 3 : 7.* [IAS 74]

Solution: Let h be the height of the cone of vertical angle 90°.

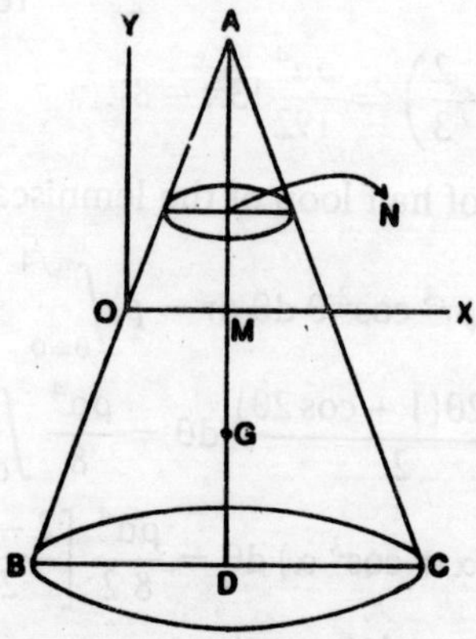

FIG 1.49

Let the generator AB be the principal axis of the cone at the point O. ΔABC is a section of the cone through the generator AB and the axis of the cone, taking OX and OY the axis of x and axis of y, perpendicular to AD and parallel to AD, respectively in this section and a line perpendicular to this section through O as z-axis.

Since the z coordinate of C.G. is zero, therefore $D = 0$, $E = 0$ and as such z-axis is the principal axes at O. The other two principal axis at O are the generator AB and the line through O and perpendicular to generator AB in the above section of the cone.

Since AB makes an angle $45°$ with OX, i.e., $\theta = 45°$, $\tan 2\theta = \infty$; $\frac{2F}{B-A} = \infty$, $B - A = 0$. $\therefore\ B = A$.

We consider an elementary circular disc of width δx at a distance x from the vertex A and perpendicular to the axis AD, i.e., $AN = x$.

$\therefore$ radius of the disc $= PN = x\tan 45° = x$.

Mass of the elementary disc $= \pi\rho x^2\,\delta x$

Now A = moment of inertia of the disc about OX

$$\begin{aligned}
&= \text{M.I. of the disc about parallel radius through N} \\
&\quad + \text{M.I. of the total mass at N} \\
&= \tfrac{1}{4}PN^2\pi\rho x^2\,\delta x + MN^2\rho\pi x^2\,\delta x \\
&= \left[\tfrac{1}{4}x^2 + (AM - AN)^2\right]\pi\rho x^2\,\delta x \\
&= \left[\tfrac{1}{4}x^2 + (AM - x)^2\right]\pi\rho x^2\,\delta x.
\end{aligned}$$

$\therefore$ A = moment of inertia of the cone about OX

$$= \int_0^h \left[\frac{1}{4}x^2 + (AM - x)^2\right]\pi\rho x^2\,dx = \pi\rho\int_0^h \left[\frac{5}{4}x^2 + AM^2 - 2AMx\right]x^2\,dx$$

$$= \pi\rho\int_0^h \left[\frac{5}{4}x^4 + AM^2x^2 - 2AMx^3\right]dx = \pi\rho\left[\frac{5}{4}\frac{h^5}{5} + \frac{AM^2h^3}{3} - \frac{2AMh^4}{4}\right]$$

$$= \pi\rho h^3\left[\frac{h^2}{4} + \frac{AM^2}{3} - \frac{h}{2}AM\right] = \frac{\pi\rho h^3}{12}\left[3h^2 - 6hAM + 4AM^2\right].$$

Also moment of inertia of the disc about OY

$$\begin{aligned}
&= \tfrac{1}{2}PN^2\pi\rho x^2\,\delta x + OM^2\pi\rho x^2\,\delta x \\
&= \left(\tfrac{1}{2}x^2 + AM^2\right)\pi\rho x^2\,\delta x \quad (\because\ OM = AM).
\end{aligned}$$

$\therefore$ B = moment of inertia of the cone about OY

$$= \int_0^h \left(\frac{x^2}{2} + AM^2\right)\pi\rho x^2\,dx = \pi\rho\int_0^h \left(\frac{x^4}{2} + AM^2x^2\right)dx$$

$$= \pi\rho\left[\frac{h^5}{10} + \frac{AM^2h^3}{3}\right] = \frac{\pi\rho h^3}{30}\left[3h^2 + 10AM^2\right].$$

But A = B

$$\frac{\pi\rho h^3}{12}(3h^2 - 6hAM + 4AM^2) = \frac{\pi\rho h^3}{30}(3h^2 + 10AM^2),$$

$$\text{or,}\quad 5(3h^2 - 6hAM + 4AM^2) = 2(3h^2 + 10AM^2),$$

$$\text{or,}\quad 9h^2 = 30hAM, \text{ or, } AM = \frac{3}{10}h.$$

From similar triangles AOM and ABD,

$$\frac{AO}{AM} = \frac{AB}{AD}, \quad \text{or,} \quad \frac{AO}{\frac{3h}{10}} = \frac{AB}{h},$$

$$\text{or,}\quad 10AO = 3AB,$$

$$\text{or,}\quad 10AO = 3(AO + OB),$$

$$\text{or,}\quad 7AO = 3OB, \quad \therefore \frac{AO}{OB} = \frac{3}{7}.$$

Problem 31. *A uniform lamina is bounded by a parabolic arc, of latus rectum 4a, and a double ordinate at a distance b from the vertex. If $b = \frac{a}{3}(7 + 4\sqrt{7})$, show that two of the principal axes at the end of a latus rectum are the tangent and normal there.*

Solution: Let the equation of the parabola be $y^2 = 4ax$.

Let LSL′ be its latus rectum. Thus coordinates of L are $(a, 2a)$.

Hence equation of tangent and normal at L are given by

$$y \cdot 2a = 2a(x + a), \quad \text{or,} \quad y - x - a = 0 \tag{1.50}$$

$$\text{and } y - 2a = -\frac{2a}{2a}(x - a), \quad \text{or,} \quad y - 2a = -x + a,$$

$$\text{or,}\quad x + y = 3a,$$

$$\text{or,}\quad x + y - 3a = 0. \tag{1.51}$$

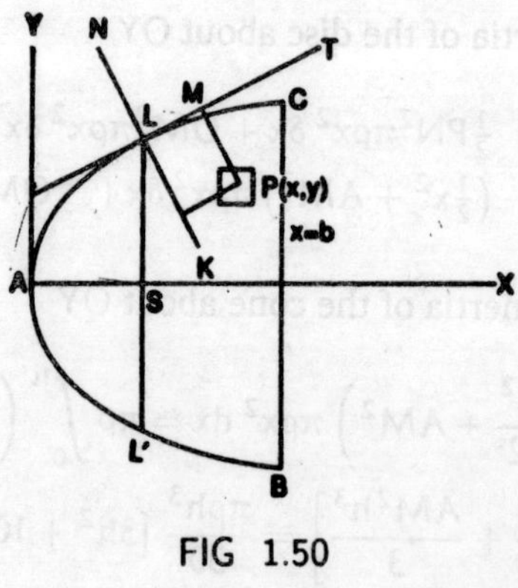

FIG 1.50

We take an elementary area $\delta x\, \delta y$ at the point $P(x, y)$ of the lamina.

$$\begin{aligned} PM &= \text{Length of perpendicular from P to equation (1.50)} \\ &= \frac{y - x - a}{\sqrt{2}} \\ PK &= \text{Length of perpendicular from P to equation (1.51)} \\ &= \frac{x + y - 3a}{\sqrt{2}}. \end{aligned}$$

Hence P.I. of the element about tangent and normal at L

$$\begin{aligned} &= \frac{(\rho\, \delta x\, \delta y)(y - x - a)(y + x - 3a)}{\sqrt{2}\sqrt{2}} \\ &= \frac{\rho}{2}\left[y^2 + yx - 3ay - xy - x^2 + 3ax - ay - ax + 3a^2\right] \delta x\, \delta y \\ &= \frac{\rho}{2}(y^2 - x^2 + 2ax - 4ay + 3a^2)\, \delta x\, \delta y. \end{aligned}$$

Hence P.I. of the lamina about the tangent and normal at L

$$\begin{aligned} &= \frac{\rho}{2}\int_{x=0}^{b}\int_{y=-2\sqrt{ax}}^{2\sqrt{ax}} \left[(y^2 - 4ay) + (3a^2 + 2ax - x^2)\right] dx\, dy \\ &= \frac{\rho}{2}\int_0^b \left[\frac{y^3}{3} - 2ay^2 + (3a^2 + 2ax - x^2 y)\right]_{-2\sqrt{ax}}^{2\sqrt{ax}} dx \\ &= \frac{\rho}{2}\int_0^b \left[\frac{2}{3}8ax\sqrt{ax} - 0 + (3a^2 + 2ax - x^2 y)4\sqrt{ax}\right] dx \\ &= \frac{\rho}{2}4\int_0^b \left[\frac{4ax\sqrt{ax}}{3} + (3a^2 + 2ax - x^2)\sqrt{ax}\right] dx \\ &= 2\rho\int_0^b \left[\frac{4a\sqrt{a}}{3}x^{3/2} + \sqrt{a}(3a^2\sqrt{x} + 2ax^{3/2} - x^{5/2}\right] dx \\ &= 2\rho\left[\frac{4a\sqrt{a}}{3}\frac{2}{5}x^{5/2} + 3a^2\sqrt{a}\frac{2}{3}x^{3/2} + 2a\sqrt{a}\frac{2}{5}x^{5/2} - \sqrt{a}\frac{2}{7}x^{7/2}\right]_0^b \\ &= 2\rho\left[\frac{8a\sqrt{a}}{15}b^{5/2} + 2a^2\sqrt{a}b^{3/2} + \frac{4a\sqrt{a}}{5}b^{5/2} - \frac{2}{7}\sqrt{a}b^{7/2}\right] \\ &= 2\rho\sqrt{a}b^{3/2}\left[\frac{8ab}{15} + 2a^2 + \frac{4ab}{5} - \frac{2}{7}b^2\right] \\ &= 4\rho\sqrt{a}b^{3/2}\left[\frac{4ab}{15} + a^2 + \frac{2ab}{5} - \frac{b^2}{7}\right] \\ &= \frac{4\rho\sqrt{a}b^{3/2}}{105}\left[28ab + 105a^2 + 42ab - 15b^2\right] \\ &= \frac{4\rho\sqrt{a}b^{3/2}}{105}\left[105a^2 + 70ab - 15b^2\right] \\ &= \frac{4\rho\sqrt{a}b^{3/2}}{21}\left[21a^2 + 14ab - 3b^2\right]. \qquad (1.52) \end{aligned}$$

The tangent and normal at L will be the principal axes of the area ABC. If P.I. of the lamina about these vanishes, then from equation (1.52) $21a^2 + 14ab - 3b^2 = 0$,

$$a = \frac{-14b \pm \sqrt{196b^2 + 252b^2}}{42} = \frac{-14b \pm b\sqrt{448}}{42} = \frac{-14b \pm b\sqrt{64 \times 7}}{42}.$$

$$\therefore\ a = \frac{(-14 \pm 8\sqrt{7})b}{42}.$$

Neglecting negative sign as b cannot be negative,

$$\frac{a}{b} = \frac{8\sqrt{7} - 14}{42},$$

$$\text{or,}\quad \frac{b}{a} = \frac{42}{2(4\sqrt{7} - 7)} = \frac{21}{4\sqrt{7} - 7} \times \frac{4\sqrt{7} + 7}{4\sqrt{7} + 7}$$

$$= \frac{21(4\sqrt{7} + 7)}{112 - 49} = \frac{21(4\sqrt{7} + 7)}{63} = \frac{(4\sqrt{7} + 7)}{3}.$$

Type VII: Problems based on Principal Moments

Problem 32. *Show that for a thin hemispherical shell of radius a and mass M, the principal moments of inertia at the centre of gravity are* $\frac{83}{320}Ma^2$, $\frac{83}{320}Ma^2$, $\frac{2}{3}Ma^2$.

Solution: Let G be the centre of gravity of a hemispherical solid of radius a and mass M.

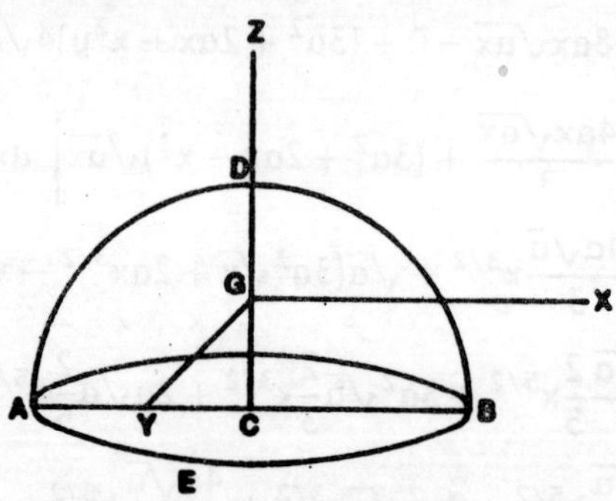

FIG 1.51

If C is the centre and CD the central radius of the hemisphere, then $CG = \frac{3a}{8}$.

We take GX and GY, the axes through G and parallel to the plane base as x-axis and y-axis, respectively and GZ, the central radius as z-axis, then

$$A = \text{M.I. about GX} = \text{M.I. about AB} - M.CG^2$$

$$= \frac{2}{5}Ma^2 - M\left(\frac{3}{8}a\right)^2 = Ma^2\left[\frac{2}{5} - \frac{9}{64}\right]$$

$$= Ma^2\frac{(128 - 45)}{320} = \frac{Ma^2 \cdot 83}{320} = \frac{83}{320}Ma^2.$$

Similarly,

$$B = \text{M.I. about } GY = \frac{83}{320}Ma^2; \ C = \text{M.I. about } CZ = \frac{2}{3}Ma^2.$$

Now coordinates of C are $(0, 0, -\frac{3a}{8})$

$$\begin{aligned} D &= \text{P.I. about } GY, GZ = \text{P.I. about parallel lines } CB, CE \\ &\quad - \text{P.I. of } M \text{ at } C \text{ about } GY, GZ \\ &= 0 - M.0\left(-\frac{3a}{8}\right) = 0. \end{aligned}$$

Similarly, $E = 0$, $F = 0$.

Hence $D = E = F = 0$.

Therefore, GX, GY, GZ are the principal axes at G and principal moments are

$$\frac{83}{320}Ma^2, \ \frac{83}{320}Ma^2, \ \frac{2}{3}Ma^2.$$

Problem 33. *Three rods AB, BC, CD each of mass m and length $2a$ are such that each is perpendicular to the other two. Show that the principal moments of inertia at the centre of mass are ma^2, $\frac{11}{3}ma^2$ and $4ma^2$.* [IAS 78]

Solution: Let BY be a line parallel to CD. Let AB, BY, and BC be taken as the axes of x, y, z, respectively.

The coordinates of middle points L, M, N of rods AB, BC, CD are $(a, 0, 0)$, $(0, 0, a)$ and $(0, a, 2a)$, respectively.

If $(\bar{x}, \bar{y}, \bar{z})$ be the coordinates of C.G. at G of the rods AB, BC, CD each of mass m, then

$$\bar{x} = \frac{m \cdot a + m \cdot 0 + m \cdot 0}{m + m + m} = \frac{a}{3}$$

$$\bar{y} = \frac{m \cdot 0 + m \cdot 0 + m \cdot a}{m + m + m} = \frac{a}{3}$$

$$\text{and } \bar{z} = \frac{m \cdot 0 + m \cdot a + m \cdot 2a}{m + m + m} = \frac{3am}{3m} = a.$$

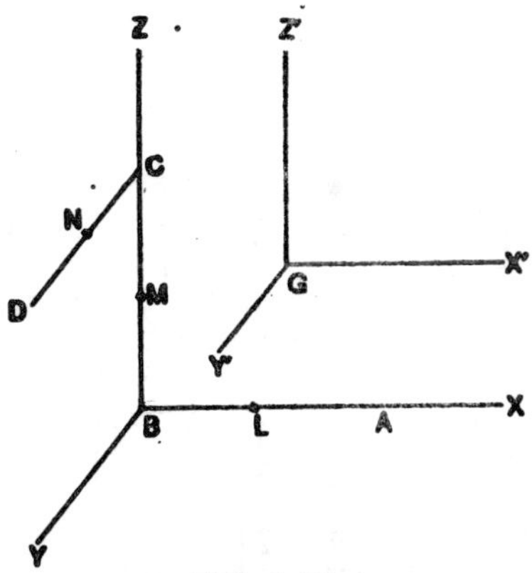

FIG 1.52

Let GX′, GY′, GZ′ be the axes parallel to BA, BY and BC.

Hence the coordinates of L, M, N referred to G as origin are

$$\left(a-\frac{a}{3},0-\frac{a}{3},0-a\right)=\left(\frac{2}{3}a,-\frac{a}{3},-a\right),$$

$$\left(0-\frac{a}{3},0-\frac{a}{3},a-a\right)=\left(-\frac{a}{3},-\frac{a}{3},0\right)$$

and $$\left(0-\frac{a}{3},a-\frac{a}{3},2a-a\right)=\left(-\frac{a}{3},\frac{2a}{3},a\right).$$

Hence

A_1 = M.I. of the three rods about GX′

= M.I. of AB + M.I. of BC + M.I. of CD about GX′

$$=\left[m\cdot 0+m\left(-\frac{a}{3}\right)^2+(-a)^2\right]+\left[\frac{m}{3}a^2+m\left\{\left(-\frac{a}{3}\right)^2+0^2\right\}\right]$$

$$+\left[\frac{m}{3}a^2+\left\{\left(\frac{2a}{3}\right)^2+a^2\right\}m\right]$$

$$=ma^2\left[\frac{1}{9}+1+\frac{1}{3}+\frac{1}{9}+\frac{1}{3}+\frac{4}{9}+1\right]$$

$$=ma^2\left[\frac{6}{9}+2+\frac{2}{3}\right]=ma^2\left(\frac{2}{3}+2+\frac{2}{3}\right)$$

$$=ma^2\left(\frac{4}{3}+2\right)=ma^2\frac{10}{3}=\frac{10}{3}ma^2.$$

B_1 = M.I. of the three rods about GY′

$$=\left[\frac{1}{3}ma^2+m\left\{(-a)^2+\left(\frac{2}{3}a\right)^2\right\}\right]+\left[\frac{1}{3}ma^2+m\left\{0+\left(-\frac{1}{3}a\right)^2\right\}\right]$$

$$+m\left[a^2+\left(-\frac{a}{3}\right)^2\right]$$

$$=ma^2\left[\frac{1}{3}+1+\frac{4}{9}+\frac{1}{3}+\frac{1}{9}+1+\frac{1}{9}\right]=\frac{10}{3}ma^2.$$

C_1 = M.I. of the three rods about GZ′

$$=\left[\frac{1}{3}ma^2+m\left\{\left(\frac{2}{3}a\right)^2+\left(-\frac{a}{3}\right)^2\right\}\right]+\left[m\left\{\left(-\frac{a}{3}\right)^2+\left(-\frac{a}{3}\right)^2\right\}\right]$$

$$+\left[\frac{1}{3}ma^2+m\left\{\left(-\frac{1}{3}a\right)^2+\left(\frac{2}{3}a\right)^2\right\}\right]$$

$$=ma^2\left[\frac{1}{3}+\frac{4}{9}+\frac{1}{9}+\frac{1}{9}+\frac{1}{9}+\frac{1}{3}+\frac{1}{9}+\frac{4}{9}\right]=ma^2\left[\frac{4}{9}+\frac{8}{9}+\frac{2}{3}\right]$$

$$=ma^2\left[\frac{12}{9}+\frac{2}{3}\right]=ma^2\left[\frac{4}{3}+\frac{2}{3}\right]=2ma^2.$$

$$
\begin{aligned}
D_1 &= \text{P.I. about } GY', GZ' = \Sigma m y_1 z_1 \\
&= m\left(-\frac{a}{3}\right)(-a) + m\left(-\frac{a}{3}\right)\cdot 0 + m\left(\frac{2}{3}a\right)a \\
&= ma^2\left[\frac{1}{3}+\frac{2}{3}\right] = ma^2. \\
E_1 &= \text{P.I. about } GZ', GX' = \Sigma m z_1 x_1 \\
&= m(-a)\left(\frac{2}{3}a\right) + m(0)\left(-\frac{a}{3}\right) + m\cdot a\cdot\left(-\frac{a}{3}\right) \\
&= ma^2\left(-\frac{2}{3}-\frac{1}{3}\right) = -ma^2. \\
F_1 &= \text{P.I. about } GX', GY' = \Sigma m x_1 y_1 \\
&= m\left(\frac{2}{3}a\right)\left(-\frac{a}{3}\right) + m\left(-\frac{a}{3}\right)\left(-\frac{a}{3}\right) + m\left(-\frac{a}{3}\right)\left(\frac{2}{3}a\right) \\
&= ma^2\left(-\frac{2}{9}+\frac{1}{9}-\frac{2}{9}\right) = ma^2\left(-\frac{3}{9}\right) = -\frac{1}{3}ma^2.
\end{aligned}
$$

Hence the equation of momental ellipsoid at G is

$$
\begin{aligned}
& A_1x^2 + B_1y^2 + C_1z^2 - 2D_1yz - 2E_1zx - 2F_1xy = 3mk^4, \\
\text{or,}\quad & \frac{10}{3}ma^2x^2 + \frac{10}{3}ma^2y^2 + 2ma^2z^2 - 2ma^2yz + 2ma^2zx \\
& \qquad + \frac{2}{3}ma^2xy = 3mk^4, \\
\text{or,}\quad & \frac{ma^2}{3}\left[10x^2 + 10y^2 + 6z^2 - 6yz + 6zx + 2xy\right] = 3mk^4.
\end{aligned}
$$

Reducing it to the form free from the terms of yz, zx and xy, we get

$$
\tfrac{1}{3}ma^2(\lambda_1x^2 + \lambda_2y^2 + \lambda_3z^2) = 3mk^4, \tag{1.53}
$$

where $\lambda_1, \lambda_2, \lambda_3$ are the roots of λ cubic

$$
\begin{aligned}
\lambda^3 - \lambda^2(a+b+c) &+ \lambda(ab+bc+ca-f^2-g^2-h^2) \\
&- (abc + 2fgh - af^2 - bg^2 - ch^2) = 0.
\end{aligned} \tag{1.54}
$$

Here $a = 10$, $b = 10$, $c = .6$, $h = 1$, $f = -3$, $g = 3$.
Hence equation (1.53) becomes.

$$
\begin{aligned}
& \lambda^3 - 26\lambda^2 + (100 + 60 + 60 - 9 - 9 - 1)\lambda \\
& \quad -(600 - 18 - 90 - 90 - 6) = 0, \\
\text{or,}\quad & \lambda^3 - 26\lambda^2 + 201\lambda - 396 = 0, \\
\text{or,}\quad & \lambda^3 - 3\lambda^2 - 23\lambda^2 + 69\lambda + 132\lambda - 396 = 0, \\
\text{or,}\quad & \lambda^2(\lambda - 3) - 23\lambda(\lambda - 3) + 132(\lambda - 3) = 0, \\
\text{or,}\quad & (\lambda - 3)(\lambda^2 - 23\lambda + 132) = 0, \\
\text{or,}\quad & (\lambda - 3)[\lambda^2 - 11\lambda - 12\lambda + 132] = 0,
\end{aligned}
$$

$$\text{or,}\quad (\lambda-3)[\lambda(\lambda-11)-12(\lambda-11)]=0,$$
$$\text{or,}\quad (\lambda-3)(\lambda-11)(\lambda-12)=0,$$
$$\lambda=3,11,12.$$

Putting these values in equation (1.53), it becomes

$$\frac{ma^2}{3}[3x^2+11y^2+12z^2]=3mk^4,$$
$$\text{or,}\quad ma^2x^2+\frac{11}{3}ma^2y^2+4ma^2z^2=3mk^4.$$

Hence the principal moments are ma^2, $\frac{11}{3}ma^2$, $4ma^2$.

Problem 34. *Show that the principal radii of gyration at the C.G. of a triangle are the roots of the equation*

$$x^4-\frac{a^2+b^2+c^2}{36}x^2+\frac{\Delta^2}{108}=0,$$

where $\triangle$ is the area of the triangle. **[IAS 52, 86]**

Solution: Let ABC be the triangle of mass M. Let us choose G, the C.G. of the $\triangle ABC$ as origin and the principal axes through G as the axes. Let the coordinates of the vertices A, B, C be (x_1,y_1), (x_2,y_2) and (x_3,y_3), respectively.

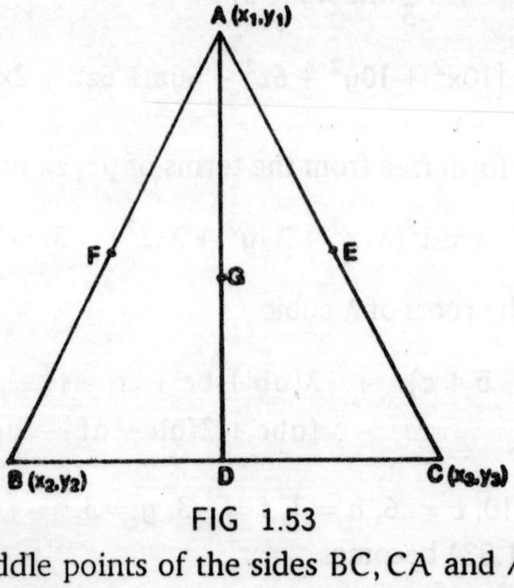

FIG 1.53

Let D, E, F be the middle points of the sides BC, CA and AB of $\triangle ABC$. Hence coordinates of D, E and F are

$$\left(\frac{x_2+x_3}{2},\frac{y_2+y_3}{2}\right);\ \left(\frac{x_3+x_1}{2},\frac{y_3+y_1}{2}\right)\ \text{and}\ \left(\frac{x_1+x_2}{2},\frac{y_1+y_2}{2}\right),$$

respectively.

Also coordinates of G are

$$\left(\frac{x_1+x_2+x_3}{3},\frac{y_1+y_2+y_3}{3}\right).$$

Since C.G. is taken as origin, we have

$$\left.\begin{array}{r} x_1 + x_2 + x_3 = 0 \\ \text{and} \quad y_1 + y_2 + y_3 = 0 \end{array}\right\}. \tag{1.55}$$

Thus

$$(x_1 + x_2 + x_3)^2 = 0,$$

$$\text{or,} \quad x_1^2 + x_2^2 + x_3^2 = -2(x_1x_2 + x_2x_3 + x_3x_1), \tag{1.56}$$

$$\text{or,} \quad y_1^2 + y_2^2 + y_3^2 = -2(y_1y_2 + y_2y_3 + y_3y_1). \tag{1.57}$$

Now

$$\begin{aligned} BC^2 &= a^2 = (x_2 - x_3)^2 + (y_2 - y_3)^2 \\ CA^2 &= b^2 = (x_3 - x_1)^2 + (y_3 - y_1)^2 \\ AB^2 &= c^2 = (x_1 - x_2)^2 + (y_1 - y_2)^2. \end{aligned}$$

$$\begin{aligned} \therefore\ a^2 + b^2 + c^2 &= 2(x_1^2 + x_2^2 + x_3^2 + y_1^2 + y_2^2 + y_3^2) - 2[(x_2x_3 + x_3x_1 + x_1x_2) \\ &\quad + (y_2y_3 + y_3y_1 + y_1y_2)] \\ &= 3(x_1^2 + x_2^2 + x_3^2 + y_1^2 + y_2^2 + y_3^2) \qquad (1.58) \end{aligned}$$

[From equations (1.56) and (1.57)]

Now we replace the triangle by three particles, each of mass $\frac{M}{3}$ placed at the middle points D, E, F.

$$\begin{aligned} A &= \text{M.I. of the } \Delta \text{ about GX} \\ &= \text{M.I. of the three particles placed at the middle points D, E, F} \\ &= \frac{M}{3}\left[\frac{y_1^2}{4} + \frac{y_2^2}{4} + \frac{y_3^2}{4}\right] = \frac{M}{12}(y_1^2 + y_2^2 + y_3^2). \end{aligned}$$

Similarly,

$$B = \text{M.I. of the } \Delta \text{ about GY} = \frac{M}{12}(x_1^2 + x_2^2 + x_3^2).$$

Hence

$$\begin{aligned} A + B &= \frac{M}{12}(x_1^2 + x_2^2 + x_3^2 + y_1^2 + y_2^2 + y_3^2) \\ &= \frac{M}{12}\frac{(a^2 + b^2 + c^2)}{3} \quad \text{[From equation (1.58)]} \end{aligned}$$

Again the products of inertia about GX and GY the principal axes vanishes as such the product of inertia of the three particles at D, E, F each of mass $\frac{M}{3}$ about GX and GY vanishes.

$$\therefore \quad \frac{M}{3}\left[\frac{x_2 + x_3}{2}\frac{y_2 + y_3}{2} + \frac{x_3 + x_1}{2}\frac{y_3 + y_1}{2} + \frac{x_1 + x_2}{2}\frac{y_1 + y_2}{2}\right] = 0,$$

$$\text{or,} \quad \frac{M}{12}[(-x_1)(-y_1) + (-x_2)(-y_2) + (-x_3)(-y_3)] = 0 \text{ [From equation (1.55)]}$$

$$\text{or,} \quad x_1y_1 + x_2y_2 + x_3y_3 = 0. \tag{1.59}$$

Also

$$AB = \frac{M^2}{144}(x_1^2 + x_2^2 + x_3^2)(y_1^2 + y_2^2 + y_3^2)$$
$$= \frac{M^2}{144}[(x_1y_1 + x_2y_2 + x_3y_3)^2 + (x_1y_2 - x_2y_1)^2$$
$$+(x_2y_3 - x_3y_2)^2 + (x_3y_1 - y_3x_1)^2].$$

Again

$$2\Delta = x_1(y_2 - y_3) + x_2(y_3 - y_1) + x_3(y_1 - y_2)$$
$$= x_1(y_1 + y_2 + y_2) + x_2(-y_1 - y_2 - y_1) + (-x_1 - x_2)(y_1 - y_2)$$

[From equation (1.55)]

$$= x_1y_1 + 2x_1y_2 - 2x_2y_1 - x_2y_2 - x_2y_1 + x_2y_2 - x_1y_1 + x_1y_2$$
$$= 3(x_1y_2 - x_2y_1), \text{ or, } x_1y_2 - x_2y_1 = \frac{2\Delta}{3}.$$

Similarly

$$x_2y_3 - x_3y_2 = \frac{2\Delta}{3};\ x_3y_1 - x_1y_3 = \frac{2\Delta}{3}.$$

Hence

$$AB = \frac{M^2}{144}\left[0 + \left(\frac{2}{3}\Delta\right)^2 + \left(\frac{2}{3}\Delta\right)^2 + \left(\frac{2}{3}\Delta\right)^2\right] = \frac{M^2}{144}\cdot 3\cdot\frac{4}{9}\Delta^2 = \frac{M^2\Delta^2}{108}.$$

Now of k_1 and k_2 be the principal radii of gyration at the C.G., then $A = Mk_1^2$ and $B = Mk_2^2$.

$$\therefore\ k_1^2 + k_2^2 = \frac{A+B}{M} = \frac{a^2+b^2+c^2}{36} \text{ and } k_1^2k_2^2 = \frac{AB}{M^2} = \frac{\Delta^2}{108}.$$

Hence k_1^2 and k_2^2 are the roots of the equation

$$x^4 - (k_1^2 + k_2^2)x^2 + k_1^2k_2^2 = 0;\ \therefore\ x^4 - \frac{a^2+b^2+c^2}{36}x^2 + \frac{\Delta^2}{108} = 0.$$

1.18 Exercise 1

Type I

1. Show that the moment of inertia of a semi-circular lamina about a tangent parallel to the bounding diameter is $Ma^2\left(\frac{5}{4} - \frac{8}{3\pi}\right)$, where a is the radius and M is the mass of the lamina.

2. Find the moment of inertia of the area of the lemniscate $r^2 = a^2\cos 2\theta$ about:

   (a) a line through the origin in its plane and perpendicular to its axis

   [*Ans.* $\frac{Ma^2}{48}(3\pi + 8)$, where M is the mass of the curve.]

(b) a line through the origin and perpendicular its plane. [*Ans.* $\frac{\pi M a^2}{8}$]

3. Find the moment of inertia of a hollow sphere about a diameter, its external and internal radii being a and b, respectively.

[*Ans.* $\frac{2M}{5}\frac{a^5-b^5}{a^3-b^3}$, where M is the mass of the sphere]

4. From a uniform sphere of radius a, spherical sector of vertical angle 2α is removed. Show that the moment of inertia of the remainder of mass M about the axis of symmetry is $\frac{1}{5}Ma^2(1+\cos\alpha)(2-\cos\alpha)$.

5. Prove that the moment of inertia about an axis through the centre and perpendicular to the plane of a circular ring whose outer and inner radii are a and b is $\frac{M}{2}(a^2+b^2)$, where M denotes the mass of the ring.

6. Find the P.I. of:

(a) an elliptic quadrant about its axes.

[*Ans.* $\frac{Mab}{2\pi}$, where M is the mass of the ellipse]

(b) an octant of an ellipsoid.

[*Ans.* $\frac{2Mab}{5\pi}$, where M is the mass of the ellipsoid]

Type II

7. The moment of inertia about its axis of a solid rubber tyre of mass M and circular cross-section of radius a is $\frac{M}{4}(4b^2+3a^2)$, where b is the radius of the cone. If the tyre be hollow and of small uniform thickness, show that corresponding result is $\frac{M}{2}(2b^2+3a^2)$.

Type III

8. Show that the moment of inertia of an ellipse of mass M and semi-axes a and b about a tangent is $\frac{5M}{4}p^2$, where p is the perpendicular from the centre on the tangent. **[LNMU 94H, MU 93H]**

9. Show that the sum of moment of inertia of an elliptic area about any two tangents at right angles is always the same.

10. Show that the moment of inertia of an elliptic area of mass M and equation $ax^2+2hxy+by^2+2gx+2fy+c=0$ about a diameter parallel to the axis of x is $-\frac{aM\Delta}{4(ab-h^2)^2}$, where $\Delta = abc+2fgh-af^2-bg^2-ch^2$.

11. Show that the moment of inertia of a right solid cone whose height is h and radius of whose base is a is

$$\frac{3Ma^2}{20}\frac{6h^2+a^2}{h^2+a^2}$$

about a slant side and

$$\frac{3M}{80}(h^2+4a^2)$$

about a line through the C.G. of the cone perpendicular to its axes.

[Agra 82, 76; Raj 79, 82; BPSC 83, MU 91H]

12. Show that for a thin hemispherical shell of mass M and radius a, the moment of inertia about any line through the vertex is $\frac{2}{3}Ma^2$.

Type IV

13. Show that the momental ellipsoid at a point on the edge of a circular base of a thin hemispherical shell is $2x^2 + 5(y^2 + z^2) - 3zx = \text{constant}$.

14. Prove that the equation of momental ellipsoid at a point on the circular edge of a solid cone is
$(3a^2 + 2h^2)x^2 + (23a^2 + 2h^2)y^2 + 26a^2z^2 - 10ahxz = \text{constant}$,
where h is the height and a the radius of the base.

15. Find the momental ellipsoid at any point O of a material straight rod AB of mass M and length $2a$. [MU 05H]
[*Ans.* $y^2 + z^2 = \text{constant}$]

16. The principal axes at the centre of gravity being the axes of reference, prove that the momental ellipsoid at the point (p, q, r) is
$\left(\frac{A}{M} + q^2 + r^2\right)x^2 + \left(\frac{B}{M} + r^2 + p^2\right)y^2 + \left(\frac{C}{M} + p^2 + q^2\right)z^2 - 2qryz - 2rpzx - 2pqxy = \text{constant}$.

Type V

17. Show that a uniform triangular lamina of mass m is equi-momental with three particles each of mass $\frac{1}{12}m$ placed at the angular points and a particle of mass $\frac{3}{4}m$ placed at the centre of inertia of the triangle. [CU 94H, 2000H]

18. ABCD is a uniform parallelogram of mass M; at the middle points of the four sides are placed particles each equal to $\frac{M}{6}$ and at the intersection of the diagonals a particle of mass $\frac{M}{3}$. Show that these five particles and the parallelogram are equi-momental systems.

19. Show that a uniform rod of mass m is kinetically equivalent to three particles, rigidly connected and situated one at each end of the rod and one at its middle point, the masses of the particles being $\frac{m}{6}$, $\frac{m}{6}$ and $\frac{2}{3}m$.

20. Show that there is a momental ellipse at an angular point of a triangular area which touches the opposite side at its middle point and bisects the adjacent sides.

21. Show that any lamina is dynamically equivalent to the three particles, each one-third of the mass of the lamina, placed at the corners of a maximum triangle inscribed on the ellipse, whose equation referred to the principal axes at the centre of inertia is $\frac{x^2}{B^2} + \frac{y^2}{A^2} = 2$, where mA and mB are the principal moment of inertia's about OX and OY and m is the mass.

Type VI

22. At the vertex C of a triangle ABC which is right angled at C, the principal axes are a perpendicular to the plane and two others inclined to the sides at an angle $\frac{1}{2}\tan^{-1}\frac{ab}{a^2 - b^2}$.

23. A uniform square lamina is bounded by the axes of x and y and the lines $x = 2c$, $y = 2c$ and a corner is cut off by the line $\frac{x}{a} + \frac{y}{b} = 2$. Show that the principal axes at the centre of the square are inclined to the axes of x at angles is given by $\tan 2\theta = \frac{ab-2(a+b)c+3c^2}{(a-b)(a+b-2c)}$.

24. A uniform rectangular plate whose sides are of length $2a, 2b$ has a portion cut out in the form of a square whose centre is the centre of the rectangle and whose mass is half the mass of the plate. Show that the axes of greatest and least moment of inertia at a corner of the rectangle makes angle θ, $\frac{\pi}{2} + \theta$ with a side, where $\tan 2\theta = \frac{6}{5}\frac{ab}{a^2-b^2}$.

25. Show that at an extremity of the bounding diameter of a semi-circular lamina, the principal axes makes an angle $\frac{1}{2}\tan^{-1}\left(\frac{8}{3\pi}\right)$ to the diameter.

26. Find the principal axes of a right-circular cone at a point on the circumference of the base and show that one of them will pass through its C.G. if the vertical angle of the cone is $2\tan^{-1}\frac{1}{2}$. [MU 94, 99H, 02H]

[*Ans.* y-axis at O]

27. Find the principal axes of an elliptic area at any point of its bounding arc. [BU 95H]

[*Ans.* If principal axis makes angle θ with x-axis, then $\theta = \frac{1}{2}\tan^{-1}\frac{8ab\tan\phi}{(5a^2-b^2)} + (a^2-5b^2)\tan^2\phi$, where symbols have usual meanings.]

28. Two particles each of mass m are placed at the extremities of the minor axis of an elliptic area of mass M. Prove that the principal axes at any point of the circumference of the ellipse will be the tangent and normal to the ellipse of $\frac{m}{M} = \frac{5}{8}\frac{e^2}{1-2e^2}$, where e is the eccentricity of the ellipse.

29. ABC is a triangular area and AD is perpendicular to BC, and AE is a median, O is the middle point of DE. Show that BC is a principal axis of the triangle at O.

Type VII

30. Show that for a thin hemispherical solid of radius a and mass M, the principal moment of inertia's at the centre of gravity are

$\frac{83}{320}Ma^2$, $\frac{83}{320}Ma^2$, $\frac{2}{5}Ma^2$.

31. A uniform tetrahedron of mass M is bounded by the planes of reference and the plane $x + y + z = a$. Show that the principal moment of inertia's at the origin are $\frac{1}{4}Ma^2$, $\frac{1}{4}Ma^2$, $\frac{1}{10}Ma^2$.

Chapter 2

D' Alembert's Principle

2.1 Dynamics of a Particle

For a single particle we know from Newton's second law of motion, the dynamical equations are

$$m\frac{d^2x}{dt^2} = X,\ m\frac{d^2y}{dt^2} = Y,\ m\frac{d^2z}{dt^2} = Z,$$

where m is the mass of the particle and (x, y, z) the coordinates of the particle at time t and X, Y, Z the components of external forces parallel to coordinate axes, respectively.

2.2 Dynamics of Rigid Body

Rigid body is considered as a collection of several particles so that we have to consider the mutual action between these particles in the equations of motion in 2.1. For any two particles the mutual action is along the line joining them and the action and reaction between them are equal and opposite. This mutual action is called the *force due to rest of the body.*

D' Alembert's proposed a method by which all the necessary equations may be obtained without writing down the equations of motions of all particles and without considering the unknown inner forces. This important principle is based on the following rule which is a natural consequence of Newton's third law of motion.

The internal actions and reactions of any system of rigid bodies in motion are in equilibrium amongst themselves.

2.3 Definitions

Impressed Forces

The external forces acting on a body are called *impressed force. For example,* weight of a body is the impressed force on the body. In case a body is tied to a string, then the tension in the string is also an impressed force on the body.

Effective Forces

The effective force on a particle is defined as the product of its mass m and its acceleration f. If a particle of mass m is situated at the point (x, y, z) at time t, then the effective forces on this particle at this time t are

$$m\frac{d^2x}{dt^2},\ m\frac{d^2y}{dt^2} \text{ and } m\frac{d^2z}{dt^2}$$

parallel to the axes.

Also

$$m\frac{d^2s}{dt^2},\ \frac{mv^2}{\rho}$$

are the effective forces on the particle along the tangent and normal.

2.4 D' Alembert's Principle

[MU 91H, 94H; LNMU 94H; TMBBU 93H, 95H, 98H, 00H, 01H, 03H; MU 96H, 99H, 01H, 04H; CU 91H, 94H, 96H, 00H, 02H, 04H; BBABU 96H, 99H, 02H, 04H]

Statement. *The reversed effective forces acting on each particle of the body and the external forces of the system are in equilibrium.*

Proof. Let m be the mass of a particle at (x, y, z) of a rigid body at any time t. It f is the resultant of component accelerations $\ddot{x}, \ddot{y}, \ddot{z}$, then the effective force on the particle is mf. Let F denote the resultant of the impressed forces and R the resultant of the internal forces (mutual actions) on the particle. Then by Newton's second law of forces, mf is the resultant of F and R.

Thus $-mf$ (reversed effective force), F and R are inequilibrium. This holds good for every particle of the body.

Hence $\Sigma(-mf)$, ΣF and ΣR are in equilibrium, i.e.,

$$\Sigma F + \Sigma R + \Sigma(-mf) = 0 \tag{2.1}$$

But the internal actions and reactions of a body are in equilibrium.

So $\Sigma R = 0$.

Thus equation (2.1) gives $\Sigma F - \Sigma mf = 0$.

Therefore, $\Sigma(-mf)$ and ΣF are in equilibrium.

Hence the reversed effective forces acting at each particle of the body and the impressed (external) forces on the system are in equilibrium.

Vector Method. We consider a rigid body in motion. Let at any time t, $\vec{r}$ be the position vector of a particle of mass m, and $\vec{F}$ and $\vec{R}$ the external and internal forces, respectively acting on it.

Now by Newton's second law

$$m\frac{d^2\vec{r}}{dt^2} = \vec{F} + \vec{R}, \text{ or, } \vec{F} + \vec{R} - m\frac{d^2\vec{r}}{dt^2} = \vec{0},$$

i.e., the forces $\vec{F}, \vec{R}, -m\frac{d^2\vec{r}}{dt^2}$ acting on a particle of mass m are in equilibrium.

Now applying the same argument to every particle of the rigid body, the forces $\Sigma\vec{F}, \Sigma\vec{R}$ and $\Sigma\left(-m\frac{d^2\vec{r}}{dt^2}\right)$ are in equilibrium, where the summation extends to all particles.

Since the internal forces acting on a rigid body form pairs of equal and opposite forces, thus $\Sigma\vec{R} = \vec{0}$.

Hence the forces $\Sigma\vec{F}$ and $\Sigma\left(-m\frac{d^2\vec{r}}{dt^2}\right)$ are in equilibrium,

$$\text{i.e., } \Sigma\vec{F} + \Sigma\left(-m\frac{d^2\vec{r}}{dt^2}\right) = \vec{0}.$$

This proves the D' Alembert's Principle.

Note 1. ***This principle reduces to the problem of dynamics to the problem of statics. Thus we mark all the external forces of the system and mark the effective forces in opposite directions. Then we solve the problem as a problem of statics by equating to zero the resolved parts of all these forces in two mutually perpendicular directions and taking moments about suitable point on the line.***

2.5 General Equations of Motion of a Body

[MU 93H; LNMU 94H; TMBU 93H, 95H, 98H, 00H, 01H, 03H; BBABU 96H, 99H, 02H, 04H]

To deduce the general equation of motion of a rigid body from D' Alembert's principle.

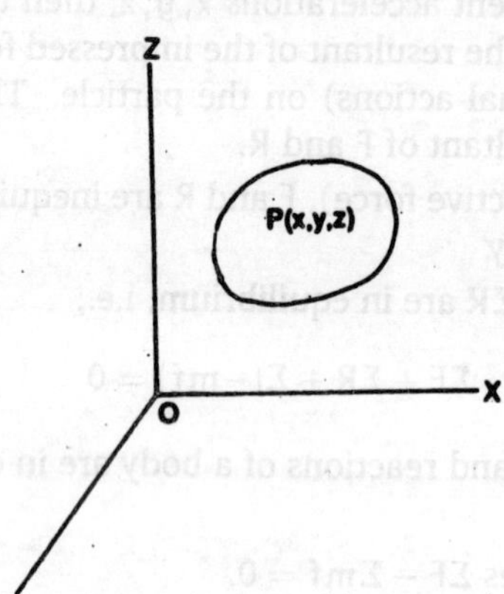

FIG 2.1

Let (x, y, z) be the coordinates of a particle of mass m in a continuous rigid body at time t, $\ddot{x}, \ddot{y}, \ddot{z}$ and X, Y, Z are accelerations and components of the external forces acting on the particle parallel to the axes.

By D' Alembert's principle, the forces $X - m\ddot{x}$, $Y - m\ddot{y}$, $Z - m\ddot{z}$ at the point (x, y, z) together with similar forces at all such points of the body form a system in equilibrium.

Therefore, by the principle of statics, we find the six conditions of equilibrium as

$$\Sigma(X - m\ddot{x}) = 0,\ \Sigma(Y - m\ddot{y}) = 0,\ \Sigma(Z - m\ddot{z}) = 0$$
$$\Sigma\left[y(Z - m\ddot{z}) - z(Y - m\ddot{y})\right] = 0$$
$$\Sigma\left[z(X - m\ddot{x}) - x(Z - m\ddot{z})\right] = 0$$
$$\Sigma\left[x(Y - m\ddot{y}) - y(X - m\ddot{x})\right] = 0,$$

where Σ denotes the summation over all the particles of the body.

These six equations can be written as

$$\begin{aligned}
\Sigma m\ddot{x} &= \Sigma X && (2.2)\\
\Sigma m\ddot{y} &= \Sigma Y && (2.3)\\
\Sigma m\ddot{z} &= \Sigma Z && (2.4)\\
\Sigma m(y\ddot{z} - z\ddot{y}) &= \Sigma(yZ - zY) && (2.5)\\
\Sigma m(z\ddot{x} - x\ddot{z}) &= \Sigma(zX - xZ) && (2.6)\\
\text{and } \Sigma m(x\ddot{y} - y\ddot{x}) &= \Sigma(xY - yX). && (2.7)
\end{aligned}$$

These are the six equations of motion of any rigid body.

Equations (2.2), (2.3) and (2.4) show that the sums of all the components of effective forces along the coordinate axes are respectively equal to the sum of the components of external impressed forces along the axes.

Equations (2.5), (2.6) and (2.7) show that the sums of the moments about the axes of effective forces are respectively equal to the sum of the moments of the external forces about the same axes.

Vector Method. We consider a rigid body in motion. At time t let $\vec{r}$ be the position vector of a particle of mass m and $\vec{F}$ the external force acting on it.

Then by D' Alembert's principle, we have

$$\Sigma\vec{F} + \Sigma\left(-m\frac{d^2\vec{r}}{dt^2}\right) = 0, \text{ or, } \Sigma m\frac{d^2\vec{r}}{dt^2} = \Sigma\vec{F}. \qquad (2.8)$$

Taking cross product by $\vec{r}$, we have

$$\Sigma m\left(\vec{r} \times \frac{d^2\vec{r}}{dt^2}\right) = \Sigma\left(\vec{r} \times \vec{F}\right). \qquad (2.9)$$

Equations (2.8) and (2.9) are in general vector equations of motions of a rigid body.

Again

$$\begin{aligned}
\vec{r} &= x\vec{i} + y\vec{j} + z\vec{k}\\
\vec{F} &= X\vec{i} + Y\vec{j} + Z\vec{k},
\end{aligned}$$

where X, Y, Z are the components of $\vec{F}$.

Hence

$$\frac{d^2\vec{r}}{dt^2} = \frac{d^2x}{dt^2}\vec{i} + \frac{d^2y}{dt^2}\vec{j} + \frac{d^2z}{dt^2}\vec{k}.$$

Now

$$\begin{aligned}
\vec{r} \times \frac{d^2\vec{r}}{dt^2} &= \begin{vmatrix} \vec{i} & \vec{j} & \vec{k} \\ x & y & z \\ \frac{d^2x}{dt^2} & \frac{d^2y}{dt^2} & \frac{d^2z}{dt^2} \end{vmatrix}\\
&= \vec{i}\left(y\frac{d^2z}{dt^2} - z\frac{d^2y}{dt^2}\right) + \vec{j}\left(z\frac{d^2x}{dt^2} - x\frac{d^2z}{dt^2}\right) + \vec{k}\left(x\frac{d^2y}{dt^2} - y\frac{d^2x}{dt^2}\right).
\end{aligned}$$

Also

$$\vec{r}\times\vec{F} = \begin{vmatrix} \vec{i} & \vec{j} & \vec{k} \\ x & y & z \\ X & Y & Z \end{vmatrix}$$
$$= \vec{i}(yZ - zY) - \vec{j}(xZ - zX) + \vec{k}(xY - yX).$$

With these values equation (2.9) becomes

$$\Sigma m\left[\vec{i}\left(y\frac{d^2z}{dt^2} - z\frac{d^2y}{dt^2}\right) + \vec{j}\left(z\frac{d^2x}{dt^2} - x\frac{d^2z}{dt^2}\right) + \vec{k}\left(x\frac{d^2y}{dt^2} - y\frac{d^2x}{dt^2}\right)\right]$$
$$= \Sigma\left[\vec{i}(yZ - zY) + \vec{j}(zX - xZ) + \vec{k}(xY - yX)\right].$$

Equating the coefficients of $\vec{i}, \vec{j}, \vec{k}$ on both the sides, we get

$$\Sigma m\left(y\frac{d^2z}{dt^2} - z\frac{d^2y}{dt^2}\right) = \Sigma(yZ - zY)$$
$$\Sigma m\left(z\frac{d^2x}{dt^2} - x\frac{d^2z}{dt^2}\right) = \Sigma(zX - xZ)$$
$$\Sigma m\left(x\frac{d^2y}{dt^2} - y\frac{d^2x}{dt^2}\right) = \Sigma(xY - yX).$$

Again putting the values of $\frac{d^2\vec{r}}{dt^2}$ and $\vec{F}$ in equation (2.8) we get

$$\Sigma m\left(\frac{d^2x}{dt^2}\vec{i} + \frac{d^2y}{dt^2}\vec{j} + \frac{d^2z}{dt^2}\vec{k}\right) = \Sigma(X\vec{i} + Y\vec{j} + Z\vec{k}).$$

Equating the coefficient of $\vec{i}, \vec{j}, \vec{k}$ on both the sides we get

$$\Sigma m\frac{d^2x}{dt^2} = \Sigma X,\ \Sigma m\frac{d^2y}{dt^2} = \Sigma Y,\ \Sigma m\frac{d^2z}{dt^2} = \Sigma Z.$$

2.6 Independence of the Motion of Centre of Inertia Equation or Motion of the Centre of Inertia [CU 91H, 94H, 96H, 2000H, 02H, 04H; TMBU 92H]

To prove that the centre of inertia of a body moves as if all the mass of a body moves, as if all the mass of the body were collected at it and as if all the external forces were acting at it in the directions parallel to these in which they act.

Let $(\bar{x}, \bar{y}, \bar{z})$ be the coordinates of the centre of inertia of a body of mass M. Then

$$\bar{x} = \frac{\Sigma mx}{\Sigma m} = \frac{\Sigma mx}{M},\ \Sigma mx = M\bar{x}.$$

Similarly,

$$\Sigma my = M\bar{y},\ \Sigma mz = M\bar{z}.$$

Since $\Sigma mx = M\bar{x}$,

$$\therefore \Sigma m\frac{d^2x}{dt^2} = M\frac{d^2\bar{x}}{dt^2}. \tag{2.10}$$

Also from the general equation of motion of a body we have

$$\Sigma m \frac{d^2x}{dt^2} = \Sigma X.$$

Hence from equation (2.10)

$$\left.\begin{aligned} M\frac{d^2\bar{x}}{dt^2} &= \Sigma X. \\ M\frac{d^2\bar{y}}{dt^2} &= \Sigma Y \\ M\frac{d^2\bar{z}}{dt^2} &= \Sigma Z \end{aligned}\right\}. \qquad (2.11)$$

Similarly,

These are the equations of motion of a particle of mass M placed at the centre of inertia of the body; and acted on by forces $\Sigma X, \Sigma Y, \Sigma Z$ parallel to the original directions of the forces acting on the different points of the body. This proves the theorem.

Vector Method. We consider a rigid body in motion. At time t let $\vec{r}$ be the position vector of a particle m of the body and $\vec{F}$ the external force acting on it. Then the equation of motion of the body is

$$\Sigma m \frac{d^2\vec{r}}{dt^2} = \Sigma F. \qquad (2.12)$$

Let $\vec{\bar{r}}$ be the position vector of the centre of inertia of the body. Then we have

$$\vec{\bar{r}} = \frac{\Sigma m\vec{r}}{\Sigma m} = \frac{\Sigma m\vec{r}}{M}$$

$$\Sigma m\vec{r} = M\vec{\bar{r}}$$

$$\therefore \quad \Sigma m \frac{d^2\vec{r}}{dt^2} = M\frac{d^2\vec{\bar{r}}}{dt^2}. \qquad (2.13)$$

From equations (2.12) and (2.13), we have

$$M\frac{d^2\vec{\bar{r}}}{dt^2} = \Sigma F. \qquad (2.14)$$

If the components of $\vec{\bar{r}}$ and $\vec{F}$ be $(\bar{x}, \bar{y}, \bar{z})$ and (X, Y, Z), then

$$\vec{\bar{r}} = \bar{x}\vec{i} + \bar{y}\vec{j} + \bar{z}\vec{k}$$

$$\text{and} \quad \vec{F} = X\vec{i} + Y\vec{j} + Z\vec{k}.$$

Hence from equation (2.14), we get

$$M\left(\frac{d^2\bar{x}}{dt^2}\vec{i} + \frac{d^2\bar{y}}{dt^2}\vec{j} + \frac{d^2\bar{z}}{dt^2}\vec{k}\right) = \Sigma(X\vec{i} + Y\vec{j} + Z\vec{k}).$$

Equating coefficient of $\vec{i}, \vec{j}, \vec{k}$ on both the sides, we get

$$M\frac{d^2\bar{x}}{dt^2} = \Sigma X, \; M\frac{d^2\bar{y}}{dt^2} = \Sigma Y, \; M\frac{d^2\bar{z}}{dt^2} = \Sigma Z.$$

This proves the theorem.

2.7 Motion Relative to the Centre of Inertia

[CU 91H, 94H, 96H, 00H, 02H, 04H; BBABU 99H, MU 95H; TMBU 91H, 96H, 99H]

To prove that the motion of a body about its centre of inertia is the same as it would be if the centre of inertia were fixed and the same forces acted on the body.

Let $(\bar{x}, \bar{y}, \bar{z})$ be the coordinates of centre of inertia 'G' of the body referred to rectangular axes OX, OY, OZ through a fixed point O. Let GX′, GY′, GZ′ be the axes through G parallel to the axes OX, OY, OZ, respectively.

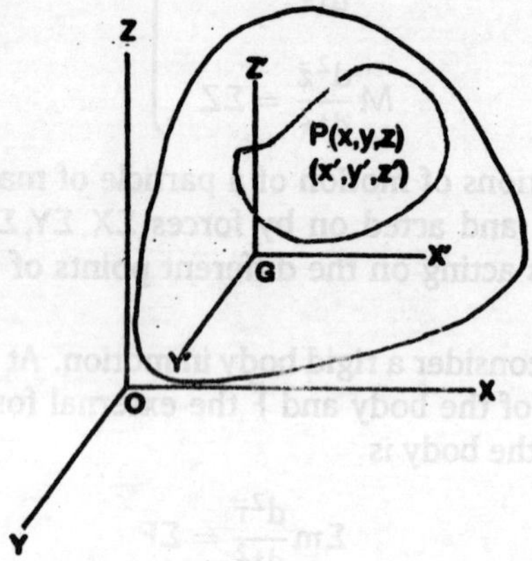

FIG 2.2

Let (x, y, z) and (x', y', z') be the coordinates of a particle of mass m at P referred to coordinate axes OX, OY, OZ and parallel axes GX′, GY′, GZ′, respectively. Then

$$x = \bar{x} + x',\ y = \bar{y} + y',\ z = \bar{z} + z'$$

$$\therefore\ \frac{d^2x}{dt^2} = \frac{d^2\bar{x}}{dt^2} + \frac{d^2x'}{dt^2},\ \frac{d^2y}{dt^2} = \frac{d^2\bar{y}}{dt^2} + \frac{d^2y'}{dt^2},\ \frac{d^2z}{dt^2} = \frac{d^2\bar{z}}{dt^2} + \frac{d^2z'}{dt^2}.$$

Hence the equation

$$\Sigma m\left(y\frac{d^2z}{dt^2} - z\frac{d^2y}{dt^2}\right) = \Sigma(yZ - zY)$$

becomes

$$\Sigma m\left[(\bar{y} + y')\left(\frac{d^2\bar{z}}{dt^2} + \frac{d^2z'}{dt^2}\right) - (\bar{z} + z')\left(\frac{d^2\bar{y}}{dt^2} + \frac{d^2y'}{dt^2}\right)\right]$$

$$= \Sigma\left[(\bar{y} + y')Z - (\bar{z} + z')Y\right]$$

or,

$$\Sigma m\left[\bar{y}\frac{d^2\bar{z}}{dt^2} + \bar{y}\frac{d^2z'}{dt^2} + y'\frac{d^2\bar{z}}{dt^2} + y'\frac{d^2z'}{dt^2} - \bar{z}\frac{d^2\bar{y}}{dt^2} - \bar{z}\frac{d^2y'}{dt^2} - z'\frac{d^2\bar{y}}{dt^2} - z'\frac{d^2y'}{dt^2}\right]$$

$$= \Sigma\left[(\bar{y}Z - \bar{z}Y) + (y'Z - z'Y)\right]$$

$$\text{or,} \quad \Sigma m\left(\bar{y}\frac{d^2\bar{z}}{dt^2} - \bar{z}\frac{d^2\bar{y}}{dt^2}\right) + \Sigma m\left(y'\frac{d^2z'}{dt^2} - z'\frac{d^2y'}{dt^2}\right)$$
$$+\bar{y}\Sigma m\frac{d^2z'}{dt^2} + \frac{d^2\bar{z}}{dt^2}\Sigma my' - \bar{z}\Sigma m\frac{d^2y'}{dt^2} - \frac{d^2\bar{y}}{dt^2}\Sigma mz'$$
$$= \Sigma\left[(\bar{y}Z - \bar{z}Y) + (y'Z - z'Y)\right]. \qquad (2.15)$$

Since (x', y', z') are the coordinates of P referred to G as origin

$$\frac{\Sigma my'}{\Sigma m} = 0, \text{ or, } \Sigma my' = 0. \therefore \Sigma m\frac{d^2y'}{dt^2} = 0.$$

Similarly,

$$\Sigma mz' = 0, \ \Sigma m\frac{d^2z'}{dt^2} = 0.$$

Also

$$M\frac{d^2\bar{y}}{dt^2} = \Sigma Y, \ M\frac{d^2\bar{z}}{dt^2} = \Sigma Z, \ \Sigma m = M.$$

Hence from equation (2.15), we have

$$\bar{y}M\frac{d^2\bar{z}}{dt} - \bar{z}M\frac{d^2\bar{y}}{dt^2} + \Sigma m\left(y'\frac{d^2z'}{dt^2} - z'\frac{d^2y'}{dt^2}\right)$$
$$= \Sigma\left[\bar{y}Z - \bar{z}Y + (y'Z - z'Y)\right]$$
$$\text{or,} \quad \Sigma m\left(y'\frac{d^2z'}{dt^2} - z'\frac{d^2y'}{dt^2}\right) = \Sigma(y'Z - z'Y).$$

Similarly, two other equations can be obtained.

But these equations are the same as it would have been obtained if we have regarded the centre of inertia as a fixed point.

Vector Method. We consider a rigid body in motion. At time t let $\vec{r}$ be the position vector of centre of inertia G of a rigid body of mass M. Let m be the mass of a particle whose position vector with respect to centre of inertia be $\vec{\bar{r}}$ and position vector referred to a fixed origin be $\vec{r}\,'$. Then

$$\vec{r} = \vec{\bar{r}} + \vec{r}\,', \ \therefore \frac{d^2\vec{r}}{dt^2} = \frac{d^2\vec{\bar{r}}}{dt^2} + \frac{d^2\vec{r}\,'}{dt^2}.$$

The moment vector equation of motion for a rigid body is

$$\Sigma m\vec{r} \times \frac{d^2\vec{r}}{dt^2} = \Sigma\vec{r} \times \vec{F}$$
$$\Rightarrow \Sigma m(\vec{\bar{r}} + \vec{r}\,') \times \left(\frac{d^2\vec{\bar{r}}}{dt^2} + \frac{d^2\vec{r}\,'}{dt^2}\right) = \Sigma(\vec{\bar{r}} + \vec{r}\,') \times \vec{F}$$
$$\Rightarrow \Sigma m\left[\vec{\bar{r}} \times \frac{d^2\vec{\bar{r}}}{dt^2} + \vec{\bar{r}} \times \frac{d^2\vec{r}\,'}{dt^2} + \vec{r}\,' \times \frac{d^2\vec{\bar{r}}}{dt^2} + \vec{r}\,' \times \frac{d^2\vec{r}\,'}{dt^2}\right]$$
$$= \Sigma(\vec{\bar{r}} + \vec{r}\,') \times \vec{F}. \qquad (2.16)$$

Now position vector of the centre of inertia G of the body referred to G as origin is O.

$$\therefore \quad \frac{\Sigma m\vec{r}\,'}{\Sigma m} = \vec{0}, \text{ or, } \Sigma m\vec{r}\,' = \vec{0}$$

$$\therefore \quad \Sigma m\frac{d^2\vec{r}\,'}{dt^2} = \vec{0} \text{ and } \Sigma m = M.$$

Also the equation of motion of the centre of inertia is

$$M\frac{d^2\vec{\bar{r}}}{dt^2} = \Sigma\vec{F}.$$

Hence equation (2.16) becomes

$$\Sigma m\vec{r}\,' \times \frac{d^2\vec{r}\,'}{dt^2} + \left(\vec{\bar{r}} \times \frac{d^2\vec{\bar{r}}}{dt^2}\right)\Sigma m + \vec{\bar{r}}\,' \times \left(\Sigma m\frac{d^2\vec{r}\,'}{dt^2}\right)$$

$$+ \ \vec{\bar{r}} \times \left(\Sigma m\frac{d^2\vec{r}\,'}{dt^2}\right) = \Sigma(\vec{\bar{r}} + \vec{r}\,') \times \vec{F}$$

$$\text{or, } \Sigma m\vec{r}\,' \times \frac{d^2\vec{r}\,'}{dt^2} + \left(\vec{\bar{r}} \times \frac{d^2\vec{\bar{r}}}{dt^2}\right)M = \Sigma\vec{\bar{r}} \times \vec{F} + \Sigma\vec{r}\,' \times \vec{F}$$

$$\Rightarrow \Sigma m\vec{r}\,' \times \frac{d^2\vec{r}\,'}{dt^2} + \left(\vec{\bar{r}} \times M\frac{d^2\vec{\bar{r}}}{dt^2}\right) = \Sigma\vec{\bar{r}} \times \vec{F} + \vec{r}\,' \times \vec{F}$$

$$\Rightarrow \Sigma m\vec{r}\,' \times \frac{d^2\vec{r}\,'}{dt^2} + \left(\vec{\bar{r}} \times \Sigma\vec{F}\right) = \Sigma\vec{\bar{r}} \times \vec{F} + \Sigma\vec{r}\,' \times \vec{F}$$

$$\Rightarrow \Sigma m\vec{r}\,' \times \frac{d^2\vec{r}\,'}{dt^2} = \Sigma\vec{r}\,' \times \vec{F}, \tag{2.17}$$

which is the vector equation of motion of a rigid body when the centre of inertia is regarded as a fixed point.

If (x, y, z) and (x', y', z') are the cartesian coordinates of the particle of mass m referred to the rectangular axes through the fixed point O and the parallel axes through the centre of inertia G, respectively we have

$$\vec{r} = x\vec{i} + y\vec{j} + z\vec{k} \text{ and } \vec{r}\,' = x'\vec{i} + y'\vec{j} + z'\vec{k}.$$

Let $(\bar{x}, \bar{y}, \bar{z})$ be the coordinates of G referred to the axes through O. Then

$$\vec{\bar{r}} = \bar{x}\vec{i} + \bar{y}\vec{j} + \bar{z}\vec{k}.$$

Also if X, Y, Z be the components of external forces $\vec{F}$ parallel to the axes, then

$$\vec{F} = X\vec{i} + Y\vec{j} + Z\vec{k}.$$

Also

$$\frac{d^2\vec{r}\,'}{dt^2} = \frac{d^2x'}{dt^2}\vec{i} + \frac{d^2y'}{dt^2}\vec{j} + \frac{d^2z'}{dt^2}\vec{k}.$$

Now

$$\vec{r}' \times \frac{d^2\vec{r}'}{dt^2} = \begin{vmatrix} \vec{i} & \vec{j} & \vec{k} \\ x' & y' & z' \\ \frac{d^2x'}{dt^2} & \frac{d^2y'}{dt^2} & \frac{d^2z'}{dt^2} \end{vmatrix}$$

$$= \vec{i}\left(y'\frac{d^2z'}{dt^2} - z'\frac{d^2y'}{dt^2}\right) + \vec{j}\left(z'\frac{d^2x'}{dt^2} - x'\frac{d^2z'}{dt^2}\right) + \vec{k}\left(x'\frac{d^2y'}{dt^2} - y'\frac{d^2x'}{dt^2}\right).$$

Also

$$\vec{r}' \times \vec{F} = \begin{vmatrix} \vec{i} & \vec{j} & \vec{k} \\ x' & y' & z' \\ X & Y & Z \end{vmatrix}$$

$$= \vec{i}(y'Z - z'Y) + \vec{j}(z'X - x'Z) + \vec{k}(x'Y - y'X).$$

Hence from equation (2.17), we get

$$\Sigma m\left[\vec{i}\left(y'\frac{d^2z'}{dt^2} - z'\frac{d^2y'}{dt^2}\right) + \vec{j}\left(z'\frac{d^2x'}{dt^2} - x'\frac{d^2z'}{dt^2}\right) + \vec{k}\left(x'\frac{d^2y'}{dt^2} - y'\frac{d^2x'}{dt^2}\right)\right]$$

$$= \Sigma\left[\vec{i}(y'Z - z'Y) + \vec{j}(z'X - x'Z) + \vec{k}(x'Y - y'X)\right].$$

Equating the coefficient of $\vec{i}, \vec{j}, \vec{k}$, we get

$$\Sigma m\left(y'\frac{d^2z'}{dt^2} - z'\frac{d^2y'}{dt^2}\right) = \Sigma(y'Z - z'Y);$$

$$\Sigma m\left(z'\frac{d^2x'}{dt^2} - x'\frac{d^2z'}{dt^2}\right) = \Sigma(z'X - x'Z);$$

$$\Sigma m\left(x'\frac{d^2y'}{dt^2} - y'\frac{d^2x'}{dt^2}\right) = \Sigma(x'Y - y'X).$$

Note 1. *The two important properties discussed in 2.6 and 2.7 are called the* principle of conservation of motion of translation *and rotation respectively and together called the* principle of independence *of translation and rotation.*

Note 2. *2.6 states that the motion of the C.G. is the same as if the whole mass were collected at that point and is, therefore, independent of rotation.*

Note 3. *2.7 states that the motion round the C.G. is the same as if that point were fixed and is, therefore, independent of the motion of that point.*

2.8 Centre of Inertia in a Straight Line

To prove that if no external forces act on a system of particles the centre of inertia is either at rest or moves with constant velocity in a straight line.

Let E and I be the resolved parts of the external and internal forces acting on the particle of mass m of coordinate x along the given straight line taken as x-axis. Then

$$m\frac{d^2x}{dt^2} = E + I$$

and similarly, for all other particles.

$$\therefore \Sigma m\frac{d^2x}{dt^2} = \Sigma E + \Sigma I.$$

The summation being carried for all the particles.

But ΣI, being the resolved part of the resultant internal force, is zero and ΣE being the resolved part of the resultant external force, is zero as there is no external force acting on the system as a whole.

$$\therefore \Sigma m\frac{d^2x}{dt^2} = 0. \tag{2.18}$$

Let $\bar{x}$ be the coordinate of the C.G. of the system at time t. Then

$$\bar{x} = \frac{\Sigma mx}{\Sigma m}, \quad \therefore \Sigma m\bar{x} = (\Sigma m)\,\bar{x} = M\bar{x},$$

where $M = \Sigma m$.

$$\therefore \Sigma m\frac{d^2\bar{x}}{dt^2} = M\frac{d^2\bar{x}}{dt^2}.$$

Hence from equation (2.18)

$$M\frac{d^2\bar{x}}{dt^2} = 0, \text{ or, } \frac{d^2\bar{x}}{dt^2} = 0.$$

Integrating, we get $\frac{d\bar{x}}{dt}$ = constant = k (say).

Hence if $k = 0$, the centre of inertia of the system of particles is at rest; otherwise the centre of inertia of the system moves with uniform velocity.

2.9 Impulsive Forces

A force a of very great magnitude acting on a body for a very short time is called an impulsive force.

Impulse. If an impulsive force X is acting on a body for a very short time t, the impulse X' is defined as

$$X' = \int_0^t X\,dt.$$

It may be noted that impulse is not a force.

2.9.1 *Negligible effect of finite forces compared to impulsive forces.*

Let f be the finite force and X be the impulsive force acting on a particle of mass m in the body for a short time T. Then

$$\int_0^T m\frac{dv}{dt} = \int_0^T X\,dt + \int_0^T f\,dt.$$

$$\therefore\ m(V' - V) = X' + fT.$$

Since T is small, fT vanishes

$$\therefore\ X' - m(V' - V).$$

2.9.2 *Application of D' Alembert's principle to impulsive forces, general equation of motion.*

Let u, v, w and u', v', w' be the velocities parallel to the axes, respectively before and after the action of impulsive forces on the particle of mass m. If X', Y', Z' are the resolved parts of the total impulse on m parallel to the axes, then

$$\Sigma m(u' - u) = \Sigma \int_0^T X\,dt = \Sigma X' \text{ or, } \Sigma mu' - \Sigma mu = \Sigma X'. \tag{2.19}$$

Similarly,

$$\Sigma mv' - \Sigma mv = \Sigma Y' \tag{2.20}$$
$$\Sigma mw' - \Sigma mw = \Sigma Z'. \tag{2.21}$$

Thus the change in the momentum parallel to any of the axes is equal to the total impulse of the external forces parallel to the corresponding axis.

Hence the change in momentum parallel to any of the axes of the whole mass M, supposed collected at the centre of inertia and moving with it, is equal to the impulse of the external force parallel to the corresponding axis.

Again we have the equation

$$\Sigma m(y\ddot{z} - z\ddot{y}) = \Sigma m(yZ - zY)$$
$$\text{or, } \frac{d}{dt}\Sigma m(y\dot{z} - z\dot{y}) = \Sigma m(yZ - zY).$$

Integrating this, we have

$$[\Sigma m(y\dot{z} - z\dot{y})]_0^T = \Sigma m\left[y\int_0^T Z dt - z\int_0^T Y dt\right].$$

Since the time interval T is so small that the body has not moved during this interval, we may take x, y, z as constants. Thus the above equation becomes

$$\Sigma m\{y(w' - w) - z(v' - v)\} = \Sigma(yZ' - zY')$$
$$\text{or, } \Sigma m(yw' - zv') - \Sigma m(yw - zv) = \Sigma(yZ' - zY').$$

Similarly,

$$\Sigma m(xv' - yu') - \Sigma m(xv - yu) = \Sigma(xY' - yX')$$
$$\text{and}\quad \Sigma m(zu' - xw') - \Sigma m(zu - xw) = \Sigma(zX' - xZ').$$

Hence the change in the moment of momentum about any of the axes is equal to the moment about that axis of the impulses of the external forces.

Vector Method. Let I and I′ be the resultant external and internal impulses acting on the particle of mass m at P. Also let the velocity of m change from v_1 to v_2. Then Impulse = change in momentum.

$$\therefore \vec{I} + \vec{I}' = m(\vec{V}_2 - \vec{V}_1).$$

$$\Sigma\vec{I} + \Sigma\vec{I}' = \Sigma m\vec{V}_2 - \Sigma m\vec{V}_1. \qquad (2.22)$$

But $\Sigma\vec{I}' = \vec{0}$ by Newton's third law.
We get

$$\Sigma\vec{I} = \Sigma m\vec{V}_2 - \Sigma m\vec{V}_1,$$

i.e.,*the total external impulse applied to the system of particles is equal to the change of linear momentum produced.*

Now let $\overrightarrow{OP} = \vec{r}$, then from equation (2.22) we get

$$\Sigma\vec{r} \times (\vec{I} + \vec{I}') = \Sigma\vec{r} \times m(\vec{V}_2 - \vec{V}_1).$$

$$\Sigma\vec{r} \times \vec{I} = \Sigma\vec{r} \times m\vec{V}_2 - \Sigma\vec{r} \times m\vec{V}_1 \left(\because \Sigma\vec{r} \times \vec{I}' = 0\right).$$

Hence, the total vector sum of moments of the external impulses about any point O is equal to the increase in the angular momentum produced about the same point.

2.10 Worked out Examples

Problem 1. *A plank of mass M, and length 2a, is initially at rest along a line of greatest slope, of a smooth plane inclined at an angle α to the horizon, and a man of mass M′, starting from the upper end walks down the plank so that it does not move. Show that he will reach the other end in time* $\sqrt{\frac{2M'a}{(M+M')g\sin\alpha}}$, *where a is the length of the plank.*

[LNMU 89H, 93H, 77H, 84H; BU 72H; IAS 55; MU 75, 80, 84, 87; Raj 77; BU 93H; CU 88H; 91H, 02H, 04H, IAS 2000]

Solution: Let AB be the plank of mass M and length a resting along the line of greatest slope of a smooth plane inclined at an angle α to the horizon.

Let G be the C.G. of the plank, where its weight Mg acts and AG = GB = a/2.

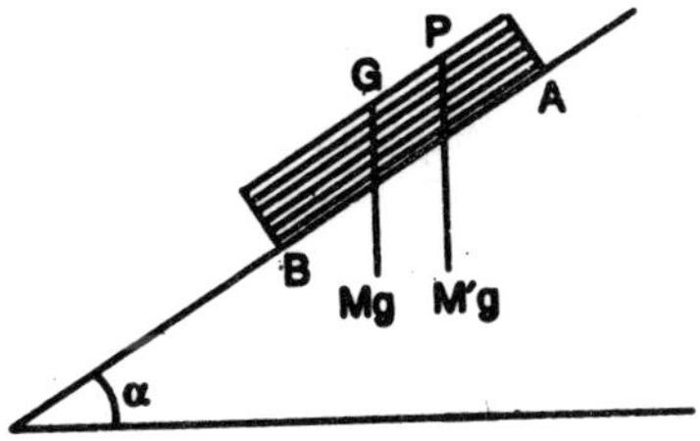

FIG 2.3

A man of mass M starts moving from the end A and let at any instant of time t, the man be at P, where $AP = x$. Let the distance of the combined C.G. of the man of mass M' at P and the mass of the plank M at G be $\bar{x}$ from A. Then

$$\bar{x} = \frac{M \cdot \frac{a}{2} + M'x}{M + M'}.$$

Differentiating it twice with respect to t we have

$$\frac{d^2\bar{x}}{dt^2} = \frac{M'}{M+M'}\frac{d^2x}{dt^2}. \tag{2.23}$$

Now the total weight $(M+M')g$ will be vertically downwards at the C.G. of the system.

$\therefore$ the equation of motion of the C.G. of the system is given by

$$(M+M')\frac{d^2\bar{x}}{dt^2} = Mg\sin\alpha + M'g\sin\alpha$$

or, $$(M+M')\frac{M'}{M+M'}\frac{d^2x}{dt^2} = (M+M')g\sin\alpha \text{ [From equation (2.23)]}$$

or, $$\frac{d^2x}{dt^2} = \frac{(M+M')}{M'}g\sin\alpha.$$

Integrating, we get

$$\frac{dx}{dt} = \frac{M+M'}{M'}g\sin\alpha t + C.$$

Initially when $t = 0$, $\frac{dx}{dt} = 0$.

$$\therefore\ C = 0.$$

Hence

$$\frac{dx}{dt} = \frac{M+M'}{M'}g\sin\alpha t.$$

Integrating again we get

$$x = \frac{M+M'}{M'}g\sin\alpha\frac{t^2}{2} + C_1.$$

Initially when $t = 0$, $x = 0$, so that $C_1 = 0$.

Hence

$$x = \frac{M+M'}{M'} g \sin\alpha \frac{t^2}{2}, \text{ or, } t = \sqrt{\frac{2M'x}{(M+M')g\sin\alpha}}.$$

Putting $x = a$, the required time to reach the other end B of the plank is given by

$$t = \sqrt{\frac{2M'a}{(M+M')g\sin\alpha}}.$$

Problem 2. *A uniform rod OA, of length 2a, free to turn about its end O, revolves with uniform, angular velocity ω about the vertical OZ through O, and is inclined at a constant angle α to OZ. Show that the value of α is either zero or* $\cos^{-1}\left(\frac{3g}{4a\omega^2}\right)$.

[LNMU 87H, 94H; CU 87H; VU 92H; BU 68H; RU 66H; PU 67H; BPSC 82, 84; CU 93H, 01H, IAS 96]

Solution: Let m be the mass of the rod. We consider an element $PQ = \delta x$ of the rod at a distance x from O.

This element PQ will make a circle in the horizontal plane with radius PN and centre at N.

Since the rod revolves with uniform angular velocity, the only effective force on this element is

$$\frac{m}{2a}\delta x\, x \sin\alpha\omega^2 \text{ along NP.}$$

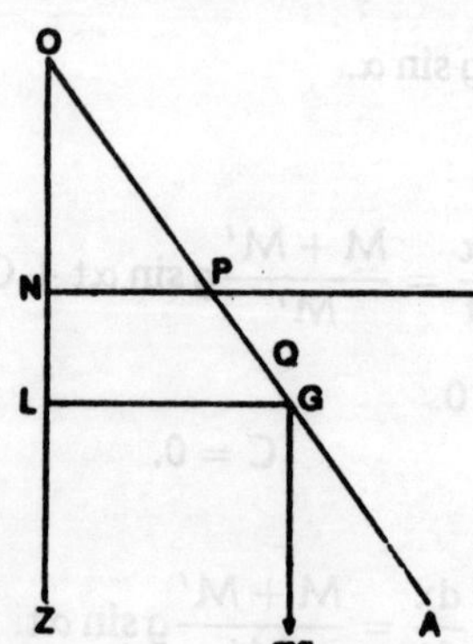

FIG 2.4

The external forces on the rod are its weight mg and reaction at O.

Now by D' Alembert's principle all the reversed effective forces acting at different points of the rod, and the external forces, weight mg and reaction at O are in equilibrium.

To avoid reaction at O, taking moment about O, we get

$$\Sigma\left(\frac{m}{2a}\delta x\omega^2 x\sin\alpha\right)x\cos\alpha - mg.GL = 0$$

$$\Rightarrow \int_0^{2a}\frac{m}{2a}\omega^2x^2\sin\alpha\cos\alpha - mga\sin\alpha = 0$$

$$\Rightarrow \frac{m\omega^2\sin\alpha\cos\alpha}{2a}\frac{8a^3}{3} - mga\sin\alpha = 0$$

$$\Rightarrow mga\sin\alpha\left(\frac{4a\omega^2\cos\alpha}{3g} - 1\right) = 0.$$

Either

$$\sin\alpha = 0 \quad \text{or,} \quad \frac{4a\omega^2\cos\alpha}{3g} - 1 = 0 \;\therefore \alpha = 0$$

$$\text{or,} \quad \omega^2 = \frac{3g}{4a\cos\alpha}; \quad \cos\alpha = \frac{3g}{4a\omega^2}.$$

If $3g > 4a\omega^2$, then $\cos\alpha > 1$ which is impossible. Hence only solution in this case is $\alpha = 0$, i.e., the rod hangs vertically.

If $3g < 4a\omega^2$, then $\alpha = \cos^{-1}\frac{3g}{4a\omega^2}$.

Problem 3. *A rod of length 2a, is suspended by a string of length l, attached to one end. If the string and rod revolve about the vertical with uniform angular velocity, and their inclinations to the vertical be θ and φ, respectively, show that* $3l/a = \frac{(4\tan\theta - 3\tan\phi)\sin\phi}{(\tan\phi - \tan\theta)\sin\theta}$. [BU 65H; RU 65H; MU 80, 81, 82, 83; VU 87H]

Solution: Let the rod AB of length 2a and mass m be suspended by a string OA of length l. Let θ and φ be the inclinations of the string and rod to the vertical, respectively.

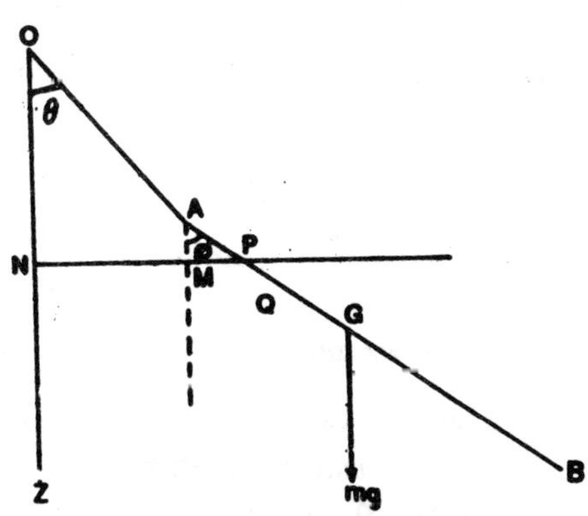

FIG 2.5

We consider an element $PQ = \delta x$ of the rod at a distance x from A, then mass of this element is $\frac{m}{2a}\delta x$, where m is the mass of AB.

From P we draw PN perpendicular to the vertical through O.

$$PN = PM + MN = x\sin\phi + l\sin\theta.$$

As the rod revolves with uniform angular velocity ω about the vertical OZ, the element δx will describe a circle of radius PN in the horizontal plane.

The reversed effective force on element δx is

$$\frac{m}{2a}\delta x\omega^2 PN = \frac{m}{2a}\delta x\omega^2(l\sin\theta + x\sin\phi) \text{ along NP}$$

The external forces acting on the rod are

1. tension T at A along AO; and
2. its weight mg acting vertically downwards at its middle point G.

Resolving horizontally and vertically the forces acting on the rod, we get

$$\begin{aligned}
T\sin\theta &= \Sigma\frac{m}{2a}\delta x\omega^2(l\sin\theta + x\sin\phi)\\
\Rightarrow T\sin\theta &= \frac{m}{2a}\omega^2\int_0^{2a}(l\sin\theta + x\sin\phi)dx\\
\Rightarrow T\sin\theta &= \frac{m\omega^2}{2a}\left[l\sin\theta 2a + 2a^2\sin\phi\right]\\
\Rightarrow T\sin\theta &= m\omega^2(l\sin\theta + a\sin\phi) \qquad (2.24)\\
\text{and } T\cos\theta &= mg. \qquad (2.25)
\end{aligned}$$

Again taking moment about A, we have

$$\begin{aligned}
mga\sin\phi &= \frac{m}{2a}\omega^2\int_0^{2a}(l\sin\theta + x\sin\phi)x\cos\phi dx\\
ga\sin\phi &= \frac{\omega^2}{2a}\left[l2a^2\sin\theta + \frac{8a^3}{3}\sin\phi\right]\cos\phi\\
&= \frac{\omega^2}{2a}2a^2\left[l\sin\theta + \frac{4a}{3}\sin\phi\right]\cos\phi\\
ga\sin\phi &= \omega^2 a\left(l\sin\theta + \frac{4a}{3}\sin\phi\right)\cos\phi. \qquad (2.26)
\end{aligned}$$

From equations (2.24) and (2.25), we get

$$\tan\theta = \frac{\omega^2(l\sin\theta + a\sin\phi)}{g}, \text{ or, } \omega^2 = \frac{g\tan\theta}{l\sin\theta + a\sin\phi}.$$

Hence from equation (2.26), we get

$$ga\sin\phi = \frac{g\tan\theta\, a\left(l\sin\theta + \frac{4a}{3}\sin\phi\right)\cos\phi}{l\sin\theta + a\sin\phi}$$

$$\text{or,}\quad \tan\phi = \frac{\tan\theta(3l\sin\theta + 4a\sin\phi)}{3(l\sin\theta + a\sin\phi)}$$

$$\text{or,}\quad 3l\sin\theta\tan\phi + 3a\sin\phi\tan\phi = 3l\sin\theta\tan\theta + 4a\sin\phi\tan\theta$$

$$\text{or,}\quad 3l\sin\theta(\tan\phi - \tan\theta) = a\sin\phi(4\tan\theta - 3\tan\phi)$$

$$\therefore\quad \frac{3l}{a} = \frac{\sin\phi(4\tan\theta - 3\tan\phi)}{\sin\theta(\tan\phi - \tan\theta)}.$$

Problem 4. *A rod revolving on a smooth horizontal plane about one end which is fixed, breaks into two parts. What is the subsequent motion of the two parts?*

[LNMU 76H, 88H]

Solution: Let the rod OA revolving round O break into two parts OB and AB. Let C be the middle point of AB.

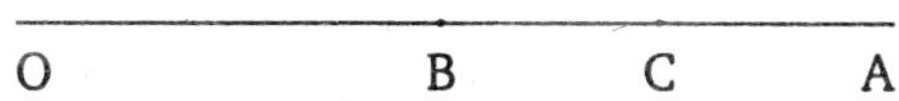

FIG 2.6

Now the part OB will continue to rotate about O with the same angular velocity.

The part AB whose C.G. is at C has a linear velocity. Therefore, this part will fly off in a tangent line at C to the circle with O as centre and OC as radius. Again, we know that the motion of a body about its centre of inertia is the same as if the centre of inertia were fixed and the same forces acted on it. Hence the part AB will continue to rotate about C with the same angular velocity.

Problem 5. *A cannon of mass M, resting on a rough horizontal plane of coefficient of friction μ is fired with such a charge that the relative velocity of the ball and cannon, at the moment when it leaves the cannon is u. Show that the cannon will recoil a distance $\left(\frac{mu}{M+m}\right)^2 \frac{1}{2\mu g}$ along the plane, m being the mass of the ball.*

Solution: Let I be the impulse between the cannon and the ball. If v is the velocity of the ball and V the velocity of cannon in opposite direction, then the relative velocity of the ball and cannon, at the moment ball leaves the cannon is

$$v + V = u \text{ (given)}. \tag{2.27}$$

Also since, impulse = change in momentum,

$$\begin{aligned}
\therefore\ I &= m(v-0) \text{ (for the ball)} \\
I &= mv \\
\text{and } I &= M(V-0) \text{ (for the cannon)} \\
I &= MV \\
\therefore\ mv &= MV, \text{ or, } v = \frac{MV}{m}.
\end{aligned} \tag{2.28}$$

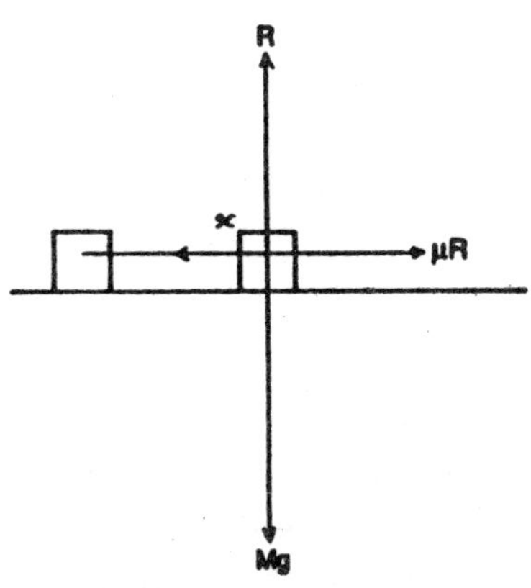

FIG 2.7

Hence from equation (2.27), we get

$$\frac{MV}{m} + V = u, \text{ or, } V(M+m) = mu$$

$$V = \frac{mu}{M+m}. \tag{2.29}$$

If the cannon moves through a distance x in the direction opposite to the direction of motion of the ball in time t, then on the rough plane, for the cannon the equation of motion is

$$M\frac{d^2x}{dt^2} = -\mu R = -\mu Mg, \text{ or, } \frac{d^2x}{dt^2} = -\mu g.$$

Multiplying both the sides by $2\frac{dx}{dt}$, we get

$$2\frac{dx}{dt}\frac{d^2x}{dt^2} = -\mu g 2\frac{dx}{dt}, \text{ or, } \frac{d}{dt}\left(\frac{dx}{dt}\right)^2 = -2\mu g\frac{dx}{dt}.$$

Integrating we get

$$\left(\frac{dx}{dt}\right)^2 = -2\mu g x + C.$$

But initially, when $x = 0$, $\frac{dx}{dt} = V$.

$$\therefore V^2 = C.$$

Hence

$$\left(\frac{dx}{dt}\right)^2 = V^2 - 2\mu g x.$$

When the cannon comes to rest, $\frac{dx}{dt} = 0$.

$$\therefore x = \frac{V^2}{2\mu g} = \left(\frac{mu}{M+m}\right)^2 \frac{1}{2\mu g} \text{ [From equation (2.29)].}$$

2.11 Exercise 2

1. A rough uniform board of mass m and length $2a$, rests on a smooth horizontal plane and a man of mass M walks on it from one end to the other. Find the distance through which the board moves in this time.

[MU 76, 78, 80, 82; Kanpur 81, 83; MU 04H]

[*Ans.* $\frac{2aM}{m+M}$]

2. A rod of length $2a$, revolves with uniform angular velocity ω about a vertical axis through a smooth joint at one extremity of the rod so that it describes a cone of semi-vertical angle α, show that $\omega^2 = \frac{3g}{4a\cos\alpha}$.

   Prove also that the direction of the reaction at the hinge makes with the vertical an angle $\tan^{-1}\left(\frac{3}{4}\tan\alpha\right)$. **[CU 81H, 95H]**

3. A thin circular disc of mass M and radius a can turn freely about a thin axis OA, which is perpendicular to its plane and passes through a point O of its circumference. The axis OA is compelled to move in a horizontal plane with angular velocity ω about its end A. Show that the inclination θ to the vertical of the radius of the disc through O is $\cos^{-1}\left(\frac{g}{a\omega^2}\right)$ unless $\omega^2 < \frac{g}{a}$ and then θ is zero.

4. Two uniform spheres, each of mass M and radius a, are firmly fixed to the ends of two uniform thin rods, each of mass m and length l, and the other ends of the rods are freely hinged to a point O. The whole system revolves; as in the governer of a steam engine about a vertical line through O with angular velocity ω. Show that, when the motion is steady, the rods are inclined to the vertical at an angle θ given by the equation

$$\cos\theta = \frac{g}{\omega^2}\frac{M(l+a)+\frac{m}{2}l}{M(l+a)^2+\frac{m}{3}l^2}.$$

5. A circular board is placed on a smooth horizontal plane, and a boy runs round the edge of it at a uniform rate. What is the motion of the centre of the board?

   [*Ans.* The centre of the disc describes a circle round G as centre and GO as radius, where O and G are C.G.'s of the board and the system.]

6. Two persons are situated on a perfectly smooth horizontal plane at a distance a from each other. One of the persons of mass M throws a ball of mass m towards the other which reaches him in time t.

   Prove that the first person will begin to slide along the plane with velocity $\frac{ma}{Mt}$.

7. A thin heavy disc can turn freely about an axis in its own plane and this axis revolves horizontally with a uniform angular velocity ω about a fixed point on itself. Show that the inclination θ of the plane to the disc to the vertical is $\cos^{-1}\frac{gh}{\omega^2k^2}$, where h is the distance of the centre of inertia of the disc from the axis and k is the radius of gyration of the disc about the axis.

   If $\omega^2 < \frac{gh}{k^2}$, then show that the plane of the disc is vertical. [CU 87H, VU 92H]

8. A heavy uniform rigid straight rod OA of length $2a$ is free to turn about its end O. If the rod revolves uniformly with angular velocity π about the vertical through O making a constant angle $\pi/4$ with the vertical, show that $\pi^2 = \frac{39}{2\sqrt{2}a}$. [CU 98H, 02H]

   [*Hint.* See worked-out problem 2.2. Here $\alpha = \pi/4$.]

Chapter 3

Motion about a Fixed Axis

3.1 A rigid body is rotating about a fixed axis, to find the moment of effective forces about the axis of rotation [LNMU 94H, BBABU 01H]

Let the fixed axis of rotation be OZ, perpendicular to the plane of the paper, and let a fixed plane through OZ cut the paper in OA. Let a plane ZOG, through OZ and fixed in the body, make an angle θ with the fixed plane so that $\angle AOG = \theta$.

We consider an element of mass m at P of the body and let the plane ZOQ through P and OZ cut the plane of the paper along OQ. Let the plane ZOQ make an angle α to the plane ZOG. Obviously, when the body rotates about OZ, α remains constant.

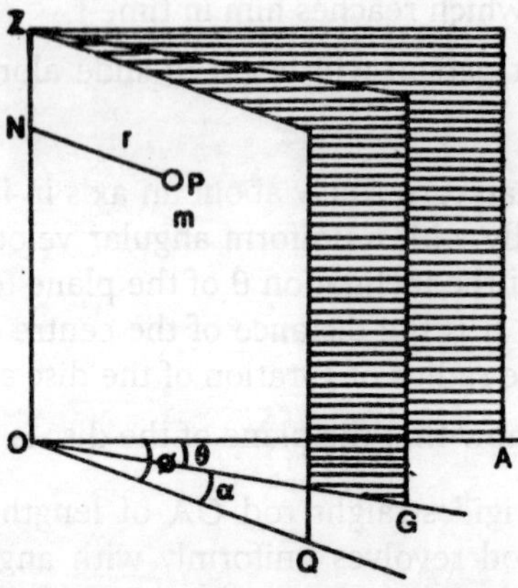

FIG 3.1

Let plane ZOQ make an angle ϕ with ZOA, i.e., $\angle AOQ = \phi$.

$$\therefore \quad \theta + \alpha = \phi.$$

$$\therefore \quad \frac{d\theta}{dt} = \frac{d\phi}{dt}, \text{ also } \frac{d^2\theta}{dt^2} = \frac{d^2\phi}{dt^2}. \tag{3.1}$$

Let PN be the perpendicular from P on OZ and PN $= r$. Then as the body rotates about OZ, the point describes a circle of radius r about N as centre.

Thus acceleration of mass m at P are $r\left(\frac{d\phi}{dt}\right)^2$ and $r\frac{d^2\phi}{dt^2}$ along and perpendicular to PN, respectively.

Therefore, effective forces along and perpendicular PN are $mr\left(\frac{d\phi}{dt}\right)^2$ and $mr\frac{d^2\phi}{dt^2}$, respectively, i.e., $mr\left(\frac{d\theta}{dt}\right)^2$ and $mr\frac{d^2\theta}{dt^2}$.

The effective force $mr\left(\frac{d\theta}{dt}\right)^2$ is along PN, thus it cuts OZ at N, so that its moment is zero about OZ.

Hence taking moment about OZ, we have moments of the effective forces of

$$m = r \times mr\frac{d^2\theta}{dt^2} = mr^2\frac{d^2\theta}{dt^2}.$$

Hence moment about OZ of all the effective forces of the body

$$= \Sigma mr^2\frac{d^2\theta}{dt^2} = \frac{d^2\theta}{dt^2}\Sigma mr^2$$

since $\frac{d^2\theta}{dt^2}$ is the same for all particles of the body

$$= \frac{d^2\theta}{dt^2}MK^2$$

since $\Sigma mr^2 = MK^2$ = moment of inertia of the body about OZ.

Hence the required moment of all the effective forces of the body about OZ

$$= MK^2\frac{d^2\theta}{dt^2},$$

where θ is the angle of any plane through the axis which is fixed in the body makes with any plane through the axis which is fixed in the space.

3.2 Equation of motion of a rigid body about a fixed axis

When a body rotates about a fixed axis OZ, the impressed forces includes, besides the external forces, the reaction on the axis of rotation OZ. The moment of the reaction on the axis OZ, about OZ will be zero.

Since by D' Alembert's principle the reversed effective forces and the impressed forces are in equilibrium, therefore, the algebraic sum of their moments about OZ will be zero.

Hence Moment of the effective force about OZ = moment of all the external forces about OZ,

$$\text{or, } MK^2\frac{d^2\theta}{dt^2} = L,$$

where L is the moment of all the external forces about the axis of rotation OZ.

This equation is called the *equation of motion of the body* about the axis of rotation.

3.3 Moment of Momentum of the body about the fixed axis

[CU 97H, TMBU 96H, 99H, BBABU 00H, 02H, 04H]

The velocity of the particle m is $r\frac{d\theta}{dt}$ in a direction perpendicular to PN (see figure of Section 3.1)

Hence momentum of $m = mr\frac{d\theta}{dt}$.

$\therefore$ moment of this momentum about OZ

$$= r \times mr\frac{d\theta}{dt} = mr^2\frac{d\theta}{dt}.$$

Hence moment of momentum of the whole body about OZ

$$= \Sigma mr^2\frac{d\theta}{dt} = \frac{d\theta}{dt}\Sigma mr^2 = MK^2\frac{d\theta}{dt}.$$

3.4 Expression for Kinetic Energy of a rigid body moving about a fixed axis

[CU 96H]

The velocity of the particle m is

$$r\frac{d\phi}{dt} = r\frac{d\theta}{dt}.$$

So its

$$\text{K.E.} = \frac{1}{2}m\left(r\frac{d\theta}{dt}\right)^2 = \frac{1}{2}mr^2\left(\frac{d\theta}{dt}\right)^2.$$

Hence the kinetic energy of the body

$$= \Sigma\frac{1}{2}mr^2\left(\frac{d\theta}{dt}\right)^2 = \frac{1}{2}\left(\frac{d\theta}{dt}\right)^2\Sigma mr^2 = \frac{1}{2}MK^2\left(\frac{d\theta}{dt}\right)^2.$$

3.5 Compound Pendulum

[CU 93H, 04H; BBABU 92H, 00H, 03H; MU 05H]

A rigid body of any shape and size which is free to turn about a fixed horizontal axis, the external forces being the gravity and the reaction of the fixed axis, is called a *compound pendulum.*

3.6 Time of a Complete Oscillation

[BBABU 2000H, 03H; MU 92H, 97H, 2000H, 03H]

To determine the motion of a rigid body acted on by the force of gravity only and moving about a fixed horizontal axis or if a rigid body swing under gravity about a fixed horizontal axis, show that the time of a complete oscillation is $2\pi\sqrt{\frac{K^2}{gh}}$, where K is the radius of gyration about the fixed axis and h is the distance between the fixed axis and centre of inertia of the body.

Let the plane of the paper be taken as the plane through the centre of gravity G of the body and perpendicular to the axis of rotation.

Let OA be the line through G and perpendicular to the axis of rotation, i.e., O is the point, where the axis of rotation meet the plane of the paper.

Let θ be the angle between the vertical OZ and the line OA, so that θ is the angle that a plane fixed in the body makes with a plane fixed in the space.

Given $OG = h$.

The impressed forces on the body are

(*a*) its weight Mg acting vertically downwards at G, and

(*b*) the reaction of the fixed axis at O.

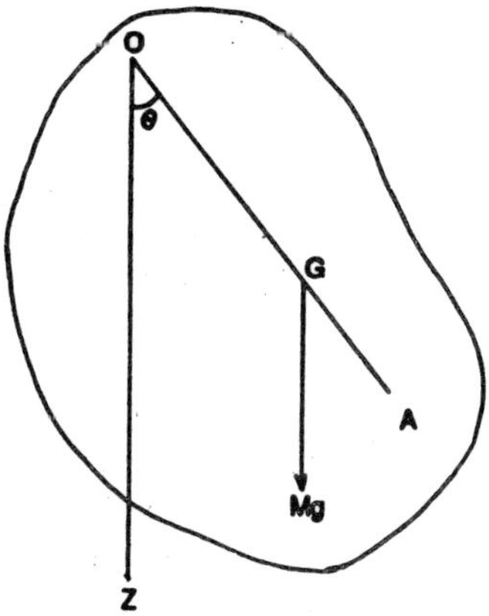

FIG 3.2

To avoid the reaction of the fixed axis at O, we take moments about the horizontal axis through O. The moment of Mg about O is $-Mgh\sin\theta$.

Also we know that equation of motion of a rigid body about a fixed axis is

$$MK^2\frac{d^2\theta}{dt^2} = L. \tag{3.2}$$

Here $L = -Mgh\sin\theta$.

Hence from equation (3.2)

$$MK^2\frac{d^2\theta}{dt^2} = -Mgh\sin\theta, \text{ or, } \frac{d^2\theta}{dt^2} = -\frac{gh}{K^2}\sin\theta.$$

Since θ is small, hence $\sin\theta = \theta$.

Thus

$$\frac{d^2\theta}{dt^2} = -\frac{gh}{K^2}\theta,$$

which shows that the motion is simple harmonic motion.

Hence the time of one complete oscillation $= 2\pi\sqrt{\frac{K^2}{gh}}$.

3.7 Simple Equivalent Pendulum [BBABU 93H, 95H, 00H; CU 04H, MU 97H]

A simple pendulum having the same periodic time as the compound pendulum is called a *simple equivalent pendulum*.

Let l be the length of a simple pendulum, then its time period $= 2\pi\sqrt{l/g}$.

The period of a compound pendulum $= 2\pi\sqrt{\frac{k^2}{gh}}$.

If this simple pendulum is equivalent to the compound pendulum, then

$$2\pi\sqrt{\frac{l}{g}} = 2\pi\sqrt{\frac{K^2}{gh}}, \therefore l = \frac{K^2}{h}.$$

Thus the length K^2/h is called the *length of simple equivalent pendulum* of a compound pendulum.

3.8 Some Definitions

Centre of Suspension [CU 97H, 00H, 03H]

The point O, where the perpendicular from the C.G. at G of the body cuts the axis of rotation; is called the *centre of suspension of the body.*

Centre of Oscillation [CU 97H, 00H, 03H]

If O is the centre of suspension and G the centre of gravity of the body, then the point O_1 on OG, produced such that $OO_1 = K^2/h$, is called the centre of oscillation of the body.

Note 1. *From the definition of centre of oscillation, we conclude that if the whole mass of the body were collected at the centre of oscillation and hanged from the centre of suspension by a string of length $l = K^2/h$, the time of oscillation of this simple pendulum will be the same as that of the body swinging about the centre of suspension.*

The time of oscillation $T = 2\pi\sqrt{l/g}$ of a simple pendulum depends upon the length l of the string and not upon the mass attached at the other end of the string.

Hence we conclude that if an additional mass is attached at the centre of oscillation of a compound pendulum, its time period is unaltered. [CU 98H]

3.9 Minimum Time of Oscillation of a Compound Pendulum

[MU 93H, 96H, 98H, 02H, 05H; BBABU 98H; TMBU 00H, 04H; CU 93H, 04H]

Let K be the radius of gyration of the body about a line through the centre of gravity G, parallel to the axis of rotation. Then

$$k^2 = K^2 + h^2,$$

where k is the radius of gyration of the body about the axis of rotation.

Let l be the length of simple equivalent pendulum, then

$$l = \frac{k^2}{h} = \frac{K^2 + h^2}{h} = \frac{K^2}{h} + h.$$

Now the time of oscillation of a compound pendulum will be least when the length l of the simple equivalent pendulum is minimum.

Now

$$\frac{dl}{dh} = 1 - \frac{K^2}{h^2}; \quad \frac{d^2l}{dh^2} = \frac{2K^2}{h^3}.$$

For maximum or minimum value of l,

$$\frac{dl}{dh} = 0, \text{ or, } 1 - \frac{k^2}{h^2} = 0, \text{ or, } h = K.$$

At $h - K$,

$$\frac{d^2l}{dh^2} = \frac{2K^2}{K^3} = \frac{2}{K},$$

which is positive.

Hence l is minimum when $h = K$.

Thus minimum value of $l = K + K = 2K$.

Thus minimum time of oscillation of a compound pendulum is given by

$$T = 2\pi\sqrt{\frac{l}{g}} = 2\pi\sqrt{\frac{2K}{g}}.$$

Note 1. *In case $h = 0$ or ∞, then $l = \infty$, i.e., if the axis of suspension either passes through the centre of gravity or be at infinite distance from the centre of gravity, the length of simple equivalent pendulum is infinite and thereby the time of oscillation is also infinite.*

3.10 To show that centre of suspension and oscillation of a compound pendulum are convertible

[BBABU 97H, 01H; MU 94H; LNMU 89H; BHU 62H, 64H, 66H; BU 59H, 62H; MU 01H, 04H, CU 91H, TMBU 96H, 02H]

Let O_1 be the centre of oscillation of a compound, pendulum when O is the centre of suspension.

$$\therefore \; OO_1 = k^2/h,$$

where k is the radius of gyration of the body about the axis of rotation through O.

Now let the body be suspended from O_1 instead of from O. Then we shall prove that the body will swing about O_1 with O as centre of oscillation.

Let K be the radius of gyration of the body about an axis through the centre of gravity G and parallel to the axis of rotation. Then by the theorem of parallel axis, moment of inertia about any axis through O = moment of inertia about parallel axis through $G + M.OG^2$.

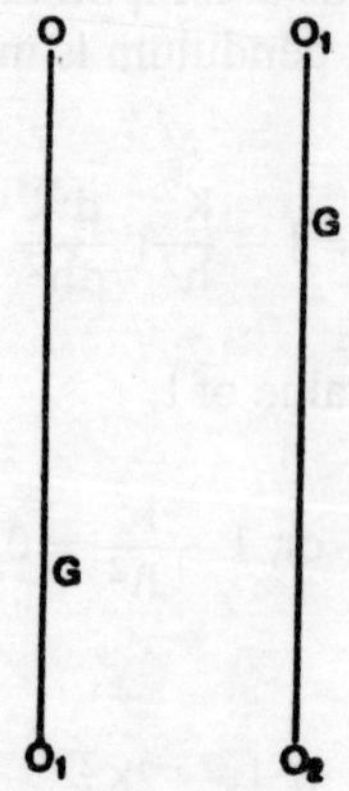

FIG 3.3

$\therefore\ Mk^2 = MK^2 + M.OG^2$, or, $k^2 = K^2 + OG^2$.

Now

$$OO_1 = \frac{k^2}{h} = \frac{K^2 + OG^2}{OG}, \quad \text{or,} \quad OO_1 \cdot OG = K^2 + OG^2,$$

$$\text{or,} \quad OO_1 \cdot OG - OG^2 = K^2,$$

$$\text{or,} \quad OG(OO_1 - OG) = K^2,$$

$$\text{or,} \quad OG \cdot O_1G = K^2. \tag{3.3}$$

Let us suppose that when O_1 is the centre of suspension of the body, then O_2 is the centre of oscillation.

Then similarly we have

$$O_1G \cdot O_2G = K^2 \tag{3.4}$$

Comparing equations (3.3) and (3.4), we find that O_2 and O are the same point, i.e., if O_1 is the centre of suspension, O is the centre of oscillation.

Hence, the centre of suspension and oscillation are interchangeable.

3.11 Reactions of the Axis of Rotation

To find the reaction of the axis of rotation when a rigid body is rotating about it, given that the forces and the body are symmetrical with respect to the plane through the centre of gravity perpendicular to the fixed axis and gravity is the only external force.

Let the plane perpendicular to the axis of rotation through the C.G. 'G' meet the axis at O, i.e., O is the centre of suspension of the body. By the symmetry of forces and the body the action on the axis of the body reduces to a single force in the plane of rotation.

Let X, Y be the components of the single force of rotation along and perpendicular to GO.

Now the C.G. at G of the body describes a circle round O as centre and h as radius.

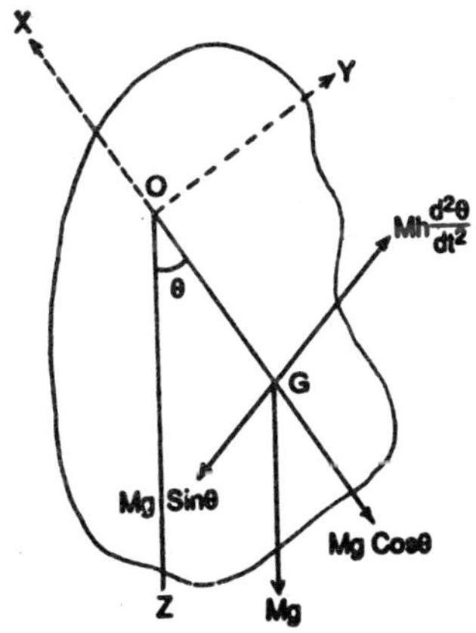

FIG 3.4

Hence its acceleration along and perpendicular to GO are

$$h\left(\frac{d\theta}{dt}\right)^2 \quad \text{and} \quad h\frac{d^2\theta}{dt^2}.$$

Now the equations of motion of the C.G. at G are

$$Mh\left(\frac{d\theta}{dt}\right)^2 = X - Mg\cos\theta \tag{3.5}$$

$$\text{and } Mh\left(\frac{d^2\theta}{dt^2}\right) = Y - Mg\sin\theta. \tag{3.6}$$

Also equation of motion gives

$$MK^2\frac{d^2\theta}{dt^2} = -Mgh\sin\theta, \tag{3.7}$$

where k is the radius of gyration about the axis of rotation.

To get Y we eliminate $\frac{d^2\theta}{dt^2}$ from equations (3.6) and (3.7).

Integrating equation (3.7) we shall get $\left(\frac{d\theta}{dt}\right)^2$ and putting in equation (3.5) we will get X.

If R be the resultant reaction, then

$$R = \sqrt{X^2 + Y^2}.$$

If the reaction is inclined at an angle ϕ to X, then $\phi = \tan^{-1}\frac{Y}{X}$.

3.12 Equation of Motion about Fixed Axis under Impulsive Forces

Let the impulsive force be applied to a rigid body movable about a fixed axis. If ω and ω' are the angular velocities of the bodies just before and after the action of the impulsive forces, then since moment of momentum about the axis of rotation is $MK^2\frac{d^2\theta}{dt^2}$ so that change in moment of momentum about the axis is $MK^2(\omega' - \omega)$.

If L be the moment of the impulsive forces about the axis of rotation, then since change in the moment of momentum of the body about the axis is equal to the moment of the impulsive forces about it,

$$\therefore\ MK^2(\omega' - \omega) = L.$$

3.13 Centre of Percussion

If a body rotating about a given axis is so struck that there is no impulsive pressure on the axis, then any point on the line of action of the force is called a centre of percussion. The line of action of the force is called the line of percussion.

If the line of action of the blow is given the axes about which the body begins to turn is called the *axis of spontaneous rotation.* This obviously coincides with the position of the fixed axis in the first case.

3.14 Centre of Percussion of a Rod

Let a rod OA of mass M and length $2a$ be suspended freely from the end A. Let a horizontal blow P be given to the rod at the point C such that $OC = x$.

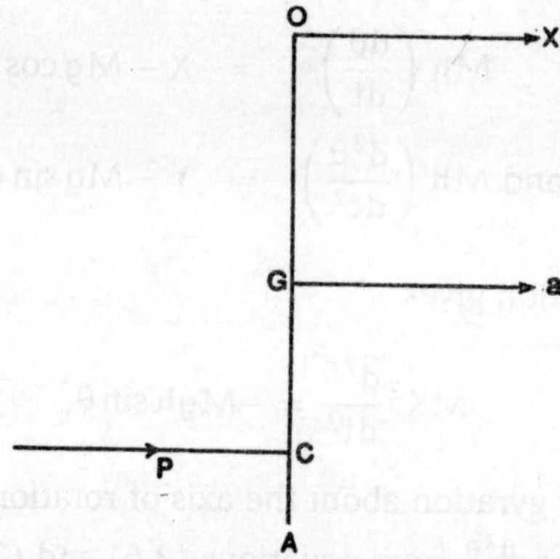

FIG 3.5

If ω be the angular velocity of the rod due to the impulse, then the centre of gravity G of the rod will move with velocity $a\omega$. Let X be the impulsive action at O.

The equation of motion is

$$MK^2\omega = Px. \tag{3.8}$$

Also change in momentum = total impulsive force in that direction.

$$\therefore\ M(a\omega - 0) = P + X. \tag{3.9}$$

If the blow is given through the centre of percussion, then $X = 0$.

$\therefore$ from equation (3.9), we get

$$Ma\omega = P. \tag{3.10}$$

Dividing equations (3.8) by (3.10), we get

$$K^2/a = x,$$

or, $x = K^2/a =$ length of the simple equivalent pendulum of the rod.

Thus the centre of percussion of a rod is at a depth $l = K^2/a$ from the centre of suspension.

Hence the centre of percussion of a rod coincides with its centre of oscillation.

3.15 Worked out Examples

Type I: Problems based on Equation of Motion of a Rigid Body about a Fixed Axis

Problem 1. *A uniform rod of mass m and length 2a can turn freely about one end which is fixed. It is started with angular velocity ω from the positions which it hangs. Then*

(a) *find its angular velocity at any instant;* [LNMU 94H; BBABU 95H, 98H; MU 05H]

(b) *find the least angular velocity with which it must begin to move so that it may perform complete revolution in a vertical plane; and* [CU 90H, 96H, 01H, 04H; MU 05H]

(c) *find the time of describing an angle θ from its vertical position in case of (b).* [BPSC 84; CU 96H]

Solution: (a) Let the uniform rod OA of length 2a and mass m turns freely about one end O which is fixed. At time t, let the rod be inclined at an angle θ to the vertical OZ.

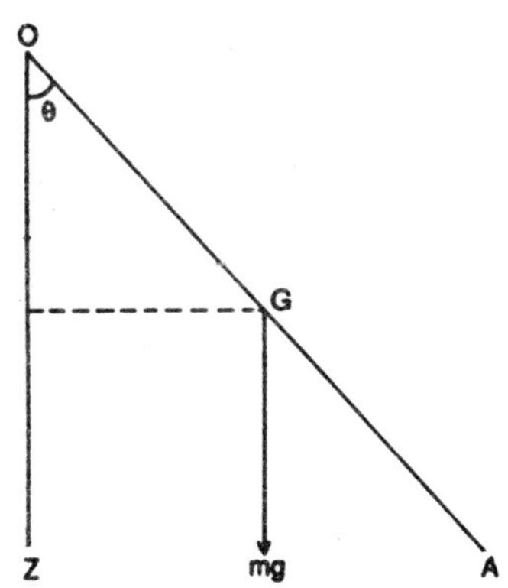

FIG 3.6

Here the axis of rotation is the horizontal line through O, OZ is the line fixed in space and OA is the line fixed in the body.

From the equation of motion of a rigid body about a fixed axis, we have

$$MK^2\frac{d^2\theta}{dt^2} = L, \tag{3.11}$$

where L is the moment of all the external forces about the axis of rotation.

Here $L = -mga\sin\theta$, $K^2 = \frac{4}{3}a^2$.

Hence from equation (3.12), we have

$$m\frac{4}{3}a^2\frac{d^2\theta}{dt^2} = -mga\sin\theta, \text{ or, } \frac{d^2\theta}{dt^2} = -\frac{3g}{4a}\sin\theta.$$

Multiplying both the sides by $2\frac{d\theta}{dt}$, we get

$$2\frac{d^2\theta}{dt^2}\frac{d\theta}{dt} = -\frac{3g}{4a}\sin\theta\, 2\frac{d\theta}{dt}, \text{ or, } \frac{d}{dt}\left\{\left(\frac{d\theta}{dt}\right)^2\right\} = -\frac{3g}{2a}\sin\theta\frac{d\theta}{dt}.$$

Integrating, we get

$$\left(\frac{d\theta}{dt}\right)^2 = \frac{3g}{2a}\cos\theta + C.$$

But when $\theta = 0$, $\frac{d\theta}{dt} = \omega$. Hence $C = \omega^2 - \frac{3g}{2a}$.

$$\therefore \left(\frac{d\theta}{dt}\right)^2 = \frac{3g}{2a}\cos\theta + \omega^2 - \frac{3g}{2a}$$

$$\Rightarrow \left(\frac{d\theta}{dt}\right)^2 = \omega^2 - \frac{3g}{2a}(1 - \cos\theta). \tag{3.12}$$

This gives the angular velocity of the rod at any instant.

(*b*) The rod will perform complete revolution in vertical plane if ω is such that $\theta = 0$ when $\theta = \pi$.

Therefore, from equation (3.12), we have,

$$0 = \omega^2 - \frac{3g}{2a}\cdot 2;\ \omega^2 = \frac{3g}{a}, \text{ or, } \omega = \sqrt{\frac{3g}{a}},$$

which is the least angular velocity with which the rod must begin to move to perform a complete revolution in the vertical plane.

(*c*) If ω be such that $\frac{d\theta}{dt} = 0$, when $\theta = \pi$, then from equation (3.12)

$$0 = \omega^2 - \frac{3g}{2a}\cdot 2.$$

Hence

$$\omega = \sqrt{\frac{3g}{a}}.$$

Therefore, from equation (3.12), we have

$$\left(\frac{d\theta}{dt}\right)^2 = \frac{3g}{a} - \frac{3g}{2a}(1-\cos\theta) = \frac{3g}{2a}(1+\cos\theta) = \frac{3g}{a}\cos^2\frac{\theta}{2},$$

$$\text{or, } \frac{d\theta}{dt} = \sqrt{\frac{3g}{a}}\cos\frac{\theta}{2}.$$

If t be the time of describing an angle θ, then

$$dt = \int \sqrt{\frac{a}{3g}} \sec\frac{\theta}{2}\, d\theta,$$

$$\Rightarrow \quad \int_0^t dt = \int_0^\theta \sqrt{\frac{a}{3g}} \sec\frac{\theta}{2}\, d\theta,$$

$$\Rightarrow \quad t = 2\sqrt{\frac{a}{3g}} \left[\log\tan\left(\frac{\theta}{4}+\frac{\pi}{4}\right)\right]_0^\theta;$$

$$\Rightarrow \quad t = 2\sqrt{\frac{a}{3g}} \log\tan\left(\frac{\theta}{4}+\frac{\pi}{4}\right).$$

Problem 2. *A uniform rod AB is freely moveable on a rough inclined plane, where inclination to the horizon is i and whose coefficient of friction is μ. About a smooth pin fixed through the end A, the bar is held in the horizontal position in the plane and allowed to falls from this position. If θ be the angle through which it falls from rest, show that* $\frac{\sin\theta}{\theta} = \mu \cot i$. [CU 94H; VU 93H]

Solution: Let AB be the rod movable on a rough inclined plane about the end A. Let at any time t, the rod fall through an angle θ from its original horizontal position.

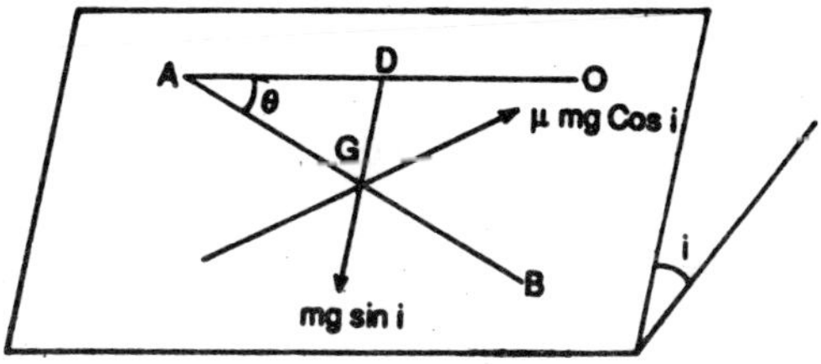

FIG 3.7

If R is the reaction of the plane, then

$$R = mg\cos i. \tag{3.13}$$

The external forces on the rod in the plane are

1. the resolved part of its weight $mg\sin i$, acting downwards along the line of greatest slope;
2. the frictional force μR, acting perpendicular to AB at its middle point G and
3. the reaction at A.

Now the equation of motion of a rigid body about a fixed axis is

$$MK^2\frac{d^2\theta}{dt^2} = L, \tag{3.14}$$

where L is the moment of external forces about the axis of rotation.

Hence from equation (3.14),

$$mK^2\frac{d^2\theta}{dt^2} = mg\sin i\, AD - \mu RAG.$$

Hence from equation (3.13),

$$mK^2\frac{d^2\theta}{dt^2} = mg\sin i\, a\cos\theta - \mu mg\cos i\, a,$$

$$\Rightarrow K^2\frac{d^2\theta}{dt^2} = ga(\sin i\cos\theta - \mu\cos i).$$

Multiplying both the sides by $2\frac{d\theta}{dt}$, we get

$$2K^2\frac{d^2\theta}{dt^2}\left(\frac{d\theta}{dt}\right) = 2ga(\sin i\cos\theta - \mu\cos i)\frac{d\theta}{dt},$$

$$\text{or,}\quad \frac{K^2 d}{dt}\left\{\left(\frac{d\theta}{dt}\right)^2\right\} = 2ag(\sin i\cos\theta - \mu\cos i)\frac{d\theta}{dt}.$$

Integrating both the sides, we get

$$K^2\left(\frac{d\theta}{dt}\right)^2 = 2ag(\sin i\sin\theta - \mu\theta\cos i) + C.$$

Initially when $\theta = 0$, $\frac{d\theta}{dt} = 0$, $\therefore\ C = 0$.
Hence

$$K^2\left(\frac{d\theta}{dt}\right)^2 = 2ag(\sin i\sin\theta - \mu\theta\cos i).$$

Now the rod will fall till $\frac{d\theta}{dt} = 0$.
Thus

$$\sin i\sin\theta - \mu\theta\cos i = 0 \text{ or, } \frac{\sin\theta}{\theta} = \mu\cot i.$$

Problem 3. *A fine string has two masses M and M′ tied to its ends and passes over a rough pulley of mass m, whose centre is fixed. If the string does not slip over the pulley, show that M will descend with acceleration*

$$\frac{(M - M')g}{M + M' + \frac{mk^2}{a^2}},$$

where a is the radius and K the radius of gyration of the pulley. [CU 92H]

Solution: Let x be the distances advanced in time t by the masses M and M′ attached at the ends C and D, respectively. Let θ be the angle turned by the pulley during this time.

Then $x = a\theta$

$$\Rightarrow \dot{x} = a\dot{\theta} \quad\text{and}\quad \ddot{x} = a\dot{\theta} \qquad (3.15)$$

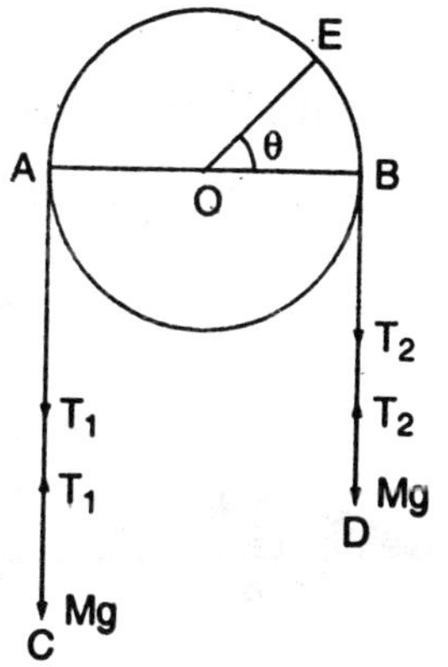

FIG 3.8

Let T_1 and T_2 be the tensions in the strings AC and BD, respectively.
$\therefore$ the equations of motion of masses M and M′ are

$$M\ddot{x} = Mg - T_1 \tag{3.16}$$

$$\text{and} \quad M'\ddot{x} = T_2 - M'g. \tag{3.17}$$

Also the equation of motion of the pulley is

$$I\ddot{\theta} = T_1 \cdot a - T_2 \cdot a.$$

where I is the moment of inertia of the pulley about its axis, i.e., $I = mk^2$.

$$\therefore \quad mk^2\ddot{\theta} = T_1 \cdot a - T_2 \cdot a$$

$$\Rightarrow \quad mk^2\left(\frac{\ddot{x}}{a}\right) = (Mg - M\ddot{x})a - (M'\ddot{x} + M'g)a$$

[From equations (3.15), (3.16) and (3.17)]

$$\Rightarrow \quad (mk^2 + Ma^2 + M'a^2)\ddot{x} = (Mg - M'g)a^2$$

$$\Rightarrow \quad \ddot{x} = \frac{(M - M')ga^2}{a^2\left(\frac{mk^2}{a^2} + M + M'\right)}$$

$$\Rightarrow \quad \ddot{x} = \frac{(M - M')g}{M + M' + \frac{mk^2}{a^2}}.$$

Type II: Problems based on Compound Pendulum

Problem 4. *A uniform wire in the form of an arc of a circle of given radius is swinging about a horizontal axis through the middle point of the arc perpendicular to the plane of the arc. Show that the time of a small oscillation is independent of the length of the arc, and length of the equivalent simple pendulum is equal to the diameter of the circle.* **[IAS 78; PU 49H; BU 61H; MU 63H]**

Solution: Let ACB be an arc of a radius subtending an angle 2α at its centre O swinging about a horizontal axis CN through the middle point C perpendicular to the plane of the arc.

Let K be the radius of gyration of the arc about CN. Then moment of inertia of the arc about CN

$$= MK^2 = M\frac{2a^2}{\alpha}(\alpha - \sin\alpha)$$

[See WO3 (iii) of M.I.]

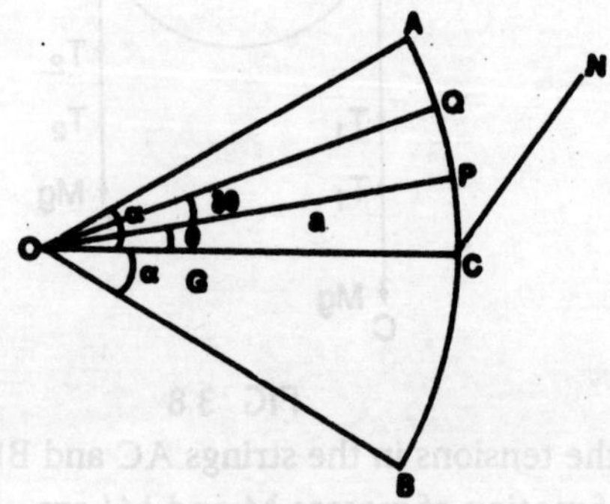

FIG 3.9

Hence

$$K^2 = \frac{2a^2}{\alpha}(\alpha - \sin\alpha)$$

and h = distance of C.G. of the arc from C.

$$\therefore\ GC = OC - OG = a - \frac{a\sin\alpha}{\alpha} = \frac{a}{\alpha}(\alpha - \sin\alpha).$$

$\therefore$ length of the simple equivalent pendulum $= l = \frac{K^2}{h} = 2a$ = diameter of the circle.

Also the time of oscillation

$$T = 2\pi\sqrt{\frac{K^2}{gh}} = 2\pi\sqrt{\frac{2a}{g}}.$$

which is independent of α, i.e., T is independent of the length of the arc.

Problem 5. *A pendulum is supported at O and P is the centre of oscillation. Show that if an additional weight is rigidly attached at P, the period of oscillation is unaltered.* **[BU 95H, 93H; IAS 84; MU 74, 77, MU 03H, 05H]**

Solution: Let m be the mass of the body forming the compound pendulum and let h be the depth of its C.G. below the point of suspension O. Let K be the radius of gyration about the horizontal axis through O. Then we have

$$OP = K^2/h.$$

So period of oscillation

$$= 2\pi\sqrt{K^2/gh} = T \text{ (say)}.$$

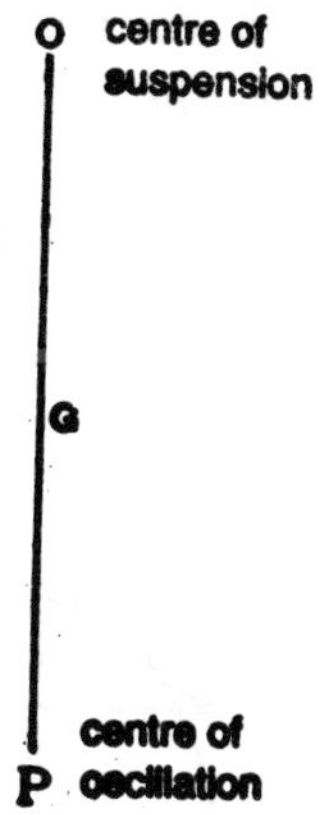

FIG 3.10

Let an additional weight M be attached at P, then if K′ be the radius of gyration about the horizontal axis through O, we have by the theorem of parallel axes on moment of inertia,

Moment of inertia about the axis of rotation = moment of inertia of the body + moment of inertia of the mass M at P about the axes of rotation.

$$\text{or, } (M+m)K'^2 = mK^2 + M\cdot OP^2,$$

$$\text{or, } (M+m)K'^2 = mK^2 + M\left(\frac{K^2}{h}\right)^2,$$

$$\text{or, } (M+m)K'^2 = K^2\left(m + M\frac{K^2}{h^2}\right). \tag{3.18}$$

If h′ be the depth of the combined C.G. below O, then we have

$$h' = \frac{mh + M\cdot OP}{m+M} = \frac{mh + M\frac{K^2}{h}}{m+M} = \frac{h\left(m+M\frac{K^2}{h^2}\right)}{m+M},$$

$$\text{or, } \quad (m+M)h' = h\left(m + M\frac{K^2}{h^2}\right). \tag{3.19}$$

Dividing equation (3.16) by equation (3.17), we get

$$\frac{K'2}{h'} = \frac{K^2}{h}. \tag{3.20}$$

Let T′ be the period of oscillation when the weight is rigidly attached at P. Then

$$T' = 2\pi\sqrt{\frac{K'^2}{gh'}} = 2\pi\sqrt{\frac{K^2}{gh}} = T \text{ [From equation (3.20)]}$$

Hence the period of oscillation is unaltered.

Problem 6. *An elliptical lamina is such that when it swings about one latus rectum as a horizontal axis, the other latus rectum passes through the centre of oscillation, prove that the eccentricity is 1/2.* **[BBABU 94H; IAS 73, TMBU 98H, 04H]**

Solution: L_1SL_2 is the latus rectum which is horizontal and its axis of rotation.

$\therefore$ S is the centre of suspension. The other latus rectum $L_1'S'L_2'$ passes through the centre of oscillation S'.

$\therefore$ SS_2' = length of simple equivalent pendulum

$$K^2/h = 2ae. \tag{3.21}$$

Also moment of inertia about OB $= \frac{Ma^2}{4}$.

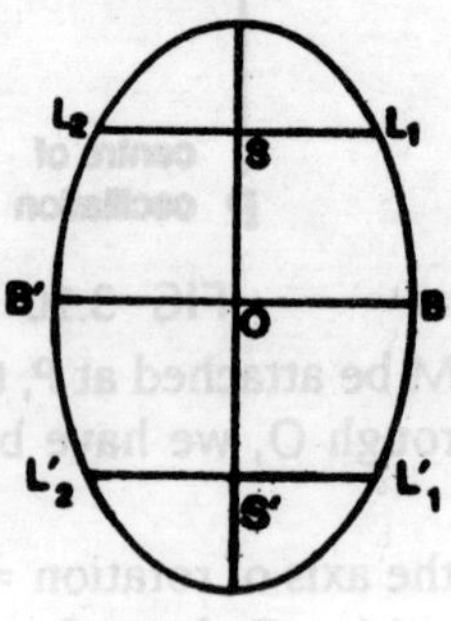

FIG 3.11

Hence be the theorem of parallel axes,

Moment of inertia about L_1L_2 = moment of inertia about OB + moment of inertia of the total mass M at O about L_1L_2.

$$MK^2 = \frac{Ma^2}{4} + M(OS)^2,$$

$$MK^2 = \frac{Ma^2}{4} + Ma^2e^2,$$

where K is the radius of gyration about L_1SL_2. Hence

$$K^2 = a^2\left(\frac{1}{4} + e^2\right).$$

If h be the depth of C.G. of the lamina below L_1SL_2, then $h = OS = ae$.

Hence from equation (3.21)

$$2ae = \frac{a^2}{h}\left(\tfrac{1}{4} + e^2\right), \text{ or, } 2ae = \frac{a^2}{ae}\left(\tfrac{1}{4} + e^2\right)$$

$$\text{or, } \quad 2e^2 = \tfrac{1}{4} + e^2, \text{ or, } e^2 = \tfrac{1}{4}, \; e = \tfrac{1}{2}.$$

Problem 7. *Find the time of oscillation of a compound pendulum consisting of a rod of mass m and length a, carrying at one end a sphere of mass m_1 and diameter 2b, the other end of the rod being fixed.*

[BHU 65H; LNMU 93H, BBABU 65H, 98H, 01H, TMBU 93H, 95H, MU 98H, 02H]

Solution: Let OA be the rod of mass m and length a attached at A to a sphere of radius b and mass m_1.

The end O of the rod is fixed and hence axis of rotation passes through O. moment of inertia of the rod OA about an axis through its extremity

$$= \frac{4}{3}m\left(\frac{a}{2}\right)^2 = \frac{ma^2}{3}.$$

Again moment of inertia of the sphere about an axis through its centre $= \frac{2}{5}m_1 b^2$.

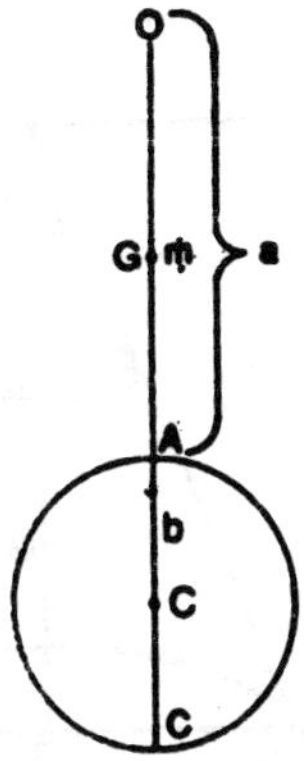

FIG 3.12

Hence by the theorem of parallel axis.

Moment of inertia of the sphere about an axis through O

= M.I. of the sphere about an axis through C
+ M.I. of the whole mass about an axis through O

$$= m_1 \tfrac{2}{5} b^2 + m_1(a+b)^2.$$

If K be the radius of gyration about an axis through O, then

$$(m+m_1)K^2 = \frac{ma^2}{3} + m_1\left[\frac{2}{5}b^2 + (a+b)^2\right]$$

$$\therefore\ K^2 = \frac{\frac{ma^2}{3} + m_1\left[\frac{2}{5}b^2 + (a+b)^2\right]}{m+m_1}. \quad (3.22)$$

Let h be the depth of C.G. of the system below O, then

$$h = \frac{m\cdot OG + m_1\cdot OC}{m+m_1} = \frac{m\cdot\frac{a}{2} + m_1(a+b)}{m+m_1}.$$

Hence the time of oscillation of compound pendulum

$$= 2\pi\sqrt{\frac{K^2}{gh}} = \frac{2\pi}{\sqrt{g}}\sqrt{\frac{\frac{ma^2}{3} + m_1\left\{\frac{2}{5}b^2 + (a+b)^2\right\}}{m\frac{a}{2} + m_1(a+b)}}.$$

Problem 8. *A solid homogeneous cone of height* h *and vertical angle* 2α*, oscillates about a horizontal axis through its vertex. Show that the length of the simple equivalent pendulum is* $\frac{h}{5}(4+\tan^2\alpha)$.

[PU 56H; BBABU 59H, 72H: RU 62H, 67H; BHU 62H; MU 74H, 79H, 81H; LNMU 88H; Meerut 76, 77, 80; BU 92H; VU 87H; BPSC 83, 82; Raj 79]

Solution: Let the cone OAB of height h and vertical angle 2α oscillate about the horizontal axis OL through the vertex O.

Let M be the mass of the cone

$$\therefore\ M = \tfrac{1}{3}\pi h^3 \tan^2 \alpha\rho, \tag{3.23}$$

where ρ is the density.

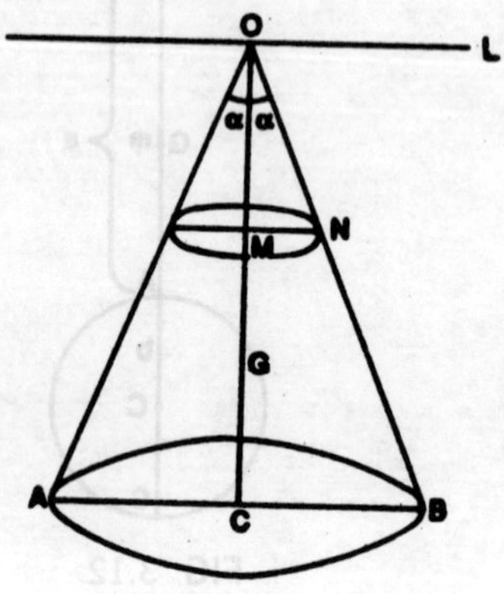

FIG 3.13

To find the moment of inertia of the cone about OL, we consider an elementary disc of width δx at a distance x from O.

Mass of the elementary disc $= \rho\pi x^2 \tan^2 \alpha\delta x$.

Hence by the theorem of parallel axes,

Moment of inertia of the disc about OL = moment of inertia of the disc about MN+ moment of inertia of the total mass about OL

$$\begin{aligned}
&= \pi x^2 \tan^2 \alpha\rho\delta x \frac{MN^2}{4} + \pi\rho x^2 \tan^2 \alpha\delta x \cdot OM^2 \\
&= \rho\pi \tan^2 \alpha \left(\frac{x^2 \tan^2 \alpha}{4} + x^2\right) \delta x\, x^2 \\
&= \rho\pi \tan^2 \alpha \left(\frac{1}{4} \tan^2 \alpha + 1\right) x^4 \delta x.
\end{aligned}$$

Hence moment of inertia of the cone about OX,

$$\begin{aligned}
MK^2 &= \int_0^h \rho\pi \tan^2 \alpha \left(\frac{1}{4} \tan^2 \alpha + 1\right) x^4\, dx \\
&= \rho\pi \tan^2 \alpha \left(\frac{1}{4} \tan^2 \alpha + 1\right) \frac{h^5}{5} \\
&= \frac{3M}{h^3} \left(\frac{1}{4} \tan^2 \alpha + 1\right) \frac{h^5}{5} \quad \text{[From equation (3.23)]} \\
&= \frac{3M}{20} (\tan^2 \alpha + 4).
\end{aligned}$$

Hence

$$K^2 = \frac{3(\tan^2\alpha+4)}{20}$$

$$h = \text{depth of C.G.} = OG = \frac{3}{4}h.$$

Therefore, length of simple equivalent pendulum

$$\frac{K^2}{h} = \frac{3(\tan^2\alpha+4)}{20}\cdot\frac{4}{3h} = \frac{1}{5}(\tan^2\alpha+4).$$

Problem 9. *A uniform triangular lamina can oscillate in its own plane about the angle A. Prove that the length of the simple equivalent pendulum is $\frac{3(b^2+c^2)-a^2}{4\sqrt{2(b^2+c^2)-a^2}}$, the axis through A being horizontal.* **[BU 70H; LNMU 86H; IAS 48, 91]**

Solution: Write the complete solution of question number 11 of worked-out examples of the 1st chapter and add the following:

Since

$$\text{M.I.} = \frac{M(3b^2+3c^2-a^2)}{12},$$

hence

$$K^2 = \frac{3b^2+3c^2-a^2}{12}.$$

Also

$$h = AG = \frac{2}{3}AD = \frac{2}{3}\frac{\sqrt{2b^2+2c^2-a^2}}{2} = \frac{1}{3}\sqrt{2b^2+2c^2-a^2}.$$

Hence length of the simple equivalent pendulum

$$l = \frac{K^2}{h} = \frac{3b^2+3c^2-a^2}{12}\cdot\frac{3}{\sqrt{2b^2+2c^2-a^2}}$$

$$= \frac{3b^2+3c^2-a^2}{4\sqrt{2b^2+2c^2-a^2}} \text{ (Proved).}$$

Problem 10. *A bent lever whose arms are of length a and b, the angle between them being α, makes small oscillation in its own plane about the fulcrum. Show that the length of the corresponding simple pendulum is $\frac{2}{3}\frac{a^3+b^3}{\sqrt{a^4+2a^2b^2\cos\alpha+b^4}}$.*

[BBABU 64H, 67H, 69H; RU 65H; BPSC 88; TMBU 99H]

Solution: Let the bent lever with arms $OA = a$ and $OB = b$, and $\angle AOB = \alpha$, oscillate its own plane about the fulcrum O. If w is the weight per unit length of the arm, then the weights aw and bw of the arms will act at their middle points G_1 and G_2, respectively.

Taking OA and a perpendicular line OY as the axes of x and y, then coordinates of G_1 and G_2 are $\left(\frac{a}{2}, 0\right)$, and $\left(\frac{b}{2}\cos\alpha, \frac{b}{2}\sin\alpha\right)$, respectively.

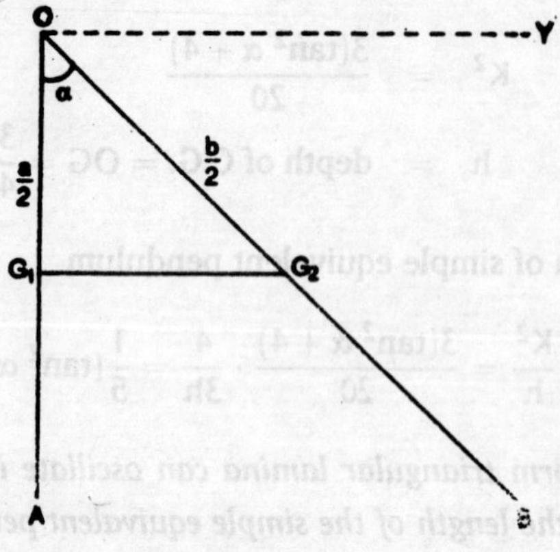

FIG 3.14

Let $(\bar{x}, \bar{y})$ be the coordinates of the C.G. of the liver. Then

$$\bar{x} = \frac{aw\frac{a}{2} + bw\frac{b}{2}\cos\alpha}{aw + bw} = \frac{a^2 + b^2\cos\alpha}{2(a+b)}$$

$$\text{and } \bar{y} = \frac{aw \cdot 0 + bw\frac{b}{2}\sin\alpha}{aw + bw} = \frac{b^2\sin\alpha}{2(a+b)}.$$

If h be the distance of C.G. of the lever from O, then

$$h = \sqrt{\bar{x}^2 + \bar{y}^2} = \frac{1}{2(a+b)}\sqrt{(a^2 + b^2\cos\alpha)^2 + b^4\sin^2\alpha}$$

$$= \frac{\sqrt{a^4 + b^4 + 2a^2b^2\cos\alpha}}{2(a+b)}.$$

If K be the radius of gyration about the axis of rotation through O, then

$$(aw + bw)K^2 = aw\frac{4}{3}\left(\frac{a}{2}\right)^2 + bw\frac{4}{3}\left(\frac{b}{2}\right)^2$$

$$\text{or, } (a+b)K^2 = \frac{1}{3}(a^3 + b^3)$$

$$\therefore K^2 = \frac{a^3 + b^3}{3(a+b)}.$$

Hence the length of corresponding simple equivalent pendulum

$$= \frac{K^2}{h} = \frac{a^3 + b^3}{3(a+b)} \cdot \frac{2(a+b)}{\sqrt{a^4 + b^4 + 2a^2b^2\cos\alpha}}$$

$$= \frac{2}{3}\frac{a^3 + b^3}{\sqrt{a^4 + b^4 + 2a^2b^2\cos\alpha}}.$$

Type III: Problems based on the Reactions of the Axis of Rotation

Problem 11. *A uniform rod of length $2a$ and weight w is turning about its end O and starts from the position in which it was vertically above G. When it has turned through an angle θ, show that the horizontal and vertical reactions at O are $\frac{3}{4}w\sin\theta$ $(2-3\cos\theta)$ and $\frac{w}{4}$ $(1-3\cos\theta)^2$.* [CU 99H]

Solution: Let OA be a uniform rod of length $2a$ and weight w be turning about O. Let G be the C.G. of the rod.

The C.G. is describing a circle of radius a about O.

If P and Q be the reactions along GO and perpendicular to GO, then equations of motion of C.G. are

$$Ma\frac{d^2\theta}{dt^2} = Q + Mg\sin\theta \tag{3.24}$$

$$\text{and} \quad Ma\left(\frac{d\theta}{dt}\right)^2 = P + Mg\cos\theta. \tag{3.25}$$

If K be the radius of gyration about the axis of rotation, then

$$MK^2 = \frac{4}{3}Ma^2, \quad \therefore\ K^2 = \frac{4}{3}a^2.$$

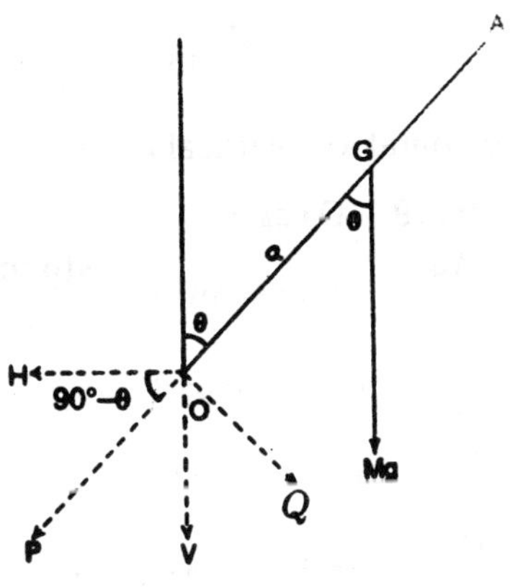

FIG 3.15

Also equation of motion is

$$MK^2\frac{d^2\theta}{dt^2} = Mg\,a\sin\theta, \text{ or, } \frac{4}{3}a^2\frac{d^2\theta}{dt^2} = ga\sin\theta$$

$$\therefore\quad \frac{d^2\theta}{dt^2} = \frac{3g}{4a}\sin\theta. \tag{3.26}$$

Hence from equations (3.24) and (3.26).

$$Ma\frac{3g}{4a}\sin\theta = Q + Mg\sin\theta$$

$$\text{or,}\quad \frac{3}{4}Mg\sin\theta - Mg\sin\theta = Q$$

$$\therefore\quad Q = -\frac{Mg\sin\theta}{4}. \tag{3.27}$$

Multiplying both the sides of equation (3.26) by $2\frac{d\theta}{dt}$.

$$2\frac{d^2\theta}{dt^2}\left(\frac{d\theta}{dt}\right) = \frac{3g}{2a}\sin\theta\frac{d\theta}{dt}, \text{ or, } \frac{d}{dt}\left[\left(\frac{d\theta}{dt}\right)^2\right] = \frac{3g}{2a}\sin\theta\frac{d\theta}{dt}.$$

Integrating, we get

$$\left(\frac{d\theta}{dt}\right)^2 = -\frac{3g}{2a}\cos\theta + C.$$

Initially, when $\theta = 0$, $\frac{d\theta}{dt} = 0$, $\therefore C = \frac{3g}{2a}$.

$$\therefore \left(\frac{d\theta}{dt}\right)^2 = \frac{3g}{2a}(1-\cos\theta).$$

Putting the value of $\left(\frac{d\theta}{dt}\right)^2$ in equation (3.24), we get

$$\begin{aligned} Ma\frac{3g}{2a}(1-\cos\theta) &= P + Mg\cos\theta \\ \therefore P &= \frac{Mg}{2}3(1-\cos\theta) - Mg\cos\theta \\ &= \frac{Mg}{2}[3 - 3\cos\theta - 2\cos\theta] \\ &= \frac{Mg}{2}(3 - 5\cos\theta). \end{aligned}$$

Let H and V be the horizontal and vertical reaction, then

$$\begin{aligned} H &= P\sin\theta - Q\cos\theta \\ &= \frac{Mg}{2}(3-5\cos\theta)\sin\theta + \frac{Mg\sin\theta}{4}\cos\theta \\ &= \frac{Mg}{4}[6\sin\theta - 10\cos\theta\sin\theta + \sin\theta\cos\theta] \\ &= \frac{Mg}{4}(6\sin\theta - 9\sin\theta\cos\theta) \\ &= \frac{3Mg}{4}(2\sin\theta - 3\sin\theta\cos\theta) \\ &= \frac{3}{4}w\sin\theta(2-3\cos\theta)\ [\because\ w = Mg] \\ \text{and } V &= P\cos\theta + Q\sin\theta \\ &= \frac{Mg}{2}(3-5\cos\theta)\cos\theta + \left(-\frac{Mg\sin\theta}{4}\right)\sin\theta \\ &= \frac{Mg}{4}[6\cos\theta - 10\cos^2\theta - \sin^2\theta] \\ &= \frac{w}{4}[6\cos\theta - 10\cos^2\theta - (1-\cos^2\theta)] \\ &= \frac{w}{4}[-9\cos^2\theta + 6\cos\theta - 1] \\ &= -\frac{w}{4}(3\cos\theta - 1)^2 \end{aligned}$$

negative sign shows that the vertical reaction is upwards.

Problem 12. *A thin uniform rod has one end attached to a smooth hinge and is allowed to fall from a horizontal position. Show that the horizontal strain on the hinge is greatest when the rod is inclined at an angle of 45° to the vertical and that the vertical strain is then $\frac{11}{8}$ times the weight of the rod.* **[CU 98H]**

Solution: Let $2a$ be the length and M be the mass of the rod OA.

Let the rod make an angle θ with vertical at any time t. The C.G. at G is describing a circle of radius a about O. Hence the equations of motion of the C.G. are

$$Ma\left(\frac{d\theta}{dt}\right)^2 = P - Mg\cos\theta \tag{3.28}$$

$$\text{and } Ma\frac{d^2\theta}{dt^2} = Q - Mg\sin\theta. \tag{3.29}$$

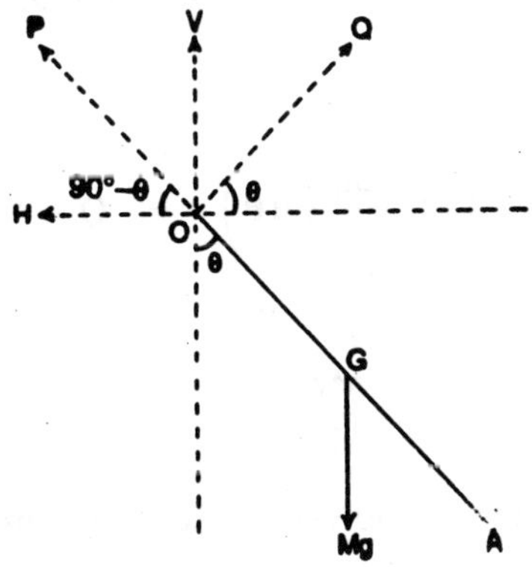

FIG 3.16

Also

$$K^2 = \frac{a^2}{3} + a^2 = \frac{4}{3}a^2.$$

Hence the equation of motion is

$$MK^2\frac{d^2\theta}{dt^2} = -Mg\,a\sin\theta$$

$$\frac{4}{3}a^2\frac{d^2\theta}{dt^2} = -ag\sin\theta$$

$$\text{or, } \frac{d^2\theta}{dt^2} = -\frac{3g}{4a}\sin\theta. \tag{3.30}$$

Multiplying both the sides by $2\frac{d\theta}{dt}$, we get

$$2\frac{d\theta}{dt}\frac{d^2\theta}{dt^2} = -\frac{3g}{2a}\sin\theta\frac{d\theta}{dt},$$

$$\text{or, } \frac{d}{dt}\left\{\left(\frac{d\theta}{dt}\right)^2\right\} = -\frac{3g}{2a}\sin\theta\frac{d\theta}{dt}.$$

Integrating, we get

$$\left(\frac{d\theta}{dt}\right)^2 = \frac{3g}{2a}\cos\theta + C.$$

Initially, when

$$\theta = \frac{\pi}{2};\ \frac{d\theta}{dt} = 0,\ C = 0.$$

Hence

$$\left(\frac{d\theta}{dt}\right)^2 = \frac{3g}{2a}\cos\theta. \tag{3.31}$$

From equations (3.28) and (3.31), we get

$$Ma\frac{3g}{2a}\cos\theta = P - Mg\cos\theta$$

$$\therefore P = \frac{5}{2}Mg\cos\theta.$$

From equations (3.29) and (3.30), we have

$$Ma\left(-\frac{3g}{4a}\sin\theta\right) = Q - Mg\sin\theta$$

$$\text{or, } Mg\sin\theta - \frac{3Mg}{4}\sin\theta = Q$$

$$\therefore \frac{Mg}{4}\sin\theta = Q.$$

If H and V be the horizontal and vertical strain on the hinge at O, then

$$H = P\sin\theta - Q\cos\theta = \frac{5}{2}Mg\cos\theta\sin\theta - \frac{Mg}{4}\sin\theta\cos\theta$$

$$= Mg\sin\theta\cos\theta\left(\frac{5}{2} - \frac{1}{4}\right) = \frac{9}{8}Mg\sin^2\theta \tag{3.32}$$

$$\text{and } V = P\cos\theta + Q\sin\theta = \frac{5}{2}Mg\cos^2\theta + \frac{Mg}{4}\sin^2\theta. \tag{3.33}$$

Now H will be maximum when $\sin 2\theta = 1$, i.e., $\theta = 45°$, i.e., when the rod is inclined at an angle of $45°$ to the vertical.

When $\theta = 45°$, then

$$V = \frac{5}{2}Mg\frac{1}{2} + \frac{Mg}{4}\frac{1}{2} = \frac{(10+1)Mg}{8} = \frac{11Mg}{8}$$

$$= \frac{11}{8}w\ (w = Mg) = \frac{11}{8}\text{ times the weight of the rod.}$$

Problem 13. *A circular area can turn freely about a horizontal axis which passes through a point O of its circumference and is perpendicular to its plane. If motion commences when the diameter through O is vertically above O, show that, when the diameter has turned through an angle θ, the components of the strain at O along and perpendicular to this diameter are respectively* $\frac{w}{3}(7\cos\theta - 4)$ *and* $\frac{w}{3}\sin\theta$.

Solution: Let a be the radius of the disc of mass M.

Let OA be the vertical diameter initially and after time t, it is making an angle θ with the vertical. Let G be the C.G. of OA.

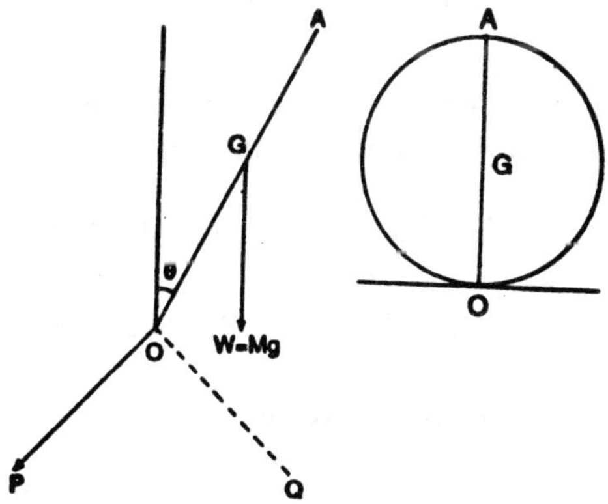

FIG 3.17

Now

$$\begin{aligned} MK^2 &= \text{M.I. of the disc about an axis through O perpendicular to its plane.} \\ &= \frac{Ma^2}{2} + Ma^2 = \frac{3}{2}Ma^2 \\ \therefore K^2 &= \frac{3}{2}a^2 \\ h &= \text{distance of C.G. of the disc from O} = a. \end{aligned}$$

Equation of motion is

$$MK^2\frac{d^2\theta}{dt^2} = Mgh\sin\theta, \quad \text{or,} \quad \frac{3}{2}a^2\frac{d^2\theta}{dt^2} = ga\sin\theta$$

$$\text{or,} \quad \frac{d^2\theta}{dt^2} = \frac{2g}{3a}\sin\theta. \tag{3.34}$$

Multiplying both the sides by $2\frac{d\theta}{dt}$, we get

$$2\frac{d\theta}{dt}\frac{d^2\theta}{dt^2} = \frac{2g}{3a}\sin\theta\frac{2d\theta}{dt}$$

$$\text{or,} \quad \frac{d}{dt}\left\{\left(\frac{d\theta}{dt}\right)^2\right\} = \frac{4g}{3a}\sin\theta\frac{d\theta}{dt}.$$

Integrating, we get

$$\left(\frac{d\theta}{dt}\right)^2 = -\frac{4g}{3a}\cos\theta + C.$$

Initially, when $\theta = 0$, $\frac{d\theta}{dt} = 0$.

$$C = \frac{4g}{3a}$$

$$\therefore \left(\frac{d\theta}{dt}\right)^2 = \frac{4g}{3a}(1 - \cos\theta). \tag{3.35}$$

Let P and Q be the components of the reaction of the axis along and perpendicular to GO. Then considering the motion of the C.G., we get

$$Ma\frac{d^2\theta}{dt^2} = Q + Mg\sin\theta \tag{3.36}$$

$$\text{and } Ma\left(\frac{d\theta}{dt}\right)^2 = P + Mg\cos\theta. \tag{3.37}$$

From equations (3.34) and (3.36), we get

$$Q = Ma\left(\frac{2g}{3a}\sin\theta\right) - Mg\sin\theta = -\frac{Mg}{3}\sin\theta = -\frac{w}{3}\sin\theta \{\because w = Mg]$$

negative sign shows that Q is in a direction opposite to that taken in the figure and its magnitude is $\frac{w}{3}\sin\theta$.

Again from equations (3.34) and (3.36), we have

$$Ma\frac{4g}{3a}(1 - \cos\theta) - Mg\cos\theta = P$$

$$\text{or, } Mg\left[\frac{4}{3} - \frac{4\cos\theta}{3} - \cos\theta\right] = P, \text{ or, } Mg\left[\frac{4}{3} - \frac{7\cos\theta}{3}\right] = P.$$

$$\therefore P = -\frac{Mg}{3}(7\cos\theta - 4) = -\frac{w}{3}(7\cos\theta - 4).$$

Here negative sign shows that O is also in opposite direction to what is shown in the figure.

Type IV: Problems based on the Impulsive Forces

Problem 14. *A uniform beam AB can turn freely about its end and is in equilibrium. Find the points of its length, where a blow must be applied to it so that the impulses at A may be in each case $\frac{1}{n}$th that of the blow.*

Solution: Let the beam AB be of mass M and length $2a$, which is free to turn about its end A. Let a blow P be applied to the beam at the point C at a distance x from A which produces an impulsive action $X = \frac{1}{n}P$ at A and an angular velocity ω.

Then the equations of motion are

$$MK^2(\omega - 0) = Px \text{ (3.38) and } Ma(W - 0) = P + \frac{P}{n} \text{ (3.39)}.$$

Dividing equation (3.35) by equation (3.36), we get

$$\frac{K^2}{a} = \frac{x}{1 + \frac{1}{n}} = \frac{nx}{n + 1}.$$

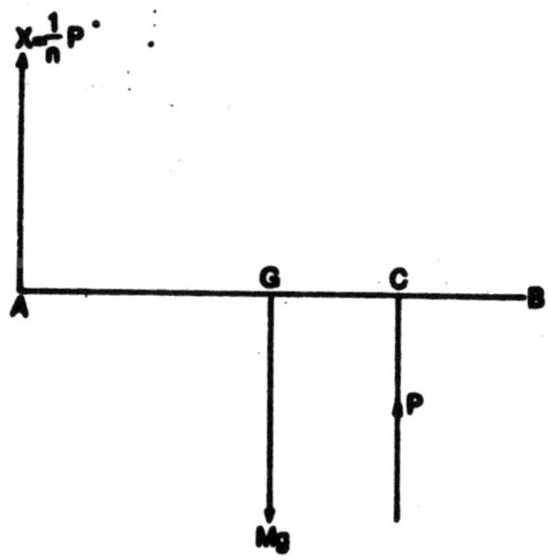

FIG 3.18

But $K^2 = \frac{4}{3}a^2$.

Hence

$$\frac{4}{3}a = \frac{nx}{n+1}; \quad nx = \frac{4a(n+1)}{3}; \quad x = \frac{4a}{3}\left(1 + \frac{1}{n}\right).$$

If X is in opposite direction, we have

$$x = \frac{4}{3}\left(\frac{n-1}{n}\right)a.$$

Problem 15. *A rod of mass* m *and length* $2a$ *which is capable of free motion about one end* A, *falls from a vertical position and when it is horizontal strikes a fixed inelastic obstacle at a distance* b *from the end* A. *Prove that the impulse of the blow is* $m\frac{2a}{b}\sqrt{\frac{2ga}{3}}$ *and that the impulse of the reaction at* A *is* $m\sqrt{\frac{3ga}{2}}\left(1 - \frac{4a}{3b}\right)$ *vertically upwards.* [CU 83H, 97H]

Solution: Let the rod AB of length $2a$ and capable of motion about the fixed end A fall from its vertical position and strike a fixed inelastic obstacle at C at a distance AC $= b$, from A.

Let ω be the angular velocity of the rod just before striking the obstacle.

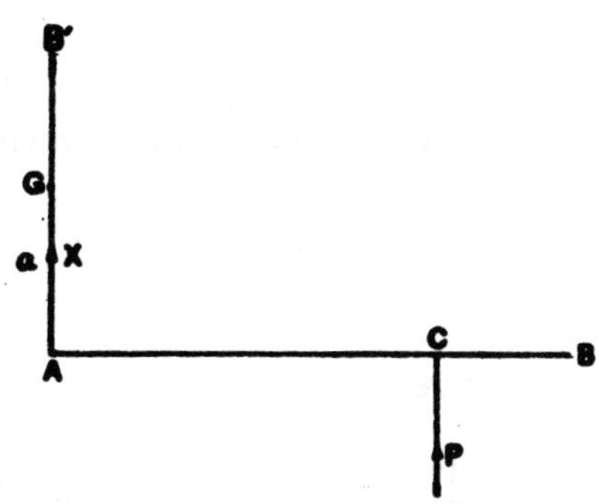

FIG 3.19

Then from energy equation, change in kinetic energy = work done, i.e.,

$$\tfrac{1}{2}mk^2\omega^2 - 0 = mg\,a.$$

But $K^2 = \frac{4}{3}a^2$.

$$\frac{m}{2}\frac{4a^2}{3}\omega^2 = mga, \text{ or, } \frac{2a}{3}\omega^2 = g.$$

$$\therefore\ \omega = \sqrt{\frac{3g}{2a}}.$$

The rod is reduced to rest after striking the obstacle at C. If P is the impulse of the obstacle and X the impulse of the reaction at A, then taking moment about A, we have

$$m \cdot \frac{4}{3}a^2(0-\omega) = -P \cdot b,$$

$$\Rightarrow\ P = \frac{4}{3}\frac{ma^2}{b}\omega = \frac{4}{3}\frac{ma^2}{b}\sqrt{\frac{3g}{2a}}$$

$$\Rightarrow\ P = \frac{m}{b}\sqrt{\frac{a^4 \cdot 16}{9}\frac{3g}{2a}} = \frac{m}{b}\sqrt{\frac{16}{9}a^4 \cdot \frac{3g}{2a}} = \frac{m}{b}\sqrt{\frac{8a^3g}{3}},$$

$$\Rightarrow\ P = \frac{2ma}{b}\sqrt{\frac{2ag}{3}}.$$

Also for C.G. G, we have

$$m(0 - a\omega) = -P - X$$

$$\Rightarrow\ -am\sqrt{\frac{3g}{2a}} = -\frac{2ma}{b}\sqrt{\frac{2ag}{3}} - X$$

$$\Rightarrow\ X = am\sqrt{\frac{3g}{2a}} - \frac{2am}{b}\sqrt{\frac{2ag}{3}} = m\sqrt{\frac{3ga}{2}}\left[1 - \frac{4a}{3b}\right].$$

Problem 16. *The door of a railway carriage stands open at right angles to the length of the train when the latter starts to move with an acceleration* f, *the door being supposed to be smoothly hinged to the carriage and to be uniform and of breadth 2a. Show that its angular velocity when it has turned through an angle θ is* $\sqrt{\left(\frac{3f}{2a}\sin\theta\right)}$.

Solution: Let AB be the axis of rotation of gate whose width BC = 2a. We consider an elementary strip PQRS of width δx. Mass of the elementary strip PQRS

$$= \frac{m}{2a}\delta x,$$

where m is the mass of the door.

Hence equation of motion gives

$$mK^2\frac{d^2\theta}{dt^2} = \int_0^{2a}\frac{m}{2a}\,dx\,f\cos\theta x.$$

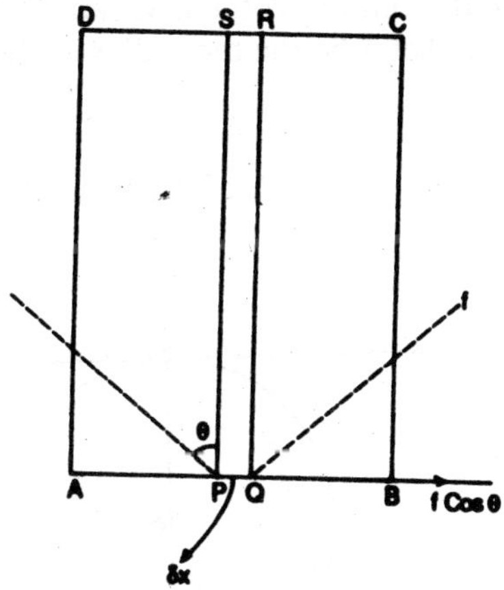

FIG 3.20

Here $K^2 = \frac{4a^2}{3}$.

$$\therefore\ m\frac{4a^2}{3}\frac{d^2\theta}{dt^2} = \frac{m}{2a}f\cos\theta\frac{4a^2}{2} = maf\cos\theta.$$

Multiplying both the sides by $2\frac{d\theta}{dt}$, we get

$$\frac{4}{3}ma^2\frac{d^2\theta}{dt^2}2\frac{d\theta}{dt} = maf\cos\theta 2\frac{d\theta}{dt}$$

$$\text{or, } \frac{d}{dt}\left[\left(\frac{d\theta}{dt}\right)^2\right] = \frac{3f}{2a}\cos\theta\frac{d\theta}{dt}.$$

Integrating, we get

$$\left(\frac{d\theta}{dt}\right)^2 = \frac{3f}{2a}\sin\theta + C.$$

Initially, when $\theta = 0$, $\frac{d\theta}{dt} = 0$.

$$\therefore\ C = 0.$$

Hence

$$\left(\frac{d\theta}{dt}\right)^2 = \frac{3f}{2a}\sin\theta. \quad \therefore\ \frac{d\theta}{dt} = \sqrt{\frac{3f}{2a}\sin\theta}.$$

Type V: Problems based on Centre of Percussion

Problem 17. *Find the position of centre of percussion in the following cases:*

(i) a uniform circular plate, its axis is a horizontal tangent;

(ii) a uniform rod with one end fixed.

Solution: (*i*) Let C be the centre, M the mass and AOB be a horizontal tangent of a uniform circular, plate. Let a be the the radius of the circular plate.

Now

$$MK^2 = \text{M.I. of the plate about AB}$$
$$= \tfrac{1}{4}Ma^2 + Ma^2 = \tfrac{5}{4}Mo^2.$$

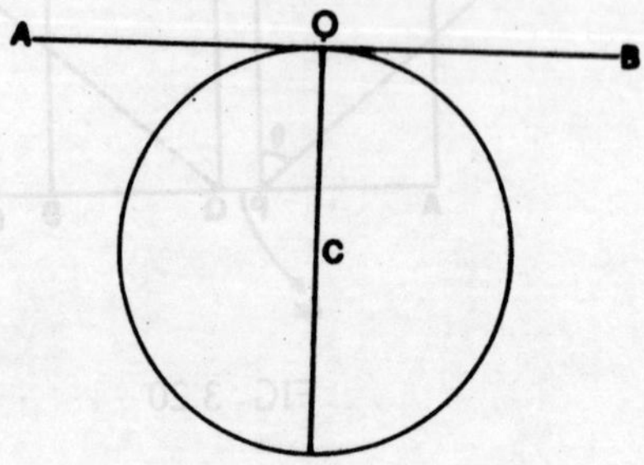

FIG 3.21

Hence

$$K^2 = \frac{5}{4}a^2 \text{ and } h = OC = a.$$

Hence distance of the centre of percussion below the highest point O

$$= \frac{K^2}{h} = \frac{5a^2}{4a} = \frac{a}{5}.$$

(*ii*) Let $2a$ be the length of the rod and M be its mass.

Then

$$h = \text{depth of its C.G. from the fixed end} = a.$$

Also

$$MK^2 = \text{M.I. about the axis of rotation through one end}$$
$$= \frac{Ma^2}{3} + Ma^2 = \frac{4}{3}Ma^2.$$

Hence

$$K^2 = \frac{4}{3}a^2.$$

Hence the distance of centre of percussion from the fixed end

$$\frac{K^2}{h} = \frac{4a^2}{3a} = \frac{4}{3}a.$$

Problem 18. *Find the centre of percussion of a triangle* ABC *which is free to move about its side* BC.

Solution: From A we draw AL perpendicular to BC. Let D be the middle point of BC. Then BC is the principal axis of the triangle at the middle point O of DL. Let AL = p.

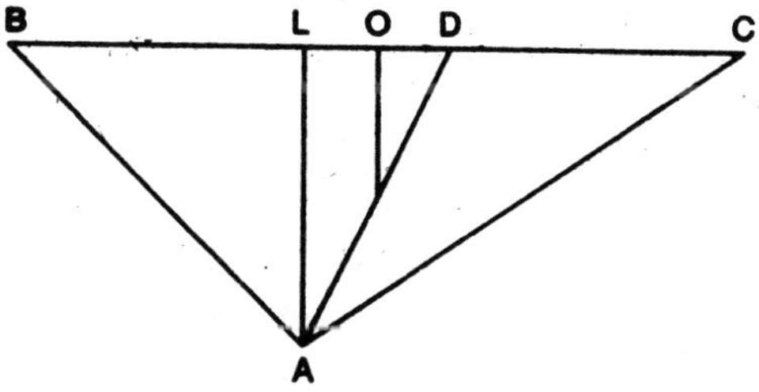

FIG 3.22

If M be the mass of the triangle, then it is kinetically equivalent to three particles, each of mass M/3 placed at the middle points D, E and F. So that

$$\begin{aligned} MK^2 &= \text{M.I. of the } \Delta ABC \text{ about BC} \\ &= \frac{M}{3}\left[\left(\frac{1}{2}p\right)^2 + \left(\frac{1}{2}p\right)^2 + 0\right] = \frac{1}{6}Mp^2. \end{aligned}$$

Hence

$$K^2 = \frac{p^2}{6}.$$

Also

$$h = \text{depth of C.G. of } \Delta ABC \text{ below BC} = p/3.$$

Therefore, depth of centre of percussion below BC along a vertical through O is

$$= \frac{K^2}{h} = \frac{p^2}{6 \cdot \frac{p}{3}} = \frac{p}{2}.$$

Clearly this point is the middle point of median AD.

Problem 19. *A pendulum is constructed of a solid sphere of mass M and radius a which is attached to the end of a rod of mass m and length b. Show that there will be no strains on the axis if the pendulum be struck at a distance* $\frac{[M\{\frac{2}{5}a^2+(a+b)^2\}+\frac{1}{3}mb^2]}{[M(a+b)+\frac{1}{2}mb^2]}$.

Solution: Let OA be the rod of mass m and length b.

Now

$$\begin{aligned} h &= \text{Depth of C.G. of the pendulum from O} \\ &= \frac{M.OC + m.OG}{M+m} = \frac{M(a+b) + m\frac{b}{2}}{M+m}. \end{aligned}$$

If K be the radius of gyration of the pendulum about the fixed axis at O, then

$$\begin{aligned} (m+M)K^2 &= M\left[\frac{2}{5}a^2 + (a+b)^2\right] + m\frac{4}{3}\left(\frac{b}{2}\right)^2 \\ \therefore K^2 &= \frac{M\left[\frac{2}{5}a^2 + (a+b)^2\right] + \frac{m}{3}b^2}{M+m}. \end{aligned}$$

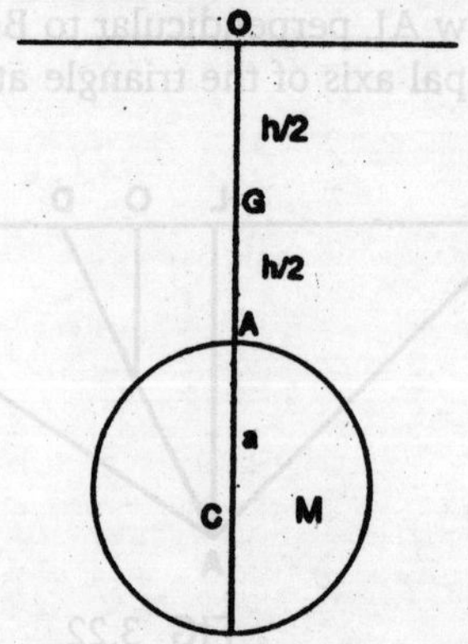

FIG 3.23

Thus the depth of the centre of percussion below O

$$= \frac{K^2}{h} = \frac{m\left[\frac{2}{5}a^2 + (a+b)^2\right] + \frac{m}{3}b^2}{M(a+b) + m\frac{b}{2}}.$$

3.16 Exercise 3

Type I

1. Two unequal masses M and M′, rest on two rough planes inclined an angles α and β to the horizon, they are connected by a fine string passing over a small pulley, of mass m and radius a, which is placed at the common vertex of the two planes, show that the acceleration of either mass is

$$\frac{g[M(\sin\alpha - \mu\cos\alpha) - M'(\sin\beta + \mu'\cos\beta)]}{M + M' + m\frac{K^2}{a^2}},$$

where μ, μ' are coefficient of friction, K is the radius of gyration of the pulley about its axis and M is the mass which moves downwards. [CU 82H]

2. A uniform disc is free to turn about a horizontal axis through its centre perpendicular to its plane. A particle of mass m is attached to a point in the edge of the disc. If the motion starts from the position in which radius to the particle makes an angle α with the upward vertical, find the angular velocity when m is in its lowest position. Take the mass of the disc as M.

[*Ans.* $2\sqrt{\frac{2mg}{a(2m+M)}}\cos\frac{\alpha}{2}$]

3. A uniform vertical circular plate of radius a, is capable of revolving about a smooth horizontal axis. Through its centre a rough perfectly flexible chain, whose mass is equal to that of the plate and whose length is equal to its circumference, hangs over its rim in equilibrium, of one end be slightly displaced. Show that the velocity of the chain when the end reaches the plate is $\sqrt{\frac{\pi a g}{6}}$. [CU 89H]

Type II

4. A weightless straight rod ACB of length $2a$ is movable about the end A which is fixed and carries two particles of the same mass, one fastened to the middle point C and the other to the end B of the rod. If the rod be held in a horizontal position and be then let go, show that the angular velocity when vertical is $\sqrt{\frac{6g}{5a}}$ and that $\frac{5}{3}a$ is the length of the simple equivalent pendulum.

[PU 62H; BU 63H; MU 64H]

5. A solid homogeneous cone of height h and semi-vertical angle α oscillates about a diameter of its base. Show that the length of the simple equivalent pendulum is $\frac{h}{5}(2+3\tan^2\alpha)$.

[CU 90H, 01H; VU 92H; BHU 63H; PU 66H; MU 81, 84, TMBU 97H]

6. An ellipse of axis $2a, 2b$ and a circle of radius b are cut from the same sheet of thin uniform metal and are superposed and fixed together with their centres coincident. The figure is free to move in its own vertical plane about one end of the major axis; show that the length of the simple equivalent pendulum is $\frac{5a^2-ab+2b^2}{4a^2}$.

7. Find the length of the simple equivalent pendulum for an elliptic lamina with the latus rectum, for its axis of rotation and the axis being horizontal.

[LNMU 80H; BU 71H]

[*Ans.* $\frac{a(1+4e^2)}{4e}$]

8. A uniform elliptic board swings about a horizontal axes at right angles to the plane of the board and passing through one focus. If the centre of oscillation beat the other focus, prove that the eccentricity of the ellipse is $\sqrt{\frac{2}{5}}$.

[CU 97H, 00H, 03H]

9. Find the length of the simple equivalent pendulum in the following cases, the axis being horizontal:

   (a) a circular disc of radius a, axis a tangent to it; [*Ans.* $\frac{5a}{4}$]

   (b) a sphere of radius a, axis a tangent at the highest point; [*Ans.* $\frac{7}{5}a$]

   (c) a circular wire of radius a, axis a tangent to it; [*Ans.* $\frac{3}{2}a$]

   (d) a hemisphere of radius a, axis a diameter of the base; [*Ans.* $\frac{16}{15}a$]

   (e) a cube of side $2a$, axes (i) a diagonal of one face, (ii) an edge.

   [*Ans.* (i) $\frac{5}{3}a$; (ii) $\frac{4}{3}\sqrt{2a}$]

10. A uniform rod of mass m and length $2a$ can oscillate about a horizontal axis through one end. A circular disc of mass $24m$ and radius $\frac{a}{3}$, can have its centre clamped to any point of the rod and its plane contains the axis of rotation. Show that for oscillations under gravity the length of simple equivalent pendulum lies between $\frac{a}{2}$ and $2a$.

11. A sphere of radius a, is suspended by a fine wire from a fixed point at a distance l from the centre. Show that the time of a small oscillation is given by $2\pi\sqrt{\frac{5l^2+2a^2}{5lg}}\left[1+\frac{1}{4}\sin^2\frac{\alpha}{2}\right]$, where α represents the amplitude of the vibration. [VU 89H; MU 80H]

12. A small weight is attached to a compound pendulum; show that the period of oscillation well increasing or diminishing according as the point of attachment is below or above the centre of oscillation. [PU 64H]

13. For a compound pendulum, show that there are three other axes of support, parallel to the original axes and intersecting the line from the centre of inertia perpendicular to the original axes, for which the time of oscillation is the same as about the original axes.

14. A triangular lamina ABC oscillates about the perpendicular from A on BC, the perpendicular being horizontal. Find the length of the simple equivalent pendulum. [*Ans.* $\frac{a^4+3(b^2-c^2)^2}{8a(b^2-c^2)}$]

15. A compound pendulum released from rest with the centre of gravity at the same level as the axis. At what inclination is the horizontal pressure on the axis greatest? [*Ans.* 45°]

16. A thin uniform rod of mass m is freely hinged at the upper end and makes n_1 small oscillations per second. A particle is then fixed at the lower end of the rod and the number of small oscillations is then formed to be n_2 per second. Show that mass of the particle is $\frac{m(n_1^2-n_2^2)}{3n_2^2-2n_1^2}$.

17. A simple circular pendulum is formed of mass M suspended from a fixed point by a weightless wire of length l. If a mass m, very small compared with M, be knotted on to the wire at a point distant a from the point of suspension, show that the time of a small vibration of the pendulum is approximately diminished by $\frac{ma}{2Ml}\left(1-\frac{a}{l}\right)$ of itself.

Type III

18. A heavy homogeneous cube of weight w, can swing about an edge which is horizontal. It starts from rest being displaced from its unstable position of equilibrium, when the perpendicular from the centre of gravity upon the edge has turned through an angle θ. Show that the components of the action at the hinge along and right angles to, this perpendicular are $\frac{w}{2}(3-5\cos\theta)$ and $\frac{w}{4}\sin\theta$.

19. A circular disc of weight w can turn freely about a horizontal axes perpendicular to its plane which passes through a point O on its circumference. If it starts from rest with the diameter vertically above O, show that the resultant pressure on the axes when that diameter is horizontal and vertically below O, are respectively $\frac{1}{3}\sqrt{17}w$ and $\frac{11}{3}w$.

    Further prove that the axis must be able to bear at least $\frac{11}{3}$ times the weight of the disc. [IAS 53]

20. A uniform semi-circular arc of mass m and radius $2a$, is fixed at its ends to two points on the same vertical line, and is rotating with constant angular velocity ω. Show that the horizontal thrust on the upper end is $\frac{m(g+\omega^2 a)}{\pi}$.

21. A right cone of angle 2α can turn freely about an axis, passing through the centre of its base and perpendicular to the axis; if the cone starts from rest with its axis horizontal, show that, when the axis is vertical, the thrust on the fixed axis is to the weight of the cone is $1 + \frac{1}{2}\cos^2\alpha$ to $1 - \frac{1}{3}\cos^2\alpha$.

Type IV

22. A rod of mass nM is lying on a horizontal table and has one end fixed; a particle of mass M is in contact with it. The rod receives a horizontal blow at its free end. Find the position of the particle so that it may start moving with the maximum velocity. In this case, show that the kinetic energies communicated to the rod and mass are equal. [*Ans.* At $2a\sqrt{\frac{n}{3}}$]

23. A uniform rod OA of mass M and length $2a$ rests on a smooth table and is free to turn about a smooth pivot at its end O. In contact with it at a distance c from O is an inelastic particle of mass m. A horizontal blow of impulse P is given to the rod at a distance x from O in a direction perpendicular to the rod. Find the resulting instantaneous angular velocity of the rod and the impulsive action at O and on the particle.

[*Ans.* $\omega = \frac{px}{\frac{4}{3}Ma^2+mb^2}, \frac{MPbx}{\frac{4}{3}Ma^2+mb^2}, P\left[1 - \frac{(Ma+mb)x}{\frac{4}{3}Ma^2+mb^2}\right]$]

24. Two wheels on spindles in fixed bearings suddenly engage so that their angular velocities become inversely proportional to their radii and in opposite directions. One wheel of radius a and moment of inertia I_1 has angular velocity ω initially, the other of radius b and moment of inertia I_2 is initially at rest. Show that their new angular velocities are $\frac{I_1 b^2}{I_1 b^2+I_2 a^3}$ and $\frac{I_1 ab}{I_1 b^2+I_2 a^2}\omega$.

Type V

25. Find how an equilateral triangular lamina must be struck so that it may commence to rotate about a side.

[*Ans.* At the middle point of the median bisecting the side about which the lamina commence to rotate.]

26. Find the position of the centre of percussion of a sector of a circle, axis in the plane of the sector, perpendicular to its symmetrical radius, and passing through the centre of the circle. [*Ans.* $\frac{3a}{8}\frac{(\alpha+\sin\alpha\cos\alpha)}{\sin\alpha}$]

Chapter 4

Motion in Two Dimensions

4.1 Equation of Motion of a Rigid Body in Two Dimensions

To determine the equations of motion in two dimensions when the forces acting on the body are finite. **[CU 04H]**

The motion of a rigid body consists of two independent motions.

(*a*) the motion of centre of gravity, and

(*b*) the motion about the centre of gravity.

Let $(\bar{x}, \bar{y})$ be the coordinates of C.G. of the body and M be the mass of the body concentrated at G. Then we have

$$M\frac{d^2\bar{x}}{dt^2} = \Sigma X \tag{4.1}$$

$$M\frac{d^2\bar{y}}{dt^2} = \Sigma Y, \tag{4.2}$$

where ΣX and ΣY are the algebraic sums of the components of the external forces parallel to the axes supposed to be acting at the C.G.

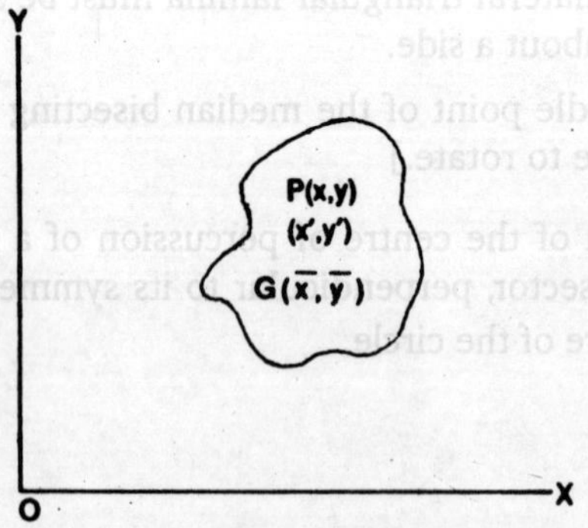

FIG 4.1

Let P be any particle of mass m whose coordinates relative to C.G. at G as origin be (x', y'). Then from D' Alembert's principle we have the following equations:

$$\Sigma m\left(x'\frac{d^2y'}{dt^2}-y'\frac{d^2x'}{dt^2}\right) = \Sigma\,(x'Y-y'X)$$

$$\text{or, } \Sigma m\frac{d}{dt}\left(x'\frac{dy'}{dt}-y'\frac{dx'}{dt}\right) = \Sigma\,(x'Y-y'X)$$

$$\text{or, } \frac{d}{dt}\left\{\Sigma m\left(x'\frac{dy'}{dt}-y'\frac{dx'}{dt}\right)\right\} = \Sigma\,(x'Y-y'X)\,. \qquad (4.3)$$

But

$$x'\frac{dy'}{dt}-y'\frac{dx'}{dt}$$

is the moment about G of the velocity m relative to G.

We determine below a simpler equivalent of this expression.

Let GP $= r$ and let GP makes an angle ϕ with a line GB fixed in the space. Also let θ be the angle that a line GA, fixed in the body makes with the line GB, fixed in the space.

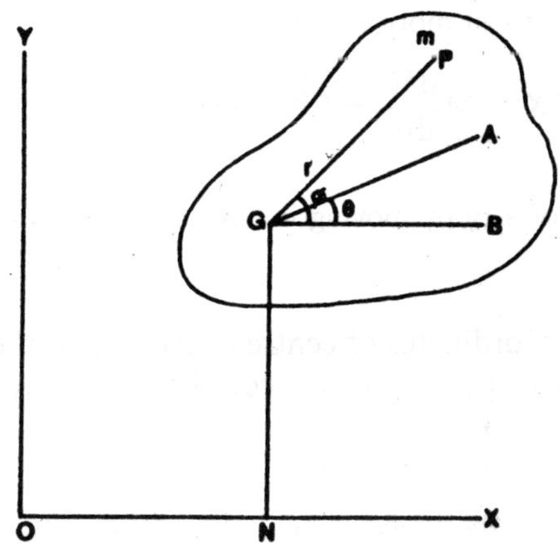

FIG 4.2

From the figure it is clear that

$$\phi = \theta + \angle AGP.$$

But $\angle AGP$ is the angle between two fixed lines GA and GP so that this angle is constant and hence we have

$$\frac{d\phi}{dt} = \frac{d\theta}{dt}. \qquad (4.4)$$

Now the velocity of particle m at P relative to G is $r\frac{d\phi}{dt}$ and hence its moment about G is

$$rX\left(r\frac{d\phi}{dt}\right) = r^2\frac{d\phi}{dt} = r^2\frac{d\theta}{dt} \quad \text{[From equation (4.4)]}$$

Hence

$$x'\frac{dy'}{dt}-y'\frac{dx'}{dt} = r^2\frac{d\theta}{dt}. \qquad (4.5)$$

Hence from equations (4.3) and (4.5) we have,

$$\frac{d}{dt}\left\{\Sigma m\left(r^2\frac{d\theta}{dt}\right)\right\} = \Sigma(x'Y-y'X)$$

$$\Rightarrow \frac{d}{dt}\left\{\Sigma mr^2\frac{d\theta}{dt}\right\} = \Sigma(x'Y-y'X)$$

$$\Rightarrow \frac{d}{dt}\left[\frac{d\theta}{dt}\Sigma mr^2\right] = \Sigma(x'Y-y'X)$$

$$\Rightarrow \frac{d}{dt}\left[\frac{d\theta}{dt}Mk^2\right] = \Sigma(x'Y-y'X),$$

where k is the radius of gyration of the body about an axis through G perpendicular to the plane of motion, i.e., xy-plane.

$$\text{or, } Mk^2\frac{d^2\theta}{dt^2} = \Sigma(x'Y-y'X) = L,$$

where L is the moment about G of all the external forces acting on the body.

Hence the equations of motion are

$$M\frac{d^2\bar{x}}{dt^2} = \Sigma X,\quad M\frac{d^2\bar{y}}{dt^2} = \Sigma Y,\quad \text{and}\quad Mk^2\frac{d^2\theta}{dt^2} = \Sigma(x'Y-y'X).$$

4.2 Kinetic Energy of a Body moving in Two Dimensions

[CU 92H, 95H, 98H, 99H, 01H, 03H]

Let $(\bar{x},\bar{y})$ be the coordinates of centre of inertia, G of the body referred to two fixed rectangular axes. Let (x,y) be the coordinates of any element of mass m. Also let (x',y') be the coordinates of this element referred to parallel axes through G.

Then $x=\bar{x}+x', y=\bar{y}+y'$ so that

$$\frac{dx}{dt} = \frac{d\bar{x}}{dt}+\frac{dx'}{dt},\quad \frac{dy}{dt} = \frac{d\bar{y}}{dt}+\frac{dy'}{dt}.$$

Hence the kinetic energy of the body

$$= \frac{1}{2}\Sigma m\left[\left(\frac{dx}{dt}\right)^2+\left(\frac{dy}{dt}\right)^2\right]$$

$$= \frac{1}{2}\Sigma m\left[\left(\frac{d\bar{x}}{dt}+\frac{dx'}{dt}\right)^2+\left(\frac{d\bar{y}}{dt}+\frac{dy'}{dt}\right)^2\right]$$

$$= \frac{1}{2}\Sigma m\left[\left(\frac{d\bar{x}}{dt}\right)^2+\left(\frac{d\bar{y}}{dt}\right)^2\right]+\frac{1}{2}\Sigma m\left[\left(\frac{dx'}{dt}\right)^2+\left(\frac{dy'}{dt}\right)^2\right]$$
$$+\Sigma m\frac{d\bar{x}}{dt}\frac{dx'}{dt}+\Sigma m\frac{d\bar{y}}{dt}\frac{dy'}{dt}$$

$$= \frac{1}{2}\Sigma m\left[\left(\frac{d\bar{x}}{dt}\right)^2+\left(\frac{d\bar{y}}{dt}\right)^2\right]+\frac{1}{2}\Sigma m\left[\left(\frac{dx'}{dt}\right)^2+\left(\frac{dy'}{dt}\right)^2\right]$$
$$+\frac{d\bar{x}}{dt}\Sigma m\frac{dx'}{dt}+\frac{d\bar{y}}{dt}\Sigma m\frac{dy'}{dt}. \tag{4.6}$$

But $\frac{\Sigma m x'}{\Sigma m} = 0$ being x coordinate of G referred to G as origin.

$\therefore\ \Sigma m x' = 0.$

Hence

$$\Sigma m \frac{dx'}{dt} = 0.$$

Similarly,

$$\Sigma m \frac{dy'}{dt} = 0.$$

Hence equation (4.6) gives

$$\text{K.E.} = \frac{1}{2} M \left[\left(\frac{d\bar{x}}{dt} \right)^2 + \left(\frac{d\bar{y}}{dt} \right)^2 \right] + \frac{1}{2} \Sigma m \left[\left(\frac{dx'}{dt} \right)^2 + \left(\frac{dy'}{dt} \right)^2 \right].$$

Now

$$\frac{1}{2} M \left[\left(\frac{d\bar{x}}{dt} \right)^2 + \left(\frac{d\bar{y}}{dt} \right)^2 \right]$$

is the kinetic energy of a particle of the mass M placed at the centre of gravity $G = \frac{1}{2} M v^2$, where V is the velocity of the centre of inertia G and

$$\frac{1}{2} \Sigma m \left[\left(\frac{dx'}{dt} \right)^2 + \left(\frac{dy'}{dt} \right)^2 \right]$$

is the kinetic energy of the body relative to the centre of inertia

$$\begin{aligned} G &- \frac{1}{2} \Sigma m \left(r \frac{d\phi}{dt} \right)^2 \quad \left[\because r \frac{d\phi}{dt} \text{ is the velocity of m relative to G} \right] \\ &= \frac{1}{2} \Sigma m \left(r \frac{d\theta}{dt} \right)^2 \quad \left[\because \frac{d\phi}{dt} = \frac{d\theta}{dt} \right] \\ &= \frac{1}{2} \left(\frac{d\theta}{dt} \right)^2 \Sigma m r^2 = \frac{1}{2} M k^2 \dot{\theta}^2, \end{aligned}$$

where θ is the angle that any body fixed in the body makes with a line fixed in the space, and k is the radius of gyration of the body about a line through G perpendicular to the motion.

Hence the required kinetic energy

$$\begin{aligned} &= \tfrac{1}{2} M v^2 + \tfrac{1}{2} M k^2 \dot{\theta}^2 \\ &= \text{K.E. of a particle of mass M placed at G and moving with it + K.E. of the body relative to the centre of inertia} \end{aligned}$$

or, K.E. $=$ K.E. due to translation + K.E. due to rotation.

4.3 Moment of Momentum [MU 92H; LNMU 93H, CU 93H]

Moment of momentum or Angular momentum about the origin O of a body moving in two dimensions.

Let $(\bar{x}, \bar{y})$ be the coordinates of centre of inertia G referred to fixed rectangular axes through O, and (x, y) the coordinates of any particle of mass m. Let (x', y') be the coordinates of this particle referred to parallel axes through G.

Then $x = \bar{x} + x'$, $y = \bar{y} + y'$.

Now moment of momentum of the body about the origin O

$$\begin{aligned}
&= \Sigma m\left(x\frac{dy}{dy} - y\frac{dx}{dt}\right) \\
&= \Sigma m\left[(\bar{x}+x')\left(\frac{d\bar{y}}{dt}+\frac{dy'}{dt}\right) - (\bar{y}+y')\left(\frac{d\bar{x}}{dt}+\frac{dx'}{dt}\right)\right] \\
&= \Sigma m\left(\bar{x}\frac{d\bar{y}}{dt} - \bar{y}\frac{d\bar{x}}{dt}\right) + \Sigma m\left(x'\frac{dy'}{dt} - y'\frac{dx'}{dt}\right) \\
&\quad + \Sigma m\left[\bar{x}\frac{dy'}{dt} + x'\frac{d\bar{y}}{dt} - \bar{y}\frac{dx'}{dt} - y'\frac{d\bar{x}}{dt}\right].
\end{aligned} \tag{4.7}$$

But $\frac{\Sigma m x'}{\Sigma m} = 0$ being x coordinate of the centre of inertia G referred to G as origin.

$\therefore$ $\Sigma m x' = 0$, $\Sigma m y' = 0$, so that

$$\Sigma m\frac{dx'}{dt} = 0, \ \Sigma m\frac{dy'}{dt} = 0.$$

$$\therefore \Sigma m x'\frac{d\bar{y}}{dt} = \frac{d\bar{y}}{dt}\Sigma m x' = 0$$

$$\text{and} \quad \Sigma m\bar{y}\frac{dx'}{dt} = \bar{y}\Sigma m\frac{dx'}{dt} = 0.$$

Similarly,

$$\frac{d\bar{x}}{dt}\Sigma m y' = 0, \ \bar{x}\Sigma m\frac{dy'}{dt} = 0.$$

Hence from equation (4.7), the moment of momentum about O

$$= M\left(\bar{x}\frac{d\bar{y}}{dt} - \bar{y}\frac{d\bar{x}}{dt}\right) + \Sigma m\left(x'\frac{dy'}{dt} - y'\frac{dx'}{dt}\right) \tag{4.8}$$

Now

$$M\left(\bar{x}\frac{d\bar{y}}{dt} - \bar{y}\frac{d\bar{x}}{dt}\right)$$

is the moment of momentum of mass M placed at G about O

$$= Mvp, \tag{4.9}$$

where v is the velocity of G and p is the perpendicular distance from O upon the direction of v.

Also

$$\Sigma m\left(x'\frac{dy'}{dt}-y'\frac{dx'}{dt}\right)$$

is the moment of momentum of the body about G

$$= \Sigma m\left(r\frac{d\phi}{dt}\right)\times r=\Sigma m\left(r\frac{d\theta}{dt}\right)\times r \quad \left(\because \frac{d\phi}{dt}=\frac{d\theta}{dt}\right)$$

$$= \Sigma mr^2\frac{d\theta}{dt}=\frac{d\theta}{dt}\Sigma mr^2=Mk^2\dot{\theta}, \tag{4.10}$$

where k is the radius of gyration of the body about a line through G perpendicular to the plane of rotation.

With the help of equations (4.9) and (4.10) the moment of momentum about O

$= Mvp + Mk^2\theta$

= moment of momentum of centre of inertia of the body

+ moment of moment relative to centre of inertia.

4.4 Motion of a solid sphere down an inclined plane [CU 92H, 94H, 01H]

A uniform sphere rolls down an inclined plane, rough enough to prevent any sliding, to discuss the motion.

Let the sphere be initially at rest with its point A in contact with O.

At any time t, let the centre of the sphere describe a distance x parallel to the inclined plane and θ the angle through which the sphere turns. Thus CA, a line fixed in the body, makes an angle θ with the normal to the plane, a line fixed in the space.

Let F be the frictional force and R the normal reaction, then equations of motion of centre of gravity of the body are

$$M\frac{d^2x}{dt^2}=Mg\sin\alpha-F \tag{4.11}$$

$$\text{and}\quad 0=\frac{d^2y}{dt^2}=Mg\cos\alpha-R \tag{4.12}$$

as there is no motion perpendicular to plane.

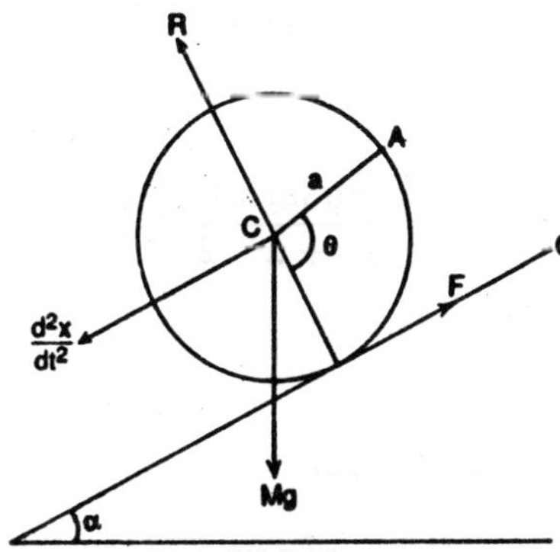

FIG 4.3

Also equation of motion about the C.G. is

$$\therefore \quad Mk^2 \frac{d^2\theta}{dt^2} = Fa. \tag{4.13}$$

Since there is no sliding, therefore, the distance CB = the arc AB, $x = a\theta$.

$$\therefore \frac{dx}{dt} = a\frac{d\theta}{dt} \text{ and } \frac{d^2x}{dt^2} = a\frac{d^2\theta}{dt^2}.$$

Hence from equation (4.13), we get

$$\frac{Mk^2}{a}\frac{d^2x}{dt^2} = Fa, \text{ or, } F = \frac{Mk^2}{a^2}\frac{d^2x}{dt^2}.$$

Putting this value of F in equation (4.11), it becomes

$$M\frac{d^2x}{dt^2} = Mg\sin\alpha - \frac{Mk^2}{a^2}\frac{d^2x}{dt^2}$$

$$\text{or,} \quad \frac{d^2x}{dt^2}\left(1 + \frac{k^2}{a^2}\right) = g\sin\alpha$$

$$\therefore \quad \frac{d^2x}{dt^2} = \frac{a^2 g\sin\alpha}{k^2 + a^2}. \tag{4.14}$$

This shows that the sphere rolls down with a constant acceleration $\frac{a^2 g\sin\alpha}{k^2+a^2}$

Integrating equation (4.14), we have

$$\frac{dx}{dt} = \frac{a^2 g\sin\alpha}{k^2 + a^2}t + C. \tag{4.15}$$

Initially, when $t = 0$, $\frac{dx}{dt} = 0$.

Hence from equation (4.15), $C = 0$.

Thus

$$\frac{dx}{dt} = \frac{a^2 g\sin\alpha}{k^2 + a^2}t.$$

Integrating again, we get

$$x = \frac{a^2 g\sin\alpha}{k^2 + a^2}\frac{t^2}{2} + C_1. \tag{4.16}$$

Initially, when $t = 0$, $x = 0$.

Thus $C_1 = 0$.

Hence from equation (4.16), we get

$$x = \frac{a^2 g\sin\alpha}{k^2 + a^2}\frac{t^2}{2}.$$

Particular Cases

1. For a solid sphere $k^2 = \frac{2}{5}a^2$.

   Hence from equation (4.14), acceleration $= \frac{a^2 g \sin\alpha}{\frac{2}{5}a^2 + a^2} = \frac{5}{7} g \sin\alpha$.

2. For a hollow sphere $k^2 = \frac{2}{3}a^2$.

   Hence from equation (4.14), acceleration $= \frac{3}{5} g \sin\alpha$.

3. For a circular disc $k^2 = \frac{a^2}{2}$.

   Hence from equation (4.14), acceleration $= \frac{2}{3} g \sin\alpha$.

4. For a circular ring $k^2 = a^2$.

   Hence from equation (4.14), acceleration $= \frac{1}{2} g \sin\alpha$.

Value of μ for Pure Rolling

For a solid sphere acceleration $= \frac{d^2x}{dt^2} = \frac{5}{7} g \sin\alpha$.

Thus from equation (4.11), we have,

$$F = Mg\sin\alpha - M\frac{d^2x}{dt^2} = Mg\sin\alpha - \frac{5M}{7}g\sin\alpha = \frac{2M}{7}g\sin\alpha.$$

Also from equation (4.12), $R = Mg\cos\alpha$.

$$\therefore \frac{F}{R} = \frac{2}{7}\tan\alpha.$$

For pure rolling,

$$F < \mu R, \text{ or, } \frac{F}{R} < \mu, \text{ or, } \mu > \frac{F}{R}.$$

Thus

$$\mu > \frac{2}{7}\tan\alpha.$$

Hence the coefficient of friction should be greater than $\frac{2}{7}\tan\alpha$ for pure rolling of a solid sphere.

In a similar manner the value of μ can be found out for other bodies.

4.5 Slipping of Rods

A uniform rod is held in vertical position with one end resting upon a perfectly rough table, and when released, rotates about the end in contact with the table. To discuss the motion. **[CU 93H]**

Let AB be a rod of mass M and length $2a$.

Let the rod which is rotating about A makes an angle θ with the vertical at any time t.

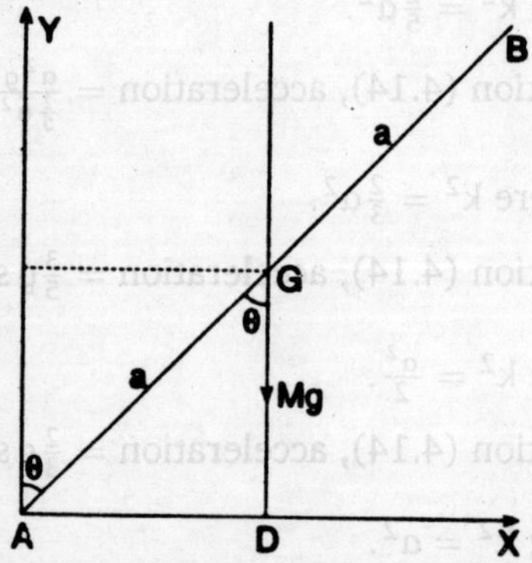

FIG 4.4

Taking A as origin, and horizontal and vertical through A as axes. Then the coordinates (x, y) of C.G. at G are given by $x = a\sin\theta$, $y = a\cos\theta$.

$$\therefore \quad \frac{dx}{dt} = a\cos\theta\frac{d\theta}{dt}, \quad \frac{dy}{dt} = -a\sin\theta\frac{d\theta}{dt}$$

$$\frac{d^2x}{dt^2} = -a\sin\theta\frac{d^2\theta}{dt^2}, \quad \frac{d^2y}{dt^2} = -a\cos\theta\left(\frac{d\theta}{dt}\right)^2.$$

Let F be the frictional force and R the normal reaction at A.

Then the equations of motion of C.G. are

$$M\frac{d^2x}{dt^2} = M\left[a\cos\theta\frac{d^2\theta}{dt^2} - a\sin\theta\left(\frac{d\theta}{dt}\right)^2\right]$$

$$= F \tag{4.17}$$

$$\text{and } M\frac{d^2y}{dt^2} = M\left[-a\sin\theta\frac{d^2\theta}{dt^2} - a\cos\theta\left(\frac{d\theta}{dt}\right)^2\right]$$

$$= R - Mg. \tag{4.18}$$

Also

$$v^2 = \dot{x}^2 + \dot{y}^2 = a^2\dot{\theta}^2$$

$$\text{and} \quad k^2 = \frac{a^2}{3}$$

and work done $= Mg(a - a\cos\theta)$.

Hence energy equation gives

$$\frac{1}{2}Mv^2 + \frac{1}{2}Mk^2\dot{\theta}^2 = Mg(a - a\cos\theta)$$

$$\Rightarrow \quad \frac{M}{2}\left[a^2\dot{\theta}^2 + \frac{a^2}{3}\dot{\theta}^2\right] = Mga(1 - \cos\theta)$$

$$\Rightarrow \quad \frac{M}{2}\frac{4}{3}a^2\dot{\theta}^2 = Mga(1 - \cos\theta)$$

$$\Rightarrow \quad \frac{2}{3}a^2\dot{\theta}^2 = ag(1 - \cos\theta)$$

$$\Rightarrow \quad \dot{\theta}^2 = \frac{2g}{3a}(1 - \cos\theta). \tag{4.19}$$

Differenciating equation (4.19) with respect to t, we get

$$2\dot{\theta}\ddot{\theta} = \frac{3g}{2a}\sin\theta\dot{\theta}, \text{ or, } \ddot{\theta} = \frac{3g}{4a}\sin\theta.$$

Hence from equation (4.17), we get

$$\begin{aligned} F &= Ma\left[\cos\theta\frac{3g}{4a}\sin\theta - \sin\theta\frac{3g}{2a}(1 - \cos\theta)\right] \\ &= \frac{Mg3\sin\theta}{4}[\cos\theta - 2(1 - \cos\theta)] \\ &= \frac{3}{4}Mg\sin\theta(3\cos\theta - 2). \end{aligned} \tag{4.20}$$

Also from equation (4.18), we have

$$\begin{aligned} R - Mg &= M\left[-a\sin\theta\frac{3g}{4a}\sin\theta - a\cos\theta\frac{3g}{2a}(1 - \cos\theta)\right] \\ &= M\frac{3g}{4}\left[-\sin^2\theta - 2\cos\theta(1 - \cos\theta)\right] \\ &= \frac{3Mg}{4}[-(1 - \cos^2\theta) - 2\cos\theta + 2\cos^2\theta] \\ &= \frac{3Mg}{4}[3\cos^2\theta - 2\cos\theta - 1] \\ \therefore R &= \frac{Mg}{4}[9\cos^2\theta - 6\cos\theta - 3 + 4] \\ &= \frac{Mg}{4}(-3\cos\theta + 1)^2. \end{aligned} \tag{4.21}$$

From equation (4.21) it is clear that R does not change its sign and vanishes when $\cos\theta = \frac{1}{2}$. *Hence the end A does not leave the plane.*

Also from equation (4.20) we see that F changes its sign as θ passes through the angle $\cos^{-1}\frac{2}{3}$, thus its direction is then reversed.

$R = 0$, when $\cos\theta = \frac{1}{3}$, hence the ratio F/R becomes infinite when $\cos\theta = \frac{1}{3}$, i.e., unless the plane be infinitely rough there will be sliding at this value of θ. In practice the end A of the rod begins to slip for some value of θ less than $\cos^{-1}\frac{1}{3}$. The end A will slip backward or forward according as the slipping takes place before or after the inclination of the rod to $\cos^{-1}\frac{2}{3}$.

4.6 Motion of a rod sliding down in a vertical plane

A uniform straight rod slides down in a vertical plane, its ends being in contact with two smooth planes, one horizontal and the other vertical. It starts from rest at an angle α with the horizontal; to discuss the motion.

Let AB be the rod of mass M and length $2a$ sliding down from rest at an angle α to the horizontal with its ends A and B on smooth horizontal and vertical planes, respectively.

At any time t, let the rod make an angle θ to the horizontal. Let R and S be the reactions at the ends A and B of the rod AB.

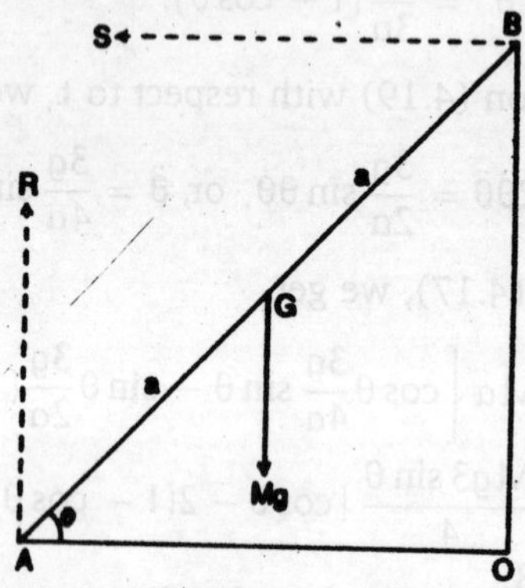

FIG 4.5

Taking O as origin, horizontal and vertical lines through O as axes, the coordinates of C.G. are given by

$$x = a\cos\theta, \qquad y = a\sin\theta,$$
$$\dot{x} = -a\sin\theta\dot{\theta}, \qquad \dot{y} = a\cos\theta\dot{\theta},$$
$$\ddot{x} = -a[\cos\theta\dot{\theta}^2 + \ddot{\theta}\sin\theta], \quad \ddot{y} = a[-\sin\theta\dot{\theta}^2 + \ddot{\theta}\cos\theta].$$

The equations of motion of the C.G. at G are given by

$$M\ddot{x} = S \text{ and } M\ddot{y} = R - Mg,$$

or, $$M(-a\cos\theta\dot{\theta}^2 - a\sin\theta\ddot{\theta}) = S \tag{4.22}$$

and $$M(-a\sin\theta\dot{\theta}^2 + a\ddot{\theta}\cos\theta) = R - Mg. \tag{4.23}$$

Also the energy equation gives

$$\frac{1}{2}M[v^2 + k^2\dot{\theta}^2] = \text{work done by the gravity.}$$

Here

$$v^2 = \dot{x}^2 + \dot{y}^2 = a^2\dot{\theta}^2.$$

Initially, when the inclination of rod was α, the height of C.G. was $a\sin\alpha$ and now it is $a\sin\theta$, so that the distance described by C.G. is $a(\sin\alpha - \sin\theta)$.

Hence the energy equation becomes

$$\frac{M}{2}\left[a^2\dot{\theta}^2+\frac{a^2}{3}\dot{\theta}^2\right]=Mga(\sin\alpha-\sin\theta),$$

$$\text{or,}\quad \frac{4}{6}a^2\dot{\theta}^2=ag(\sin\alpha-\sin\theta),$$

$$\text{or,}\quad \dot{\theta}^2=\frac{3g}{2a}(\sin\alpha-\sin\theta). \tag{4.24}$$

Differentiating it with respect to t, we have

$$2\dot{\theta}\ddot{\theta}=-\frac{3g}{2a}\cos\theta\dot{\theta}.$$

Hence

$$\ddot{\theta}=-\frac{3g}{4a}\cos\theta. \tag{4.25}$$

Substituting the values of $\dot{\theta}^2$ and $\ddot{\theta}$ in equations (4.22) and (4.23), we get

$$\begin{aligned} S &= M\left[-a\cos\theta\frac{3g}{2a}(\sin\alpha-\sin\theta)+a\sin\theta\frac{3g}{4a}\cos\theta\right] \\ &= M\frac{3g}{4}\cos\theta[\sin\theta-2(\sin\alpha-\sin\theta)] \\ &= \frac{3Mg}{4}\cos\theta[3\sin\theta-2\sin\alpha] \end{aligned} \tag{4.26}$$

$$\begin{aligned} \text{and } R &= Mg+M\left[-a\sin\theta\frac{3g}{2a}(\sin\alpha-\sin\theta)+a\cos\theta\left(-\frac{3g}{4a}\cos\theta\right)\right] \\ &= Mg+Mg\frac{3}{4}\left[-\cos^2\theta-2\sin\alpha\sin\theta+2\sin^2\theta\right] \\ &= \frac{Mg}{4}\left[4+3\left\{-(1-\sin^2\theta)-2\sin\alpha\sin\theta+2\sin^2\theta\right\}\right] \\ &= \frac{Mg}{4}\left[4+3\left\{3\sin^2\theta-2\sin\alpha\sin\theta-1\right\}\right] \\ &= \frac{Mg}{4}[9\sin^2\theta-6\sin\alpha\sin\theta+1] \\ &= \frac{Mg}{4}[9\sin^2\theta-6\sin\alpha\sin\theta+\sin^2\alpha+\cos^2\alpha] \\ &= \frac{Mg}{4}[(3\sin\theta-\sin\alpha)^2+\cos^2\alpha]. \end{aligned} \tag{4.27}$$

From equation (4.26) it is clear that $S=0$, when $\sin\theta=\frac{2}{3}\sin\alpha$, after this the value of $\dot{\theta}$ is reached, the value of R becomes negative. Hence the end B leaves the wall when $\sin\theta=\frac{2}{3}\sin\alpha$.

Again from equation (4.27), we observe that value of R is always positive, i.e., the end A never leaves the plane.

When the end B leaves the plane, then $\sin\theta=\frac{2}{3}\sin\alpha$ and $S=0$, hence the equations (4.22), (4.23), (4.24) and (4.25) do not hold good for further motion.

Putting $\sin\theta=\frac{2}{3}\sin\alpha$ in equation (4.24), the angular velocity of the rod is $\sqrt{\frac{g}{2a}}\sin\alpha$.

This will be the initial angular velocity for next part of the motion.

Second Part of the Motion

When the end B leaves the wall, let R_1 be the normal reaction at A. Let the rod have turned through an angle ϕ from the horizontal.

Now the equations of motion

$$M\frac{d^2x}{dt^2} = 0 \tag{4.28}$$

$$M\frac{d^2y}{dt^2} = R_1 - Mg \tag{4.29}$$

$$\text{and } \frac{Ma^2}{3}\ddot{\phi} = -R_1 a\cos\phi. \tag{4.30}$$

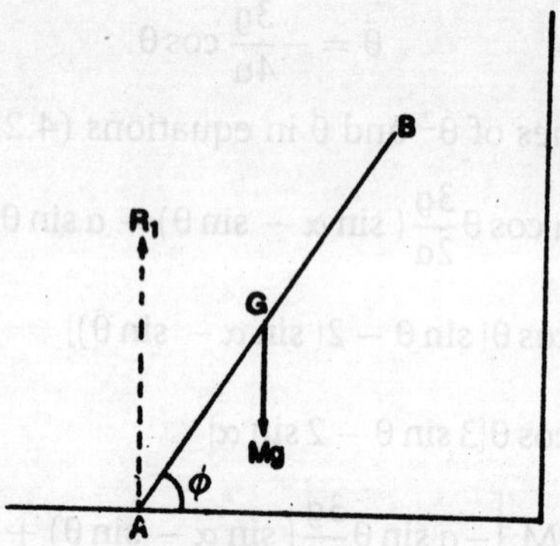

FIG 4.6

But

$$\begin{aligned} y &= a\sin\phi \\ \dot{y} &= a\cos\phi\dot{\phi} \\ \ddot{y} &= a[-\sin\phi\dot{\phi}^2 + \cos\phi\ddot{\phi}]. \end{aligned}$$

Hence from equation (4.29), we have

$$R_1 = Mg + Ma[-\sin\phi\dot{\phi}^2 + \cos\phi\ddot{\phi}].$$

Thus from equation (4.30), we get

$$\frac{Ma^2}{3}\ddot{\phi} = -a\cos\phi[Mg + Ma(-\sin\phi\dot{\phi}^2 + \cos\phi\ddot{\phi})]$$

$$\Rightarrow \quad \frac{a}{3}\ddot{\phi} = -\cos\phi[g + a(-\sin\phi\dot{\phi}^2 + \cos\phi\ddot{\phi})]$$

$$\Rightarrow \quad \frac{\ddot{\phi}}{3} = -\frac{\cos\phi}{a}[g + a(-\sin\phi\dot{\phi}^2 + \cos\phi\ddot{\phi})]$$

$$\Rightarrow \quad \ddot{\phi}\left(\frac{1}{3} + \cos^2\phi\right) - \sin\phi\cos\phi\dot{\phi}^2 = -\frac{g}{a}\cos\phi. \tag{4.31}$$

Integrating, we get

$$\dot{\phi}^2\left(\frac{1}{3} + \cos^2\phi\right) = -\frac{2g}{a}\sin\phi + C. \tag{4.32}$$

Initially, when $\sin\phi = \frac{2}{3}\sin\alpha$,

$$\frac{d\phi}{dt} = \sqrt{\frac{g}{2a}\sin\alpha}.$$

Thus from equation (4.32), we get

$$\left(\sqrt{\frac{g}{2a}\sin\alpha}\right)^2\left(\frac{1}{3}+\cos^2\phi\right) = -\frac{2g}{a}\frac{2}{3}\sin\alpha + C$$

$$\Rightarrow \quad \frac{g\sin\alpha}{2a}\left[\frac{1}{3}+\cos^2\phi\right] = -\frac{4g}{3a}\sin\alpha + C$$

$$\Rightarrow \quad \frac{g\sin\alpha}{2a}\left[\frac{1}{3}+1-\frac{4}{9}\sin^2\alpha\right] = -\frac{4g}{3a}\sin\alpha + C$$

$$\Rightarrow \quad \frac{g\sin\alpha}{2a}\left[\frac{4}{3}-\frac{4}{9}\sin^2\alpha\right] + \frac{4g}{3a}\sin\alpha = C$$

$$\Rightarrow \quad \frac{4g\sin\alpha}{3a}\left[\frac{1}{2}-\frac{1}{6}\sin^2\alpha+1\right] = C$$

$$\Rightarrow \quad \frac{4g}{3a}\sin\alpha\left[\frac{3}{2}-\frac{1}{6}\sin^2\alpha\right] = C$$

$$\Rightarrow \quad \frac{4g}{3a}\frac{3}{2}\sin\alpha\left[1-\frac{1}{9}\sin^2\alpha\right] = C$$

$$\Rightarrow \quad \frac{2g\sin\alpha}{a}\left(1-\frac{\sin^2\alpha}{9}\right) = C.$$

Thus, from equation (4.32), we get

$$\dot{\phi}^2\left(\frac{1}{3}+\cos^2\phi\right) = \frac{2g\sin\alpha}{a}\left(1-\frac{\sin^2\alpha}{9}\right) - \frac{2g}{a}\sin\phi. \tag{4.33}$$

When $\phi = 0$, i.e., when the rod becomes horizontal, its angular velocity Ω, is given by

$$\Omega^2\left(\frac{1}{3}+1\right) = \frac{2g\sin\alpha}{a}\left(1-\frac{\sin^2\alpha}{9}\right)$$

$$\Omega^2\frac{4}{3} = \frac{2g\sin\alpha}{a}\left(1-\frac{\sin^2\alpha}{9}\right)$$

$$\therefore \Omega^2 = \frac{3g\sin\alpha}{2a}\left(1-\frac{\sin^2\alpha}{9}\right).$$

4.7 Motion of a solid sphere down an inclined plane when rolling and sliding are combined

[CU 87H, 94H, 97H; VU 93H]

An imperfectly rough sphere moves from rest down a plane inclined at an angle α to the horizon, to determine the motion.

Let C be the centre of a sphere of radius a. Let CB be a radius (a line fixed in the body) which was initially normal to the plane and makes an angle θ with the normal CA at time t.

If the friction is not sufficient to produce pure rolling, then the sphere will slide as well as turn. So the maximum friction μR will act up the plane, where μ is the coefficient of friction. Let x be the distance described by the centre of gravity C parallel to the inclined plane in time t.

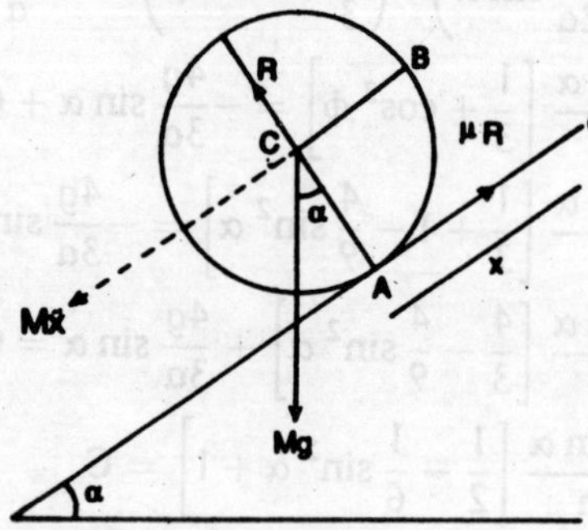

FIG 4.7

There is no motion perpendicular to the plane, so the C.G. of the sphere always moves parallel to the plane.

Hence the equations of motion are

$$M\frac{d^2x}{dt^2} = Mg\sin\alpha - \mu R \tag{4.34}$$

$$0 = R - Mg\cos\alpha \tag{4.35}$$

$$\text{and } M\frac{2}{5}a^2\ddot{\theta} = \mu R a. \tag{4.36}$$

From equation (4.35), $R = Mg\cos\alpha$.

Hence from equation (4.34),

$$M\frac{d^2x}{dt^2} = Mg\sin\alpha - \mu Mg\cos\alpha$$

$$\text{or, } \frac{d^2x}{dt^2} = g(\sin\alpha - \mu\cos\alpha).$$

Also from equation (4.36),

$$a\ddot{\theta} = \frac{5}{2}\mu g\cos\alpha.$$

Integrating these, we get

$$\frac{dx}{dt} = g(\sin\alpha - \mu\cos\alpha)t + C$$

$$\text{and } a\dot{\theta} = \frac{5}{2}\mu g t\cos\alpha + C_1.$$

Initially, when

$$t = 0,\ \frac{dx}{dt} = 0,\ \dot{\theta} = 0,\ C = 0,\ C_1 = 0.$$

Thus,

$$\frac{dx}{dt} = g(\sin\alpha - \mu\cos\alpha)t \tag{4.37}$$

$$\text{and } a\dot{\theta} = \frac{5}{2}\mu g t\cos\alpha. \tag{4.38}$$

Again integrating equations (4.37) and (4.38), we get

$$x = g(\sin\alpha - \mu\cos\alpha)\frac{t^2}{2} + C_2$$

$$\text{and } a\theta = \frac{5}{2}\mu g\cos\alpha\frac{t^2}{2} + C_3.$$

Initially, when $t = 0$, $x = 0$ and $\theta = 0$.

$$C_2 = 0,\ C_3 = 0.$$

Thus,

$$x = g(\sin\alpha - \mu\cos\alpha)\frac{t^2}{2} + C_2 \tag{4.39}$$

$$\text{and } a\theta = \frac{5}{2}\mu g\cos\alpha\frac{t^2}{2}. \tag{4.40}$$

The velocity of the point of contact A down the plane

$$\begin{aligned} &= \text{Velocity of C, the centre of sphere} \\ &\qquad + \text{Velocity of A relative to C} \\ &= \dot{x} - a\dot{\theta} \\ &= g(\sin\alpha - \mu\cos\alpha)t - \frac{5}{2}\mu g t\cos\alpha \\ &= \frac{g}{2}t[2\sin\alpha - 7\mu\cos\alpha]. \end{aligned} \tag{4.41}$$

Now equation (4.41) gives the following three cases:

Case I. *If* $2\sin\alpha - 7\mu\cos\alpha > 0$,

$$\begin{aligned} \text{i.e.,}\quad 2\sin\alpha &> 7\mu\cos\alpha, \\ \text{i.e.,}\quad \mu &< \tfrac{2}{7}\tan\alpha. \end{aligned}$$

In this case, velocity of the point of contact is positive for all values of t, i.e., it does not vanish, hence the point of contact always slides down and the maximum friction μR acts. The sphere never rolls. The equations of motion established above hold good throughout the entire motion.

Case II. *If* $2\sin\alpha = 7\mu\cos\alpha$, *i.e.*, $\mu = \frac{2}{7}\tan\alpha$.

In this case, the velocity of the point of contact is zero for all values of t and consequently the sphere always rolls throughout and the maximum force of friction μR is always exerted. The equations of motion established above hold good.

Case III. *If* $2\sin\alpha < 7\mu\cos\alpha$, *i.e.*, $\mu > \frac{2}{7}\tan\alpha$.

In this case, velocity of the point of contact is negative, i.e., if the maximum friction μR were allowed to act, the point of contact will slide up the plane which is impossible because the amount of friction will only act which is just sufficient to keep the point of contact at rest. Hence in this case the motion is of pure rolling from the very start and remains the same throughout and the maximum friction μR is not exerted. Therefore, in this case the equations of motion established above do not hold good.

If now F is the frictional force, then the equations of motion are

$$M\ddot{x} = Mg\sin\alpha - F \tag{4.42}$$

$$0 = R - Mg\cos\alpha \tag{4.43}$$

$$\text{and } M\frac{2}{5}a^2\ddot{\theta} = Fa. \tag{4.44}$$

Since the point of contact is at rest,

$$\therefore\ \dot{x} - a\dot{\theta} = 0$$

$$\dot{x} = a\dot{\theta}$$

$$\therefore\ \ddot{x} = a\ddot{\theta}.$$

From equation (4.42) and equation (4.44), we have

$$Ma\ddot{\theta} = Mg\sin\alpha - F = Mg\sin\alpha - \frac{2}{5}Ma\ddot{\theta}$$

$$\text{or, } \frac{7}{5}Ma\ddot{\theta} = Mg\sin\alpha$$

$$\text{or, } \frac{7}{5}\ddot{x} = g\sin\alpha$$

$$\therefore\ \ddot{x} = \frac{5}{7}g\sin\alpha. \tag{4.45}$$

Integrating equation (4.45), we get

$$\dot{x} = \frac{5}{7}g\sin\alpha\, t + C.$$

Initially, when $t = 0$, $\dot{x} = 0$, $C = 0$.

$$\dot{x} = \frac{5}{7}g\sin\alpha\, t.$$

Integrating again, we get

$$x = \frac{5}{7} g \sin\alpha \frac{t^2}{2} + C_1.$$

Initially, when $t = 0$, $x = 0$, $C_1 = 0$.

$$x = \frac{5}{7} g \sin\alpha \frac{t^2}{2}$$

$$\therefore\ x = a\theta = \frac{5}{14} g t^2 \sin\alpha.$$

4.8 Motion of a Circular Disc

A uniform circular disc is projected, with its plane vertical along a rough horizontal plane with a velocity V of translation, and an angular velocity ω about the centre. Find the motion.

Case I. *When V is from left to right, ω clockwise and $V > a\omega$.*

In this case, initial velocity of the point of contact P is given by $V - a\omega$, which is positive as $V > a\omega$.

Thus, the force of friction μR acts in the direction left to right.

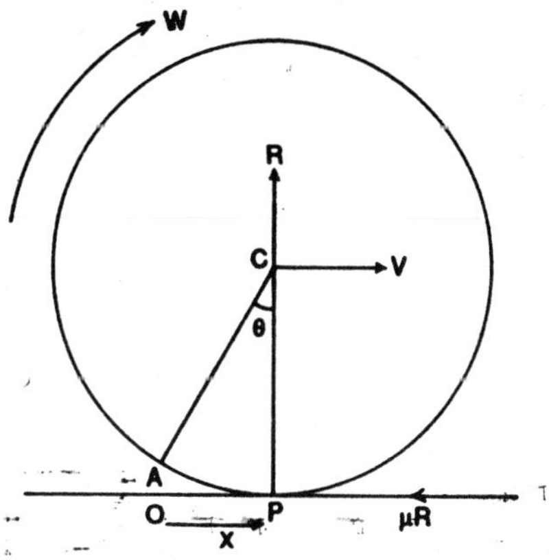

FIG 4.8

Let after time t the disc turns through an angle θ and the centre has moved through a distance x. Then the equations of motion are

$$M\ddot{x} = -\mu R = -\mu m g$$

$$\therefore\ \ddot{x} = -\mu g \tag{4.46}$$

$$\text{and } Mk^2\ddot{\theta} = M\frac{a^2}{2}\ddot{\theta} = \mu R a = \mu M g a$$

$$\text{or, } a\ddot{\theta} = 2\mu g. \tag{4.47}$$

Integrating equations (4.46) and (4.47), we get

$$\dot{x} = -\mu g t + C$$

$$\text{and } a\dot{\theta} = 2\mu g t + C_1.$$

Initially, when $t = 0$, $\dot{x} = V$, $\dot{\theta} = \omega$, $C = V$, $C_1 = a\omega$

$$\dot{x} = -\mu g t + V \tag{4.48}$$
$$\text{and } a\dot{\theta} = 2\mu g t + a\omega. \tag{4.49}$$

The rolling begins when the velocity of the point of contact P_1, i.e., $\dot{x} - a\dot{\theta} = 0$. If this happens after time t_1, then

$$\begin{aligned} \dot{x} - a\dot{\theta} &= -\mu g t_1 + V - 2\mu g t_1 - a\omega \\ 0 &= -\mu g t_1 + V - a\omega, \text{ or, } 3\mu g t_1 = V - a\omega \\ \therefore\ t_1 &= \frac{V - a\omega}{3\mu g}. \end{aligned}$$

From equation (4.48), at this time the velocity of centre is given by

$$\begin{aligned} \dot{x} &= -\mu g \frac{(V - a\omega)}{3\mu g} + V \\ &= \frac{1}{3}(3V - V + a\omega) = \frac{1}{3}(2V + a\omega). \end{aligned} \tag{4.50}$$

When rolling commences, equations of motion are

$$M\ddot{x} = -F \text{ and } Mk^2\ddot{\phi} = \frac{1}{2}Ma^2\ddot{\phi} = Fa. \tag{4.51}$$

As there is no sliding, so $\dot{x} = a\dot{\phi}$.

$$\therefore\ \ddot{x} = a\ddot{\phi}.$$

Hence from equation (4.51),

$$F = -M\ddot{x} = -Ma\ddot{\phi} = -2F, \text{ or, } 3F = 0$$
$$\therefore\ F = 0.$$

Thus no friction is required for rolling throughout the motion. Hence the equations of motion (4.51) for pure rolling reduce to

$$\left.\begin{aligned} M\ddot{x} &= 0 \\ \text{and } \tfrac{1}{2}Ma^2\ddot{\phi} &= 0, \end{aligned}\right\} \tag{4.52}$$

i.e., $\ddot{x} = 0$ and $\ddot{\phi} = 0$.

Integrating equation (4.52), we get

$$\dot{x} = \text{constant} = \frac{1}{3}(2V + a\omega) \text{ [From equation (4.50)]}$$

Hence the disc continues rolling with constant velocity $\frac{2V + a\omega}{3}$.

Case II. *When V is from left to right, ω clockwise and* $V < a\omega$.

In this case, the initial velocity of the point of contact $P = V - a\omega$ which is negative, i.e., its direction is from right to left. Therefore, the friction μR will act from left to right.

Hence the equations of motion are

$$\begin{aligned} M\ddot{x} &= \mu R = \mu M g \\ \therefore\ \ddot{x} &= \mu g \qquad (4.53) \\ \text{and } Mk^2\ddot{\theta} &= \frac{1}{2}Ma^2\ddot{\theta} = -\mu R a = -\mu M g a \\ \therefore\ a\ddot{\theta} &= -2\mu g. \qquad (4.54) \end{aligned}$$

Integrating equations (4.58) and (4.59), we get

$$\dot{x} = \mu g t + C_2 \text{ and } a\dot{\theta} = -2\mu g t + C_3.$$

Initially, when $t = 0$, $\dot{x} = V$, $\dot{\theta} = \omega$, we have

$$V = C_2,\ C_3 = a\omega.$$

Thus

$$\begin{aligned} \dot{x} &= \mu g t + V \qquad (4.55) \\ a\dot{\theta} &= -2\mu g t + a\omega. \qquad (4.56) \end{aligned}$$

Now pure rolling begins, when the velocity of the point of contact P, i.e., $\dot{x} - a\dot{\theta} = 0$. If it happens after time t_2, then

$$\dot{x} - a\dot{\theta} = \mu g t_2 + V + 2\mu g t_2 - a\omega = 0, \text{ or, } 3\mu g t_2 = a\omega - V$$

$$\therefore\ t_2 = \frac{a\omega - V}{3\mu g}.$$

From equation (4.55), we get

$$\begin{aligned} \dot{x} &= \mu g t_2 + V = \frac{a\omega - V}{3} + V \\ \dot{x} &= \frac{a\omega + 2V}{3}. \qquad (4.57) \end{aligned}$$

When rolling begins, the equations of motion are the same as in case 4.8.1. Hence $F = 0$ and the disc continues to roll with uniform velocity $\frac{1}{3}(2V + a\omega)$.

Case III. *When V is from left to right and ω anticlockwise.*

In this case, the velocity of the point of contact $P = V + a\omega$, which is positive, i.e., its direction is from left to right. Therefore, the friction μR will act from right to left.

Hence the equations of motion are

$$M\ddot{x} = -\mu R = -\mu Mg$$
$$\therefore \ddot{x} = -\mu g \tag{4.58}$$
$$\text{and } Mk^2\ddot{\theta} = \frac{1}{2}Ma^2\ddot{\theta} = -\mu Ra = -\mu Mga$$
$$\therefore a\ddot{\theta} = -2\mu g. \tag{4.59}$$

Integrating equations (4.58) and (4.59), we get

$$\dot{x} = -\mu gt + C_4, \ a\dot{\theta} = -2\mu gt + C_5.$$

Initially, when $t = 0$, $\dot{x} = V$, $\dot{\theta} = \omega$, $C_4 = V$, $C_5 = a\omega$.
Thus

$$\dot{x} = V - \mu gt \tag{4.60}$$
$$a\dot{\theta} = a\omega - 2\mu gt. \tag{4.61}$$

For pure rolling to commence, the velocity of the point of contact P is zero, i.e., $\dot{x} + a\dot{\theta} = 0$. If this happens after time t_3, then

$$\dot{x} + a\dot{\theta} = V - \mu gt_3 + a\omega - 2\mu gt_3,$$
$$\text{or, } 0 = V + a\omega - 3\mu gt_3$$
$$t_3 = \frac{V + a\omega}{3\mu g}. \tag{4.62}$$

Hence from equation (4.60),

$$\dot{x} = V - \mu gt_3 = V - \frac{(V + a\omega)}{3} = \frac{2V - a\omega}{3}. \tag{4.63}$$

If $2V > a\omega$, the velocity of the centre is from left to right. Hence as in Case 4.8.1 and 4.8.2, the motion will be of pure rolling with constant velocity $\frac{1}{3}(2V - a\omega)$.

If $2V < a\omega$, then $\dot{x}$ is negative, i.e., the direction of the velocity of the centre is from right to left, i.e., backwards. Thus when pure rolling begins, the disc rolls back towards the initial point.

From equation (4.60), we have $\dot{x} = 0$ when $t = \frac{V}{\mu g}$, then from equation (4.61) $a\dot{\theta} = a\omega - 2V$, which is positive since $2V < a\omega$.

Hence if $2V < a\omega$, the disc begins to move backwards before the pure rolling begins.

4.9 Two bodies in Contact

When two bodies are in contact, then to determine whether the relative motion involves sliding at the point of contact.

Let a moving body be placed over the other body. Let P be the point of contact of the moving body and assume that its initial velocity is zero. To find the relative motion of the bodies which is either sliding or rolling we proceed as follows.

We first assume that the body rolls and let F be the force of friction sufficient to keep the point of contact P at rest. Then we write the equations of motion and the geometrical equation to express the condition that the tangential velocity of the point P is zero. We solve these equations and then find the value of $\frac{F}{R}$. Now there are two possibilities.

Case I. *If* $\frac{F}{R} < \mu$. *In this case, the necessary friction can be called into play to keep the point* P *at rest. Thus the moving body rolls and will keep rolling so long as* $\frac{F}{R} \leq \mu$.

Case II. *If* $\frac{F}{R} > \mu$. *In this case, the point of contact will slide and the equations of motion discussed above will not hold good. Thus in this case we proceed as follows.*

We write the equations of motion, supposing that the point of contact slides. Here the frictional force is μR instead of F and there is no geometrical equation in this case.

Solving these equations we find the tangential velocity of the point of contact P. If this velocity is not zero and is in the direction opposite to the direction in which μR acts, μ has a proper sign and the body will slide at P and go on sliding so long as the velocity of the point P does not vanish. When velocity at P vanishes, we again proceed as in Case I.

4.10 Motion on a Horizontal Plane

A sphere of radius a, *whose centre of gravity* G *is at a distance* c *from its centre* C, *is placed on a rough plane so that the C.G. is horizontal. Show that it will begin to roll or slide according as the coefficient of friction* $\mu >$ *or* $< \frac{ac}{K^2+a^2}$, *where* K *is the radius of gyration about a horizontal axis through* G.

If μ *is equal to this value, what happens?*

Let M be the mass, C the centre and G the C.G. of the sphere of radius a. Initially CG is horizontal and at any time t, let it be inclined at an angle θ to the horizon.

Assuming that the sphere rolls, let OP = x be the horizontal distance moved by the point of contact P, from its initial position O, in time t.

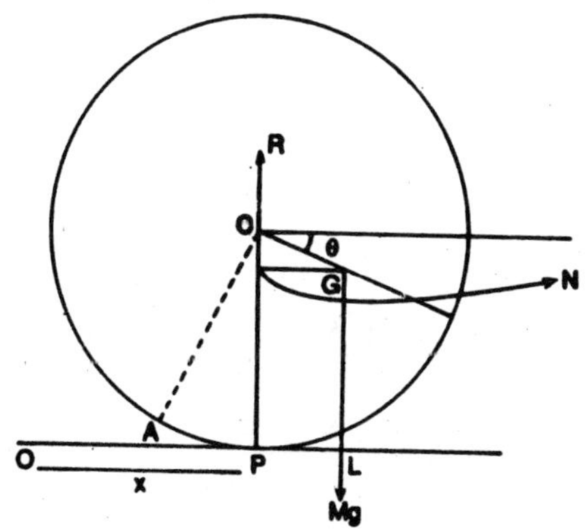

FIG 4.9

Since the motion is of pure rolling,

$\therefore\ x = a\theta$, so that $\dot{x} = a\dot{\theta}$ and $\ddot{x} = a\ddot{\theta}$.

If $(\bar{x}, \bar{y})$ be the coordinates of C.G. at G referred to O as origin and the horizontal and vertical lines through O as origin are given by

$$\bar{x} = OL = OP + PL = x + c\cos\theta$$
$$\bar{y} = GL = a - c\sin\theta.$$

Equations of motion of the sphere are

$$F = M\frac{d^2\bar{x}}{dt^2} = M\frac{d^2}{dt^2}(a\theta + c\cos\theta) = M\frac{d}{dt}[a\dot{\theta} - c\sin\theta\dot{\theta}]$$
$$= M[a\ddot{\theta} - c\{\cos\theta\dot{\theta}^2 + \sin\theta\ddot{\theta}\}]$$
$$F = M[\ddot{\theta}(a - c\sin\theta) - c\cos\theta\dot{\theta}^2] \tag{4.64}$$
$$\text{and } R - Mg = M\frac{d^2\bar{y}}{dt^2} = M\frac{d^2}{dt^2}(a - c\sin\theta)$$
$$= M\frac{d}{dt}[-c\cos\theta\dot{\theta}] = -Mc[-\sin\theta\dot{\theta}^2 + \ddot{\theta}\cos\theta].$$
$$R - Mg = M[c\sin\theta\dot{\theta}^2 - c\cos\theta\ddot{\theta}] \tag{4.65}$$
$$\text{and } MK^2\ddot{\theta} = R\cdot GN - F\cdot GL$$
$$\text{or, } MK^2\ddot{\theta} = R\cdot c\cos\theta - F(a - c\sin\theta). \tag{4.66}$$

Here we want only the initial motion when $\theta = 0$, $\dot{\theta} = 0$ but $\ddot{\theta}$ is not zero.
Hence from equations (4.64), (4.65) and (4.66), we get

$$F = Ma\ddot{\theta},\ R = Mg - cM\ddot{\theta} \text{ and } MK^2\ddot{\theta} = Rc - aF. \tag{4.67}$$

These equations determine initial values of $\ddot{\theta}$, F and R.
Eliminating F and R, from equation (4.67), we get

$$MK^2\ddot{\theta} = (Mg - cM\ddot{\theta})c - Ma^2\ddot{\theta}$$
$$\text{or, } \ddot{\theta}(K^2 + c^2 + a^2) = cg.$$
$$\therefore \ddot{\theta} = \frac{cg}{K^2 + c^2 + a^2}$$
$$\therefore F = \frac{Macg}{K^2 + c^2 + a^2}$$
$$\text{and } R = Mg - \frac{Mc\cdot cg}{K^2 + c^2 + a^2}$$
$$= \frac{Mg[K^2 + c^2 + a^2 - c^2]}{K^2 + c^2 + a^2}$$
$$= \frac{Mg(K^2 + a^2)}{K^2 + c^2 + a^2}.$$

So

$$\frac{F}{R} = \frac{ac}{K^2 + a^2}.$$

Thus the sphere will rod or slide according as

$$F < \text{or} > \mu R,$$

i.e., according as $\mu > \text{or} < \frac{F}{R}$,

i.e., according as $\mu > \text{or} < \frac{ac}{K^2+a^2}$.

When $\mu = \frac{ac}{K^2+a^2}$.

In this case, we shall consider whether $\frac{F}{R}$ is a little greater or little less than μR when θ is small but not absolutely zero.

From equations (4.64), (4.65) and (4.66), we get

$$\begin{aligned}
MK^2\theta &= [Mg + M(c\sin\theta\dot{\theta}^2 - c\cos\theta\ddot{\theta})]c\cos\theta \\
&\quad - M[a\ddot{\theta} - c\sin\theta\ddot{\theta} - c\cos\theta\dot{\theta}^2](a - c\sin\theta) \\
\Rightarrow\quad K^2\ddot{\theta} &= gc\cos\theta + c^2\sin\theta\cos\theta\dot{\theta}^2 - c^2\cos^2\theta\ddot{\theta} - a^2\ddot{\theta} + ac\sin\theta\ddot{\theta} \\
&\quad + ac\cos\theta\dot{\theta}^2 + ca\ddot{\theta}\sin\theta - c^2\sin^2\theta\ddot{\theta} - c^2\cos\theta\sin\theta\dot{\theta}^2 \\
\Rightarrow\quad K^2\ddot{\theta} &= gc\cos\theta + 2ac\sin\theta\ddot{\theta} - a^2\ddot{\theta} + ac\cos\theta\dot{\theta}^2 - c^2\ddot{\theta} \\
\Rightarrow\quad (K^2 + a^2 + c^2 - 2ac\sin\theta)\ddot{\theta} &= ac\cos\theta\dot{\theta}^2 + gc\cos\theta \\
\Rightarrow\quad \ddot{\theta}(K^2 + a^2 + c^2 - 2ac\sin\theta) - ac\cos\theta\dot{\theta}^2 &= gc\cos\theta. \qquad (4.68)
\end{aligned}$$

Integrating, we get

$$\dot{\theta}^2(K^2 + a^2 + c^2 - 2ac\sin\theta) = 2gc\sin\theta. \qquad (4.69)$$

Since θ is small, thus $\cos\theta = 1$ and $\sin\theta = \theta$. Hence from equation (4.68), we get

$$\begin{aligned}
\dot{\theta}^2(K^2 + a^2 + c^2 - 2ac\theta) &= 2gc\theta, \\
\text{or,}\quad \dot{\theta}^2(K^2 + a^2 + c^2) &= 2gc\theta \text{ neglecting } \theta\dot{\theta}^2. \qquad (4.70)
\end{aligned}$$

Then from equation (4.68), we have

$$\begin{aligned}
(K^2 + a^2 + c^2 - 2ac\theta)\ddot{\theta} &= gc\left(1 + \frac{a}{g}\dot{\theta}^2\right) \\
&= gc\left[1 + \frac{a}{g}\frac{2gc\theta}{K^2 + a^2 + c^2}\right] \quad \text{[From equation (4.70)]} \\
&= gc\left[1 + \frac{2ac\theta}{K^2 + a^2 + c^2}\right] \qquad (4.71)
\end{aligned}$$

$$\text{or,}\quad (K^2 + a^2 + c^2)\left[1 - \frac{2ac\theta}{K^2 + a^2 + c^2}\right]\ddot{\theta} = gc\left(1 + \frac{2ac\theta}{K^2 + a^2 + c^2}\right)$$

$$\text{or,}\quad (K^2 + a^2 + c^2)\ddot{\theta} = \frac{gc}{1 - \frac{2ac\theta}{K^2+a^2+c^2}}\left(1 + \frac{2ac\theta}{K^2 + a^2 + c^2}\right)$$

$$\text{or,}\quad (K^2+a^2+c^2)\ddot{\theta} = gc\left(1+\frac{2ac\theta}{K^2+a^2+c^2}\right)\left(1-\frac{2ac\theta}{K^2+a^2+c^2}\right)^{-1}$$

$$\text{or,}\quad (K^2+a^2+c^2)\ddot{\theta} = gc\left(1+\frac{2ac\theta}{K^2+a^2+c^2}\right)\left(1+\frac{2ac\theta}{K^2+a^2+c^2}\right)$$

[Expanding by Binomial Theorem and neglecting higher powers of θ]

$$= gc\left(1+\frac{4ac\theta}{K^2+a^2+c^2}\right) \text{ (neglecting } \theta^2\text{)}. \qquad (4.72)$$

Also from equations (4.64) and (4.65), we get

$$\frac{F}{R} = \frac{(a-c\sin\theta)\ddot{\theta}-c\cos\theta\dot{\theta}^2}{g-c\cos\theta\ddot{\theta}+c\sin\theta\dot{\theta}^2}$$

$$\text{or,}\quad \frac{F}{R} = \frac{(a-c\theta)\ddot{\theta}-c\dot{\theta}^2}{g-c\ddot{\theta}} \quad \text{[Since } \theta \text{ is small, neglecting } \sin\theta\dot{\theta}^2\text{]}.$$

Putting the values of $\dot{\theta}^2$ from equation (4.70) and $\ddot{\theta}$ from equation (4.72), we get

$$\frac{F}{R} = \frac{(a-c\theta)\left[\frac{gc}{K^2+a^2+c^2}\left(1+\frac{4ac\theta}{K^2+a^2+c^2}\right)\right]-\frac{c2gc\theta}{K^2+a^2+c^2}}{g-c\frac{gc}{K^2+a^2+c^2}\left(1+\frac{4ac\theta}{K^2+a^2+c^2}\right)}$$

$$= \frac{ac}{K^2+a^2}\left[1-\frac{c(3K^2-a^2)}{a(K^2+a^2)}\theta\right] \text{ [on simplification]}.$$

If $K^2 > \frac{a^2}{3}$, then

$$\frac{F}{R} < \frac{ac}{K^2+a^2} = \mu.$$

∴ in this case, sphere rolls.

If $K^2 < \frac{a^2}{3}$, then

$$\frac{F}{R} > \frac{ac}{K^2+a^2} = \mu.$$

∴ in this case, sphere slides.

4.11 Motion of one sphere over another sphere which is fixed

A solid homogeneous sphere resting on the top of another fixed sphere, is slightly displaced and begins to roll down it. Show that it will slip when the common normal makes with the vertical an angle θ given by the equation $2\sin(\theta-\lambda) = 5\sin\lambda(3\cos\theta-2)$, where λ is the angle of friction.

Let O be the centre and a the radius of the fixed sphere. Let C be the centre and b be the radius of the solid homogeneous sphere resting on this fixed sphere. The moving sphere was initially resting at the top of the fixed sphere so that BC was vertical and the point of contact B was at A. After time t let the point of contact be P and the position of CB is as shown in the figure.

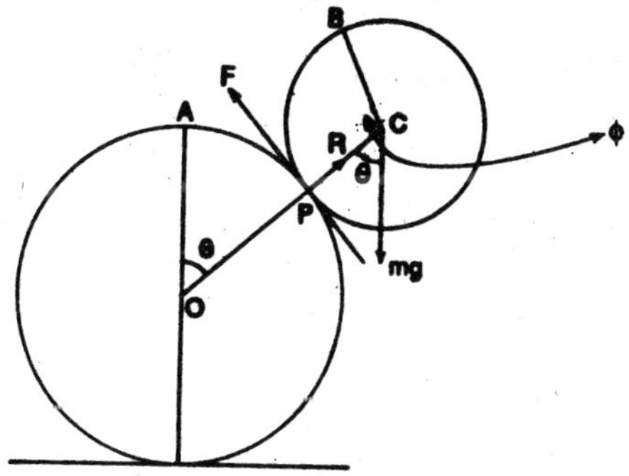

FIG 4.10

As there is pure rolling,

$$\therefore \quad \text{arc } AP = \text{arc } BP,$$

$$\text{i.e.,} \quad a\theta = b(\phi - \theta)$$

$$\text{or,} \quad b\phi = (a+b)\theta,$$

$$\text{i.e.,} \quad b\dot{\phi} = c\dot{\theta}, \text{ where } c = a + b. \tag{4.73}$$

Let R and F be the normal reaction and the friction acting on the upper sphere at P.

Since the centre C describes a circle of radius c, so that its accelerations are $c\dot{\theta}^2$ and $c\ddot{\theta}$ along and perpendicular to CO.

$\therefore$ the equations of motion of the upper sphere along and perpendicular to CO are

$$Mc\dot{\theta}^2 = -R + Mg\cos\theta \tag{4.74}$$

$$\text{and } Mc\ddot{\theta} = Mg\sin\theta - F, \tag{4.75}$$

The coordinates of the centre C referred to the horizontal and vertical lines through O as axes are $(c\sin\theta, c\cos\theta) = (x, y)$ (Let).

Then $v^2 = \dot{x}^2 + \dot{y}^2 = c^2\dot{\theta}^2$.

Also, the energy equation is

$$\frac{M}{2}(\dot{x}^2 + \dot{y}^2) + \frac{1}{2}MK^2(\dot{\phi})^2 = \text{work done by gravity}$$
$$= Mgc(1 - \cos\theta)$$

$$\text{or,} \quad c^2\dot{\theta}^2 + \frac{2}{5}b^2\dot{\phi}^2 = 2gc(1 - \cos\theta)$$

$$\text{or,} \quad c^2\dot{\theta}^2 + \frac{2c^2}{5}\dot{\theta}^2 = 2gc(1 - \cos\theta) \text{ [From equation (4.73)]}$$

$$\text{or,} \quad \frac{7}{5}c^2\dot{\theta}^2 = 2gc(1 - \cos\theta)$$

$$\text{or,} \quad c\dot{\theta}^2 = \frac{10}{7}g(1 - \cos\theta). \tag{4.76}$$

Differenciating equation (4.76) with respect to t, we get

$$2c\dot{\theta}\ddot{\theta} = \frac{10g}{7}\sin\theta\dot{\theta}, \text{ or, } c\ddot{\theta} = \frac{5g}{7}\sin\theta. \quad (4.77)$$

Now from equation (4.74), we have

$$\begin{aligned} R &= Mg\cos\theta - Mc\dot{\theta}^2 \\ &= Mg\cos\theta - M\frac{10g}{7}(1-\cos\theta) \text{ [From equation (4.76)]} \\ &= \frac{Mg}{7}[7\cos\theta - 10 + 10\cos\theta] = \frac{Mg}{7}(17\cos\theta - 10). \quad (4.78) \end{aligned}$$

Also from equation (4.75), we have

$$\begin{aligned} F &= Mg\sin\theta - Mc\ddot{\theta} \\ &= Mg\sin\theta - M\frac{5g}{7}\sin\theta \text{ [From equation (4.77)]} \\ &= \frac{Mg}{7}[7\sin\theta - 5\sin\theta] \\ F &= \frac{Mg}{7}2\sin\theta. \end{aligned}$$

Now

$$\frac{F}{R} = \frac{2\sin\theta}{17\cos\theta - 10}.$$

The sphere will slip when $\frac{F}{R} = \mu$,

$$\begin{aligned} &\Rightarrow \frac{2\sin\theta}{17\cos\theta - 10} = \mu \\ &\Rightarrow 2\sin\theta = (17\cos\theta - 10)\tan\lambda \; [\because \mu = \tan\lambda] \\ &\Rightarrow 2\sin\theta = (17\cos\theta - 10)\frac{\sin\lambda}{\cos\lambda} \\ &\Rightarrow 2\sin\theta\cos\lambda = 17\cos\theta\sin\lambda - 10\sin\lambda \\ &\Rightarrow 2[\sin\theta\cos\lambda - \cos\theta\sin\lambda] = 15\cos\theta\sin\lambda - 10\sin\lambda \\ &\Rightarrow 2\sin(\theta - \lambda) = 5\sin\lambda(3\cos\theta - 2). \end{aligned}$$

Note 1. *The upper sphere will leave the fixed sphere when* $R = 0$, *i.e.,* $17\cos\theta - 10 = 0$, *or,* $\theta = \cos^{-1}\frac{10}{17}$.

Note 2. *When the two spheres are smooth, in that case,* $F = 0$.

Hence the energy equation becomes

$$\frac{1}{2}Mc^2\dot{\theta}^2 = Mg(c - c\cos\theta), \text{ i.e., } \dot{\theta}^2 = \frac{2g}{c}(1 - \cos\theta).$$

Thus, from equation (4.74), we get

$$R = Mg\cos\theta - M2g(1 - \cos\theta) = Mg(3\cos\theta - 2).$$

The upper sphere will leave the fixed sphere when $R = 0$, i.e., $Mg(3\cos\theta - 2) = 0$, or, $\theta = \cos^{-1}\frac{2}{3}$.

4.12 Motion of a solid cylinder inside a hollow cylinder

A hollow cylinder, of radius a, *is fixed with its axis horizontal; inside it moves a solid cylinder of radius* b, *whose velocity in its lowest position is given. If the friction between the cylinders be sufficient to prevent any sliding, find the motion.* [VU 87H]

Let O be the centre and a be the radius of the fixed cylinder. Let C be the centre, M the mass and b the radius of the solid cylinder resting with its point B in contact with the lowest point A of the fixed cylinder. In time t, let the inside cylinder roll to the point P such that $\angle AOP = \theta$ and the line CB fixed in moving cylinder make an angle ϕ to the vertical. Since there is pure rolling,

$$\text{arc } AP = \text{arc } BP$$

$$\text{or,} \quad a\theta = b(\phi + \theta)$$

$$\therefore \quad b\phi = (a-b)\theta = c\theta, \quad \text{where } c = a - b$$

$$\therefore \quad b\dot{\phi} = c\dot{\theta} \tag{4.79}$$

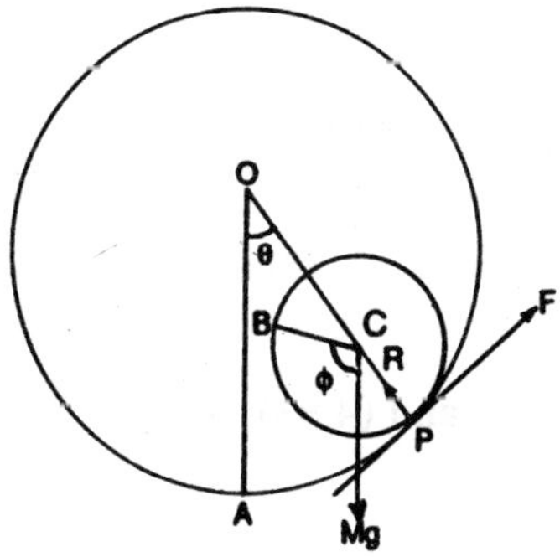

FIG 4.11

Let R and F be the normal reaction and friction at the point P. Since the centre C describes a circle of radius $OC = a - b = c$ about O. Hence its acceleration along and perpendicular to CO are $Mc\dot{\theta}^2$ and $Mc\ddot{\theta}$, respectively.

$\therefore$ the equations of motion of the moving cylinder are

$$Mc\dot{\theta}^2 = R - Mg\cos\theta \tag{4.80}$$

$$\text{and } Mc\ddot{\theta} = F - Mg\sin\theta. \tag{4.81}$$

The coordinates of C referred to the horizontal and vertical line through O as axes are $(c\sin\theta, c\cos\theta)$.

$$\therefore \; v^2 = c^2\dot{\theta}^2.$$

So kinetic energy of the moving cylinder at time t

$$\begin{aligned}
&= \frac{1}{2}MK^2\dot{\phi}^2 + \frac{1}{2}Mc^2\dot{\theta}^2 = \frac{1}{2}M\frac{b^2}{2}\dot{\phi}^2 + \frac{1}{2}Mc^2\dot{\theta}^2 \\
&= \frac{1}{4}Mc^2\dot{\theta}^2 + \frac{1}{2}Mc^2\dot{\theta}^2 \text{ [From equation (4.79)]} \\
&= \frac{3}{4}Mc^2\dot{\theta}^2.
\end{aligned}$$

If $\dot{\theta} = \omega$, is the angular initial velocity at $t = 0$, then initial kinetic energy of the moving cylinder $= \frac{3}{4}Mc^2\omega^2$.

Hence energy equation gives change in kinetic energy = work done

$$\text{or,}\quad \frac{3}{4}Mc^2\dot{\theta}^2 = \frac{3}{4}Mc^2\omega^2 = -Mg(c - c\cos\theta)$$

$$\text{or,}\quad c\dot{\theta}^2 = c\omega^2 - \frac{4}{3}g(1-\cos\theta). \tag{4.82}$$

Differentiating with respect to t, we get

$$2c\dot{\theta}\ddot{\theta} = \frac{4}{3}g\sin\theta\dot{\theta},\ \text{or,}\ c\ddot{\theta} = -\frac{2}{3}g\sin\theta. \tag{4.83}$$

From equations (4.80) and (4.82), we get

$$\begin{aligned} R &= Mg\cos\theta + M\left[c\omega^2 - \frac{4}{3}g(1-\cos\theta)\right] \\ &= Mc\omega^2 + \frac{Mg}{3}[3\cos\theta - 4 + 4\cos\theta] \\ R &= Mc\omega^2 + \frac{Mg}{3}(7\cos\theta - 4). \end{aligned} \tag{4.84}$$

Again from equations (4.81) and (4.83), we get

$$\begin{aligned} F &= Mg\sin\theta - M\frac{2}{3}g\sin\theta \\ F &= \frac{Mg}{3}\sin\theta. \end{aligned} \tag{4.85}$$

Thus equations (4.82), (4.84) and (4.85) determine motion.

Case I. *In order that the cylinder may just make complete revolution.*

In this case, $R = 0$, at the highest point, i.e., when $\theta = \pi$. Thus from equation (4.84),

$$0 = Mc\omega^2 + \frac{Mg}{3}(-7-4),\quad \text{or,}\quad \frac{11Mg}{3} = Mc\omega^2$$

$$\text{or,}\quad \omega^2 = \frac{11g}{3c}$$

$$\text{or,}\quad \omega = \sqrt{\frac{11g}{3(a-b)}}.$$

Case II. *The moving cylinder will leave the fixed cylinder.*

When $R = 0$, from equation (4.84), we get

$$0 = Mc\omega^2 + \frac{Mg}{3}(7\cos\theta - 4),$$

$$\Rightarrow \frac{Mg}{3}(4 - 7\cos\theta) = Mc\omega^2$$

$$\Rightarrow 4 - 7\cos\theta = \frac{3c\omega^2}{g}$$

$$\Rightarrow 4 - \frac{3c\omega^2}{g} = 7\cos\theta$$

$$\Rightarrow \frac{4g - 3(a-b)\omega^2}{7g} = \cos\theta. \tag{4.86}$$

Thus, the moving cylinder will leave the cylinder at an angle θ to the vertical given by equation (4.86).

Case III. *If the moving cylinder makes small oscillation about the lowest point of the fixed cylinder, then θ is always small. Thus from equation (4.83), we get* $\ddot{\theta} = -\frac{2g}{3(a-b)}\theta$.

$\therefore$ the time of small oscillation is

$$= \frac{2\pi}{\sqrt{\frac{2g}{3(a-b)}}} = 2\pi\sqrt{\frac{3(a-b)}{2g}}.$$

4.13 Worked out Examples

Type I: Problems based on Section 4.4

Problem 1. *A uniform solid cylinder is placed with its axis horizontal on a plane, whose inclination to the horizon is α. Show that the least coefficient of friction between it and the plane so that it may roll and not slide, is $\frac{1}{3}\tan\alpha$.* **[LNMU 93H; Kanpur 81]**

Solution: Let the cylinder roll down a distance x along the inclined plane in time t. If θ is the angle turned by the cylinder during this time t, then $x = a\theta$

$$\therefore\ \dot{x} = a\dot{\theta},\ \ddot{x} = a\ddot{\theta}.$$

Let R be the reaction and F be the frictional force, then equations of motion of C.G. of the cylinder are

$$M\ddot{x} = Mg\sin\alpha - F \tag{4.87}$$

$$\text{and } M\ddot{y} = 0 = Mg\cos\alpha - R. \tag{4.88}$$

Also

$$MK^2\ddot{\theta} = F\cdot a, \quad \text{or,} \quad MK^2\frac{\ddot{x}}{a} = Fa\ (\because \ddot{x} = a\ddot{\theta})$$

$$\text{or,} \quad M\ddot{x} = \frac{a^2}{K^2}F. \tag{4.89}$$

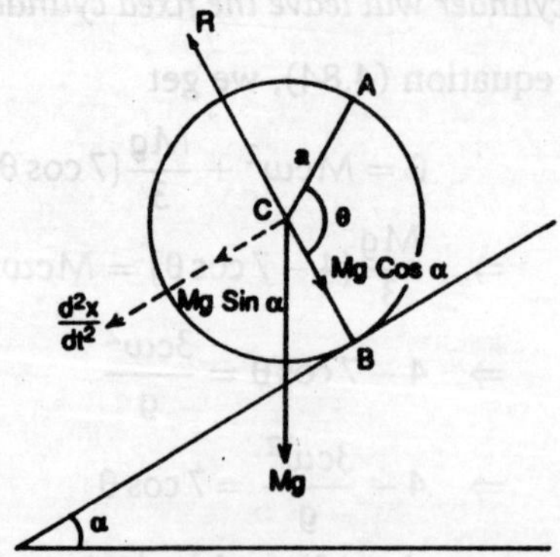

FIG 4.12

From equations (4.87) and (4.89), we have

$$F\frac{a^2}{K^2} = Mg\sin\alpha - F, \quad \text{or,} \quad F\left(\frac{a^2}{K^2}+1\right) = Mg\sin\alpha$$

$$\text{or,} \quad F\frac{(a^2+K^2)}{K^2} = Mg\sin\alpha$$

$$\therefore \quad F = \frac{K^2}{a^2+K^2}Mg\sin\alpha. \tag{4.90}$$

From equation (4.88),

$$R = Mg\cos\alpha, \ \therefore \ \frac{F}{R} = \frac{K^2}{a^2+K^2}\tan\alpha.$$

For pure rolling,

$$\mu > \frac{F}{R}, \ \therefore \ \mu > \frac{K^2}{a^2+K^2}\tan\alpha.$$

For solid cylinder, $K^2 = \frac{a^2}{2}$.

$$\therefore \ \mu > \frac{\frac{a^2}{2}}{a^2+\frac{a^2}{2}}\tan\alpha, \text{ i.e., } \mu > \frac{1}{3}\tan\alpha.$$

For hollow cylinder, $K^2 = a^2$.

$$\therefore \ \mu > \frac{a^2}{a^2+a^2}\tan\alpha, \text{ i.e., } \mu > \frac{1}{2}\tan\alpha.$$

Problem 2. *A cylinder rolls down a smooth plane whose inclination to the horizontal is α, as it goes; a fine string fixed to the highest point of the plane, find its acceleration and the tension of the string.* **[Rohilkhand 83]**

Solution: Let T be the tension in the string when the cylinder has rolled down a distance x along the inclined plane. In this time t let θ be the angle turned by the cylinder.

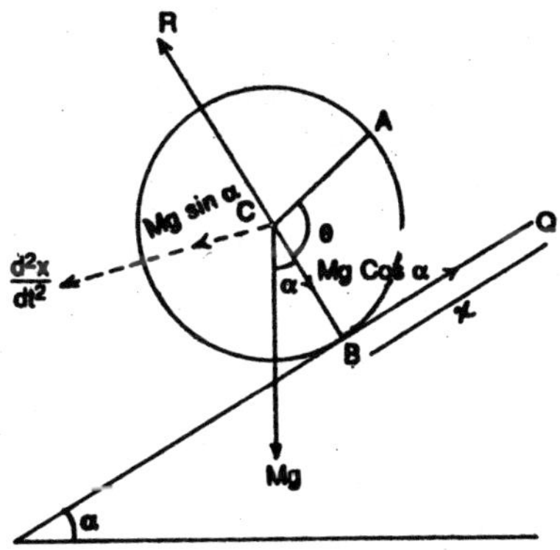

FIG 4.13

Since the string is tight, so the motion is of pure rolling.

$$\therefore \quad x = a\theta \tag{4.91}$$

$$\therefore \quad \dot{x} = a\dot{\theta} \text{ and } \ddot{x} - a\ddot{\theta}.$$

The equations of motion of the C.G. of the cylinder are

$$M\frac{d^2x}{dt^2} = Mg\sin\alpha - T \tag{4.92}$$

$$\text{and} \quad M\frac{d^2y}{dt^2} = 0 = Mg\cos\alpha - R. \tag{4.93}$$

Also

$$MK^2\ddot{\theta} = T \cdot a, \quad \text{or,} \quad M\frac{a^2}{2}\ddot{\theta} = T \cdot a$$

$$\text{or,} \quad \frac{M}{2}a\ddot{\theta} = T$$

$$\text{or,} \quad \frac{1}{2}M\ddot{x} = T \; (\because \ddot{x} = a\ddot{\theta}).$$

Putting this value of T in equation (4.92), we get

$$M\ddot{x} = Mg\sin\alpha - \frac{M}{2}\ddot{x}, \quad \text{or,} \quad \frac{3}{2}M\ddot{x} = Mg\sin\alpha$$

$$\text{or,} \quad \ddot{x} = \frac{2}{3}g\sin\alpha.$$

This gives acceleration of the cylinder.

Also

$$T = \frac{M}{2}\ddot{x} = \frac{M}{2}\frac{2}{3}g\sin\alpha = \frac{M}{3}g\sin\alpha.$$

Type II: Problems based on Section 4.5

Problem 3. *A uniform rod is held at an inclination α to the horizon with one end in contact with a horizontal table whose coefficient of friction is μ. If it be then released, show that it will commence to slide if $\mu < \frac{3\sin\alpha\cos\alpha}{1+3\sin^2\alpha}$.*

[IAS 94; Agra 81; CU 94H; VU 88H, 96H, 99H, 02H]

Solution: Let AB be the rod of mass M and length $2a$. Let F be the frictional force and R be the normal reaction.

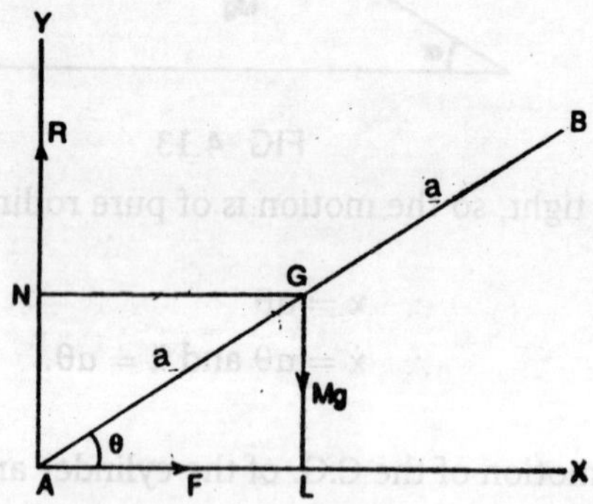

FIG 4.14

Taking the point A as origin, and the horizontal and vertical lines through A as axes, the coordinates of the C.G. at G of the rod are $(a\cos\theta, a\sin\theta)$.

Thus equations of motion of C.G. are

$$M\frac{d^2x}{dt^2} = M\left[-a\cos\theta\dot{\theta}^2 - a\sin\theta\ddot{\theta}\right] = F \tag{4.94}$$

$$\text{and } M\frac{d^2y}{dt^2} = M\left[-a\sin\theta\dot{\theta}^2 + a\cos\theta\ddot{\theta}\right] = R - Mg. \tag{4.95}$$

Initially, when the rod was inclined at an angle α to the horizontal, the coordinates of G were $(a\cos\alpha, a\sin\alpha)$.

$\therefore$ vertical downward displacement of G is $= a(\sin\alpha - \sin\theta)$.

Thus energy equation is

$$\text{K.E. at time } t = \text{work done by the gravity}$$

$$\Rightarrow \frac{M}{2}\left[(\dot{x}^2+\dot{y}^2) + K^2\dot{\theta}^2\right] = Mg(a\sin\alpha - a\sin\theta)$$

$$\Rightarrow \frac{1}{2}\left[a^2\dot{\theta}^2 + \frac{a^2}{3}\dot{\theta}^2\right] = ag(\sin\alpha - \sin\theta)$$

$$\Rightarrow \frac{1}{2}\frac{4}{3}a^2\dot{\theta}^2 = ag(\sin\alpha - \sin\theta)$$

$$\Rightarrow \dot{\theta}^2 = \frac{3g}{2a}(\sin\alpha - \sin\theta). \tag{4.96}$$

Differentiating equation (4.96) with respect to t, we get

$$2\dot{\theta}\ddot{\theta} = \frac{3g}{2a}(-\cos\theta\dot{\theta})$$

$$\ddot{\theta} = -\frac{3g}{4a}\cos\theta. \tag{4.97}$$

Putting the values of $\dot{\theta}^2$ and $\ddot{\theta}$ from equations (4.96) and (4.97) in (4.94) and (4.95), we get

$$F = M\left[-a\cos\theta\frac{3g}{2a}(\sin\alpha - \sin\theta) - a\sin\theta\left(-\frac{3g}{4a}\cos\theta\right)\right]$$

$$= \frac{M\cos\theta 3g}{4}[-2\sin\alpha + 2\sin\theta + \sin\theta]$$

$$F = \frac{3}{4}Mg\cos\theta(3\sin\theta - 2\sin\alpha)$$

$$\text{and } R = Mg + M\left[-a\sin\theta\frac{3g}{2u}(\sin\alpha - \sin\theta) + a\cos\theta\left(-\frac{3g}{4a}\cos\theta\right)\right]$$

$$= \frac{Mg}{4}\left[4 - 6\sin\alpha\sin\theta + 6\sin^2\theta - 3\cos^2\theta\right].$$

When $\theta = \alpha$.

$$F = \frac{3}{4}Mg\cos\alpha\sin\alpha$$

$$\text{and } R = \frac{Mg}{4}[4 - 6\sin^2\alpha + 6\sin^2\alpha - 3\cos^2\alpha]$$

$$= \frac{Mg}{4}(4 - 3\cos^2\alpha) = \frac{Mg}{4}[4 - 3(1 - \sin^2\alpha)]$$

$$= \frac{Mg}{4}[1 + 3\sin^2\alpha].$$

The end A will commence to slide if

$$\mu < \frac{F}{R}, \text{ or, } \mu < \frac{3\sin\alpha\cos\alpha}{1 + 3\sin^2\alpha}.$$

Problem 4. *A rough uniform rod, of length 2a, is placed on a rough table at right angles to its edge. If its centre of gravity be initially at a distance b beyond the edge, show that the rod will begin to slide, when it has turned through an angle $\tan^{-1}\frac{\mu a^2}{a^2+9b^2}$, where μ is the coefficient of friction.* **[Kanpur 82; CU 82H]**

Solution: Let AB be the rod of mass M and length 2a. Initially the rod was at right angles to the edge of the rough table.

In time t, let the rod turn through an angle θ. Let there be no sliding when the rod has turned through this angle. Let F and R be the normal reaction and force of friction on the rod. Accelerations of G along and perpendicular to GO are respectively $b\dot{\theta}^2$ and $b\ddot{\theta}$, where $OG \doteq b$.

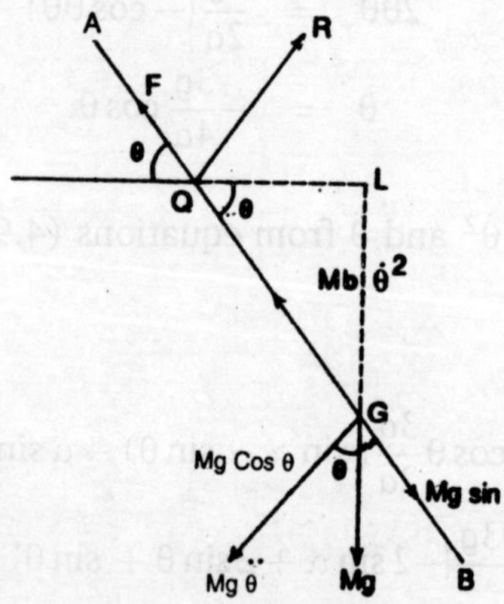

FIG 4.15

Equations of motion of C.G. at G are

$$Mb\ddot{\theta} = Mg\cos\theta - R \tag{4.98}$$

$$\text{and } Mb\dot{\theta}^2 = F - Mg\sin\theta. \tag{4.99}$$

Also energy equation gives,

$$MK^2\ddot{\theta} = Mg\,OL = Mgb\cos\theta$$

$$\text{or,}\quad M\left(b^2 + \frac{a^2}{3}\right)\ddot{\theta} = Mgb\cos\theta$$

$$\text{or,}\quad \frac{a^2+3b^2}{3}\ddot{\theta} = gb\cos\theta$$

$$\therefore\quad \ddot{\theta} = \frac{3gb}{a^2+3b^2}\cos\theta. \tag{4.100}$$

Multiplying both the sides of equation (4.100) by $2\dot{\theta}$, we get

$$2\dot{\theta}\ddot{\theta} = \frac{6gb}{a^2+3b^2}\cos\theta\dot{\theta}.$$

Integrating it, we have

$$\dot{\theta}^2 = \frac{6gb}{a^2+3b^2}\sin\theta + C.$$

Initially, when $\theta = 0$, $\dot{\theta} = 0$.

$$\therefore C = 0$$

$$\dot{\theta}^2 = \frac{6gb}{a^2+3b^2}\sin\theta. \tag{4.101}$$

Putting the values of $\ddot{\theta}$ and $\dot{\theta}^2$ in equations (4.98) and (4.99), we have,

$$\begin{aligned} R &= Mg\cos\theta - Mb\frac{3gb}{a^2+3b^2}\cos\theta \\ &= Mg\cos\theta\left[1-\frac{3b^2}{a^2+3b^2}\right] = Mg\cos\theta\frac{a^2}{a^2+3b^2} \\ \text{and } F &= Mg\sin\theta + Mb\frac{6gb}{a^2+3b^2}\sin\theta \\ &= Mg\sin\theta\left[1+\frac{6b^2}{a^2+3b^2}\right] = Mg\sin\theta\frac{a^2+9b^2}{a^2+3b^2}\end{aligned}$$

The rod will begin to slide when $F = \mu R$, i.e., when

$$Mg\sin\theta\frac{a^2+9b^2}{a^2+3b^2} = \mu Mg\frac{\cos\theta a^2}{a^2+3b^2},$$

$$\text{or,} \quad \tan\theta = \frac{\mu a^2}{a^2+9b^2}.$$

Problem 5. *A uniform rod of mass M is placed at right angles to a smooth plane of inclination α with one end in contact with it. The rod is then released. Show that when the inclination to the plane is ϕ, the reaction of the plane will be*

$$Mg\left[\frac{3(1-\sin\phi)^2+1}{(1+3\cos^2\phi)^2}\right]\cos\alpha. \qquad \text{[MU 80]}$$

Solution: Let AB be the rod of mass M and length $2a$ placed at right angles to the smooth plane of inclination α with one end A in contact with it. As there is no force acting along the inclined plane, so initially there is no motion along this plane, i.e., the C.G. at G moves perpendicular to the plane.

Let ϕ be the angle which the rod makes with the inclined plane at time T.

$\therefore \; GL = a\sin\phi = y$ (say).

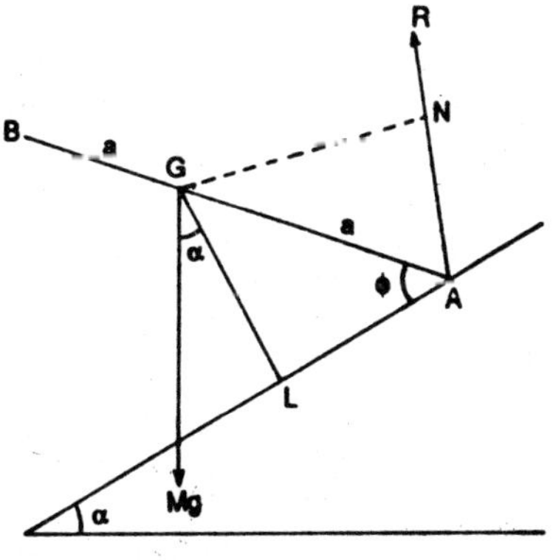

FIG 4.16

Hence the equation of motion of G, perpendicular to the inclined plane is

$$M\frac{d^2y}{dt^2} = M\frac{d^2}{dt^2}(a\sin\phi) = R - Mg\cos\alpha$$

$$\text{or,}\quad aM\frac{d}{dt}[\cos\phi\dot{\phi}] = R - Mg\cos\alpha$$

$$\text{or,}\quad aM[\cos\phi\ddot{\phi} - \sin\phi\dot{\phi}^2] = R - Mg\cos\alpha$$

$$\therefore\quad Mg\cos\alpha + aM(\cos\phi\ddot{\phi} - \sin\phi\dot{\phi}^2). \tag{4.102}$$

Also equation of motion gives

$$MK^2\ddot{\phi} = -R\cdot GN,\ \text{or, } M\frac{a^2}{3}\ddot{\phi} = -Ra\cdot\cos\phi. \tag{4.103}$$

Also the energy equation gives

$$\text{Sum of K.E.} = \frac{1}{2}Mv^2 + \frac{1}{2}MK^2\dot{\phi}^2 = \text{work done by the gravity}$$

$$\text{or,}\quad \frac{M}{2}(0^2 + \dot{y}^2) + \frac{1}{2}M\frac{a^2}{3}\dot{\phi}^2 = Mg\cos\alpha(a - a\sin\phi)$$

$$\text{or,}\quad \frac{M}{2}(a\cos\phi\dot{\phi})^2 + \frac{1}{6}Ma^2\dot{\phi}^2 = Mga\cos\alpha(1 - \sin\phi)$$

$$\text{or,}\quad \frac{Ma^2\dot{\phi}^2}{6}[3\cos^2\phi + 1] = Mga\cos\alpha(1 - \sin\phi)$$

$$\dot{\phi}^2 = \frac{6g(1-\sin\phi)}{a(1+3\cos^2\phi)}\cos\alpha. \tag{4.104}$$

Differentiating equation (4.104) with respect to t, we get

$$2\dot{\phi}\ddot{\phi} = \frac{6g\cos\alpha}{a}\left[\frac{-\cos\phi(1+3\cos^2\phi) + 6\sin\phi\cos\phi(1-\sin\phi)}{(1+3\cos^2\phi)^2}\right]\dot{\phi}$$

$$\text{or, } \ddot{\phi} = \frac{3g}{a}\cos\alpha\cos\phi\left[\frac{-1-3(1-\sin^2\phi) + 6(\sin\phi - \sin^2\phi)}{(1+3\cos^2\phi)^2}\right]$$

$$= -\frac{3g}{a}\cos\alpha\cos\phi\left[\frac{1+3-3\sin^2\phi + 6\sin^2\phi - 6\sin\phi}{(1+3\cos^2\phi)^2}\right]$$

$$= -\frac{3g}{a}\cos\alpha\cos\phi\left[\frac{3+3\sin^2\phi - 6\sin\phi + 1}{(1+3\cos^2\phi)^2}\right]$$

$$= -\frac{3g}{a}\cos\alpha\cos\phi\left[\frac{1+3(1-\sin\phi)^2}{(1+3\cos^2\phi)^2}\right].$$

$\therefore$ from equation (4.103), we get

$$\frac{Ma^2}{3}\left(-\frac{3g}{a}\cos\alpha\cos\phi\right)\left[\frac{1+3(1-\sin\phi)^2}{(1+3\cos^2\phi)^2}\right] = -Ra\cos\phi.$$

$$\therefore\ R = Mg\cos\alpha\left[\frac{1+3(1-\sin\phi)^2}{(1+3\cos^2\phi)^2}\right].$$

Type III: Problems based on Section 4.6

Problem 6. *A heavy rod, of length $2a$ is placed in a vertical plane with its ends in contact with a rough vertical wall and an equally rough horizontal plane, the coefficient of friction being* $\tan\epsilon$. *Show that it will begin to slip down if its initial inclination to the vertical is greater than 2ϵ. Prove also that the inclination θ of the rod to the vertical at any time is given by* $\ddot{\theta}(K^2 + a^2\cos 2\epsilon) - a^2\dot{\theta}^2 \sin 2\epsilon = ag\sin(\theta - 2\epsilon)$.

Solution: Let AB be the rod of mass M and length $2a$. When AB makes an angle θ with the vertical, let R and S be the resultant reactions at B and A, respectively.

Equations of motion of C.G. at G are

$$M\frac{d^2}{dt^2}(a\sin\theta) = -S\sin\epsilon + R\cos\epsilon \tag{4.105}$$

and
$$M\frac{d^2}{dt^2}(a\cos\theta) = R\sin\epsilon + S\cos\epsilon - Mg. \tag{4.106}$$

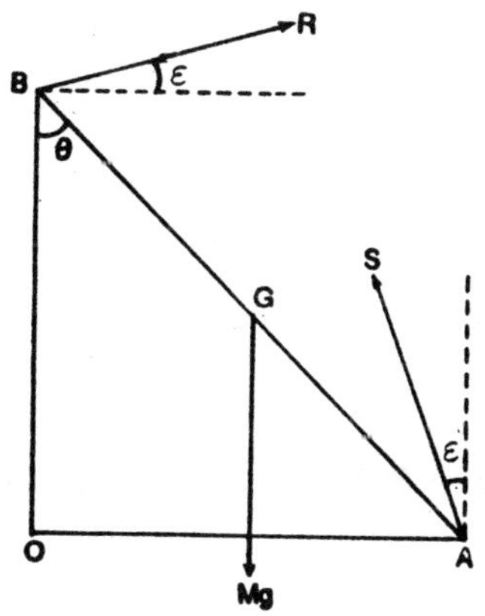

FIG 4.17

Also taking moments about G, we get

$$MK^2\ddot{\theta} = Sa\sin(\theta - \epsilon) - Ra\cos(\theta - \epsilon). \tag{4.107}$$

Equations (4.105) and (4.106) can be written as

$$M\frac{d}{dt}(a\cos\theta\dot{\theta}) = -S\sin\epsilon + R\cos\epsilon$$

or,
$$Ma[-\sin\theta\dot{\theta}^2 + \cos\theta\ddot{\theta}] = -S\sin\epsilon + R\cos\epsilon \tag{4.108}$$

and
$$Ma\frac{d}{dt}[-\sin\theta\dot{\theta}] = R\sin\epsilon + S\cos\epsilon - Mg$$

or,
$$Ma[-\{\cos\theta\dot{\theta}^2 + \sin\theta\ddot{\theta}\}] = R\sin\epsilon + S\cos\epsilon - Mg$$

or,
$$Ma[\cos\theta\dot{\theta}^2 + \sin\theta\ddot{\theta}] = Mg - R\sin\epsilon - S\cos\epsilon. \tag{4.109}$$

To solve equations (4.108) and (4.109) for R and S, equation (4.108) $\times \cos\epsilon$ gives equation (4.110) and equation (4.109) $\times \sin\epsilon$ gives equation (4.111).

$$Ma\cos\epsilon[-\sin\theta\dot{\theta}^2 + \cos\theta\ddot{\theta}] = -S\cos\epsilon\sin\epsilon + R\cos^2\epsilon. \tag{4.110}$$

$$Ma\sin\epsilon(\cos\theta\dot{\theta}^2 + \sin\theta\ddot{\theta}) = Mg\sin\epsilon - R\sin^2\epsilon - S\cos\epsilon\sin\epsilon. \quad (4.111)$$

Subtracting equations (4.111) from (4.110), we get

$$-Ma\dot{\theta}^2(\sin\theta\cos\epsilon + \cos\theta\sin\epsilon) + Ma\ddot{\theta}(\cos\epsilon\cos\theta - \sin\epsilon\sin\theta)$$
$$= R - Mg\sin\epsilon$$

$$\text{or, } R = Mg\sin\epsilon + Ma\cos(\theta+\epsilon)\ddot{\theta} - Ma\dot{\theta}^2\sin(\theta+\epsilon). \quad (4.112)$$

Similarly, we have

$$S = Mg\cos\epsilon - Ma\sin(\theta+\epsilon)\ddot{\theta} - Ma\cos(\theta+\epsilon)\dot{\theta}^2. \quad (4.113)$$

Putting these values of R and S in equation (4.107), we get

$$\begin{aligned} MK^2\ddot{\theta} &= a\sin(\theta-\epsilon)[Mg\cos\epsilon - Ma\sin(\theta+\epsilon)\ddot{\theta} \\ &\quad - Ma\cos(\theta+\epsilon)\dot{\theta}^2] - a\cos(\theta-\epsilon)[Mg\sin\epsilon \\ &\quad + Ma\cos(\theta+\epsilon)\ddot{\theta} - Ma\sin(\theta+\epsilon)\dot{\theta}^2] \\ &= Mga\sin(\theta-2\epsilon) - Ma^2\ddot{\theta}\cos 2\epsilon + Ma^2\dot{\theta}^2\sin 2\epsilon. \end{aligned}$$

$$\text{or, } \ddot{\theta}(K^2 + a^2\cos 2\epsilon) - a^2\dot{\theta}^2\sin 2\epsilon = ag\sin(\theta - 2\epsilon).$$

This gives θ.

When $\theta > 2\epsilon$, the value of $\ddot{\theta}$ is positive. Hence the rod starts slipping when $\theta > 2\epsilon$.

Type IV: Problems based on Section 4.7

Problem 7. *A sphere of radius a, is projected up an inclined plane with velocity V and angular velocity ω in the sense which would cause it to roll up. If $V > a\omega$ and the coefficient of friction μ exceeds $\frac{2}{7}\tan\alpha$, show that the sphere will cease to ascend after time $\frac{5V+2a\omega}{5g\sin\alpha}$, where α is the inclination of the plane to the horizontal.*

[IAS 95, 89; CU 89H; MU 71; Agra 77, 83]

Solution: Let C be the centre and M be the mass of the sphere. Let the radius CB which is a line fixed in the body makes an angle θ after time t with CA normal to the plane. (CA is a line fixed in the space.)

The initial velocity of the point of contact A up the plane

$$= \text{Velocity of the centre C} + \text{velocity of A relative to C.}$$
$$= V - a\omega = \text{positive as } V > a\omega.$$

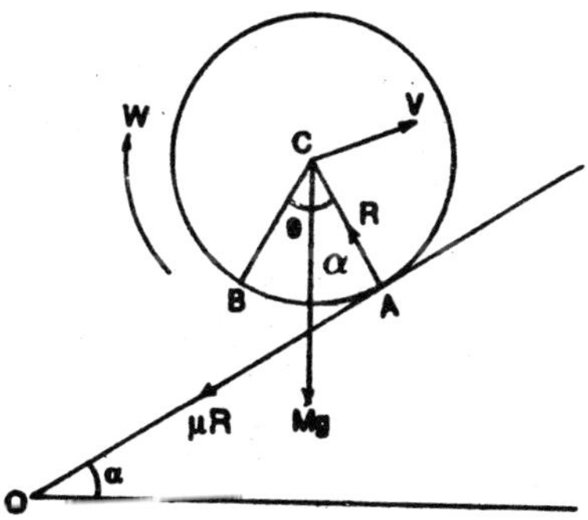

FIG 4.18

Hence the friction μR acts down the plane. Hence the sphere slides as well as turns.

The equations of motion of the sphere are

$$M\ddot{x} = -Mg\sin\alpha - \mu R \tag{4.114}$$

$$\text{and} \quad 0 = R - Mg\cos\alpha \tag{4.115}$$

$$\text{and} \quad MK^2\ddot{\theta} = M\frac{2a^2}{5}\ddot{\theta} = \mu Ra. \tag{4.116}$$

From equations (4.114) and (4.115), we have

$$M\ddot{x} = -Mg\sin\alpha - \mu Mg\cos\alpha,$$

$$\text{or,} \quad \ddot{x} = -g(\sin\alpha + \mu\cos\alpha).$$

Integrating, we get

$$\dot{x} = -g(\sin\alpha + \mu\cos\alpha)t + C_1.$$

When $t = 0$, $\dot{x} = V$

$$\therefore \quad C_1 = V$$

$$\therefore \quad \dot{x} = -g(\sin\alpha + \mu\cos\alpha)t + V. \tag{4.117}$$

From equations (4.115) and (4.116), we get

$$M\frac{2a^2}{5}\ddot{\theta} = \mu Mg\cos\alpha \cdot a, \text{ or, } a\ddot{\theta} = \tfrac{5}{2}\mu g\cos\alpha.$$

Integrating, we get

$$a\dot{\theta} = \tfrac{5}{2}\mu gt\cos\alpha + C_2.$$

But when $t = 0$, $\dot{\theta} = \omega$,

$$C_2 = a\omega$$

$$\therefore \quad a\dot{\theta} = \tfrac{5}{2}\mu gt\cos\alpha + a\omega. \tag{4.118}$$

The velocity of the point of contact $= \dot{x} - a\dot{\theta}$. Rolling commences when velocity of point of contact is zero, i.e., $\dot{x} - a\dot{\theta} = 0$, when $t = t_1$.

$$\text{or,}\quad -g(\sin\alpha + \mu\cos\alpha)t_1 + V = \tfrac{5}{2}\mu g\cos\alpha t_1 + a\omega$$

$$\text{or,}\quad V - a\omega = \frac{g}{2}[5\mu t_1\cos\alpha + 2t_1\sin\alpha + 2\mu t_1\cos\alpha]$$

$$= \frac{gt_1}{2}(7\mu\cos\alpha + 2\sin\alpha)$$

$$\therefore\quad t_1 = \frac{2V - 2a\omega}{g(7\mu\cos\alpha + 2\sin\alpha)}.$$

Putting $t = t_1$ in equations (4.117), we get

$$\dot{x} = V - \frac{g(\sin\alpha + \mu\cos\alpha)(2V - 2a\omega)}{g(7\mu\cos\alpha + 2\sin\alpha)}$$

$$= V - \frac{(\sin\alpha + \mu\cos\alpha)(2V - 2a\omega)}{7\mu\cos\alpha + 2\sin\alpha}$$

$$= \frac{7\mu V\cos\alpha + 2v\sin\alpha - 2V\sin\alpha - 2V\mu\cos\alpha + 2a\omega\sin\alpha + 2a\omega\mu\cos\alpha}{7\mu\cos\alpha + 2\sin\alpha}$$

$$\dot{x} = \frac{5V\mu\cos\alpha + 2a\omega(\sin\alpha + \mu\cos\alpha)}{7\mu\cos\alpha + 2\sin\alpha}$$

$$= V_1 \text{ (say)}.$$

Now, when rolling commences, i.e., when the point of contact has been brought to rest, let F be the friction which is sufficient for pure rolling. Because the point of contact is now at rest, therefore, friction will try to keep it at rest if it can, hence friction F acts upwards.

The equations of motion will now be

$$M\ddot{y} = -Mg\sin\alpha + F \tag{4.119}$$

$$\text{and}\quad MK^2\ddot{\phi} = M\frac{2a^2}{5}\ddot{\phi} = -Fa. \tag{4.120}$$

Since throughout the motion, the point of contact is at rest,

$$\therefore\quad \dot{y} - a\dot{\phi} = 0,$$

$$\text{or,}\quad \dot{y} = a\dot{\phi} \quad \therefore \ddot{y} = a\ddot{\phi}.$$

Thus, from equations (4.119) and (4.120), we have

$$F = M[\ddot{y} + g\sin\alpha] = M[a\ddot{\phi} + g\sin\alpha]$$

$$\text{or, } F = M\left[g\sin\alpha - \frac{5F}{2M}\right]$$

$$\text{or, } \tfrac{7}{2}F = Mg\sin\alpha$$

$$\therefore F = \tfrac{2}{7}Mg\sin\alpha. \tag{4.121}$$

Again,

$$\mu R = \mu Mg\cos\alpha > \tfrac{2}{7}\tan\alpha Mg\cos\alpha,$$
i.e., $\mu R > \tfrac{2}{7}Mg\sin\alpha$

or, $\mu R > F$ [From equation (4.121)]

Thus, $F < \mu R$.

Putting the value of F from equation (4.121) in equation (4.119), we get

$$M\ddot{y} = -Mg\sin\alpha + \frac{2}{7}Mg\sin\alpha, \text{ or, } \ddot{y} = -\frac{5}{7}g\sin\alpha.$$

Integrating we get,

$$\dot{y} = -\frac{5}{7}gt\sin\alpha + C.$$

When $t = 0$, $\dot{y} = V_1$.

$$\therefore \quad C = V_1.$$

$$\therefore \quad \dot{y} = -\frac{5}{7}gt\sin\alpha + V_1. \qquad (4.122)$$

Now the sphere will cease to ascend the plane when $\dot{y} = 0$, if this happens after time t_2, then from equation (4.122), we have

$$0 = \dot{y} = -\frac{5}{7}gt_2\sin\alpha + V_1, \text{ or, } t_2 = \frac{7V_1}{5g\sin\alpha}.$$

Hence the total time of ascent

$$\begin{aligned}
&= t_1 + t_2 \\
&= \frac{2V - 2a\omega}{g(7\mu\cos\alpha + 2\sin\alpha)} + \frac{7V_1}{5g\sin\alpha} \\
&= \frac{2V - 2a\omega}{g(7\mu\cos\alpha + 2\sin\alpha)} + \frac{7}{g5\sin\alpha}\left[\frac{5\mu V\cos\alpha + 2a\omega(\sin\alpha + \mu\cos\alpha)}{(7\mu\cos\alpha + 2\sin\alpha)}\right] \\
&= \frac{10(V - a\omega)\sin\alpha + 35\mu V\cos\alpha + 14a\omega(\sin\alpha + \mu\cos\alpha)}{5g\sin\alpha(7\mu\cos\alpha + 2\sin\alpha)} \\
&= \frac{(10V\sin\alpha + 35\mu V\cos\alpha) + 14a\omega(\sin\alpha + \mu\cos\alpha) - 10a\omega\sin\alpha}{5g\sin\alpha(7\mu\cos\alpha + 2\sin\alpha)} \\
&= \frac{5V(7\mu\cos\alpha + 2\sin\alpha) + 2a\omega(7\mu\cos\alpha + 2\sin\alpha)}{5g\sin\alpha(7\mu\cos\alpha + 2\sin\alpha)} \\
&= \frac{5V + 2a\omega}{5g\sin\alpha}.
\end{aligned}$$

Problem 8. *A sphere is projected with an underhand twist down a rough inclined plane; show that it will turn back in the course of its motion if* $2a\omega(\mu - \tan\alpha) > 5\mu u$, *where* u, ω *are the initial linear and angular velocities of the sphere,* μ *is the coefficient of friction, and* α *is the inclination of the plane.* **[VU 88H]**

Solution: Let C be the centre and M be the mass of the sphere.

Here u is in the direction ← and ω is in the sense ↙

∴ initial velocity of the point of contact A down the plane

= velocity of the centre C + velocity of A relative to C

= $u + a\omega$, which is positive.

Thus frictional force μR acts up in the plane.

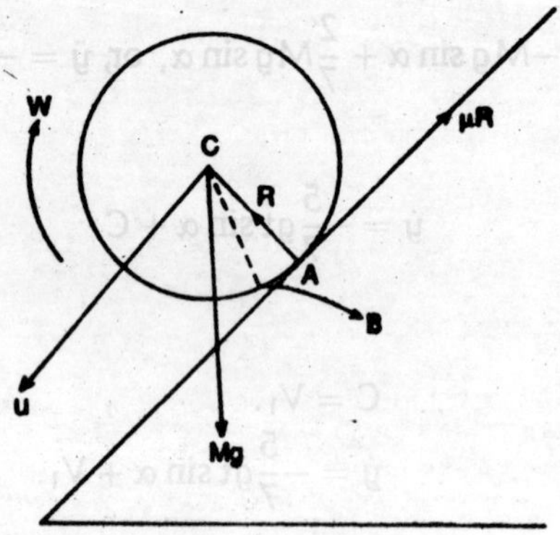

FIG 4.19

Now, equations of motion are

$$M\ddot{x} = Mg\sin\alpha - \mu R \tag{4.123}$$

$$\text{and } 0 = R - Mg\cos\alpha. \tag{4.124}$$

Also

$$MK^2\ddot{\theta} = M\frac{2}{5}a^2\ddot{\theta} = -\mu Ra. \tag{4.125}$$

From equations (4.123) and (4.124), we get

$$M\ddot{x} = Mg\sin\alpha - \mu Mg\cos\alpha$$

$$\text{or, } \ddot{x} = g(\sin\alpha - \mu\cos\alpha). \tag{4.126}$$

From equations (4.124) and (4.125), we have

$$M\frac{2}{5}a\ddot{\theta} = -\mu Mg\cos\alpha$$

$$\text{or, } a\ddot{\theta} = -\frac{5}{2}\mu g\cos\alpha. \tag{4.127}$$

Integrating equations (4.126) and (4.127), we get

$$\dot{x} = g(\sin\alpha - \mu\cos\alpha)t + C.$$

When $\dot{x} = u$, $t = 0$.

$$\therefore C = u.$$

$$\therefore \dot{x} = g(\sin\alpha - \mu\cos\alpha)t + u. \tag{4.128}$$

Also

$$a\dot{\theta} = -\tfrac{5}{2}\mu g t \cos\alpha + C_1.$$

Initially, when $t = 0$, $\dot{\theta} = \omega$.

$$a\dot{\omega} = C_1$$
$$a\dot{\theta} = a\omega - \tfrac{5}{2}\mu g t \cos\alpha.$$

The sphere will cease to go down when $\dot{x} = 0$, i.e., when $g(\sin\alpha - \mu\cos\alpha)$ $t + u = 0$.

$$\therefore\ t = \frac{u}{g(\mu\cos\alpha - \sin\alpha)}.$$

At this time

$$a\dot{\theta} = \frac{-5g\mu\cos\alpha \cdot u}{2g(\mu\cos\alpha - u\sin\alpha)} + a\omega.$$

So $a\dot{\theta}$ is positive if

$$-\frac{5\mu g u\cos\alpha}{2g(\mu\cos\alpha - u\sin\alpha)} + a\omega > 0,$$

i.e., if $-5\mu u\cos\alpha + 2a\omega(\mu\cos\alpha - \sin\alpha) > 0$, or, $2a\omega(\mu - \tan\alpha) > 5\mu u$.

Problem 9. *An inclined plane of mass M is capable of moving freely on a smooth horizontal plane. A perfectly rough sphere of mass m is placed on its inclined face and rolls down under the action of gravity. If y be the horizontal distance advanced by the inclined plane and x the part of the plane rolled over by the sphere, prove that* $(M + m)y = mx\cos\alpha$ *and* $\frac{7}{5}x - y\cos\alpha = \frac{1}{2}gt^2\sin\alpha$, *where* α *is the inclination of the plane to the horizon.*

Solution: Let C be the centre of sphere of mass m which rolls down the inclined plane of mass M and inclination α to the horizontal.

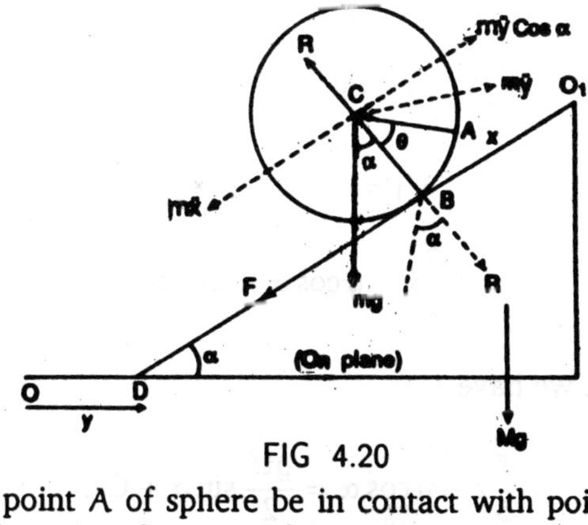

FIG 4.20

Initially, let the point A of sphere be in contact with point O of the inclined plane. If at time t, the point of contact of the sphere and the inclined plane is B and during this time the sphere turns through an angle θ, then $\angle ACB = \theta$.

If $O_1B = x$, then $x = \text{arc } AB = a\theta$.

$$\therefore\ \dot{x} = a\dot{\theta} \text{ and } \ddot{x} = a\ddot{\theta}. \qquad (4.129)$$

Let the inclined plane shift through a distance OD = y in time t.

$\therefore$ the accelerations of the centre C of the sphere are $\ddot{x}$ down the plane and parallel to it, and $\ddot{y}$ horizontally as shown in the figure.

Hence, the acceleration of the centre C parallel to the inclined plane is $\ddot{x} - \ddot{y}\cos\alpha$. (downwards).

Let F be the frictional force up the plane.

$\therefore$ the equations of motion of the sphere are

$$m(\ddot{x} - \ddot{y}\cos\alpha) = mg\sin\alpha - F \tag{4.130}$$

$$m\ddot{y}\sin\alpha = mg\cos\alpha - R \tag{4.131}$$

$$\text{and} \quad mK^2\ddot{\theta} = m\frac{2}{5}a^2\ddot{\theta} = Fa. \tag{4.132}$$

Also the equation of motion of the inclined plane is

$$M\ddot{y} = R\sin\alpha - F\cos\alpha. \tag{4.133}$$

From equations (4.129) and (4.132), we have, $\frac{2}{5}\ddot{x} = F$.

Putting this value of F in equation (4.130), we get

$$m(\ddot{x} - \ddot{y}\cos\alpha) = mg\sin\alpha - \frac{2}{5}m\ddot{x}$$

$$\text{or,} \quad \ddot{x} - \ddot{y}\cos\alpha = g\sin\alpha - \frac{2}{5}\ddot{x}$$

$$\text{or,} \quad \frac{7}{5}\ddot{x} = \ddot{y}\cos\alpha + g\sin\alpha. \tag{4.134}$$

Integrating, we get

$$\frac{7}{5}\dot{x} - \dot{y}\cos\alpha = gt\sin\alpha + C_1.$$

When $t = 0$, $\dot{x} = 0$, $\dot{y} = 0$; $\therefore$ $C_1 = 0$.

$$\therefore \quad \frac{7}{5}\dot{x} - \dot{y}\cos\alpha = gt\sin\alpha.$$

Integrating again, we have

$$\frac{7}{5}x - y\cos\alpha = \frac{gt^2}{2}\sin\alpha + C_2.$$

When $t = 0$, $x = 0$, $y = 0$, $\therefore$ $C_2 = 0$.

$$\therefore \quad \frac{7}{5}x - y\cos\alpha = \frac{gt^2}{2}\sin\alpha. \tag{4.135}$$

Putting the values of R and F from equations (4.130) and (4.131) in (4.133), we get

$$\begin{aligned} M\ddot{y} &= m(-\ddot{y}\sin\alpha + g\cos\alpha)\sin\alpha \\ &\quad -m[g\sin\alpha - (\ddot{x} - \ddot{y}\cos\alpha)]\cos\alpha \\ &= -m\ddot{y}\sin^2\alpha + mg\sin\alpha\cos\alpha - mg\sin\alpha\cos\alpha \\ &\quad +m\cos\alpha\ddot{x} - m\ddot{y}\cos^2\alpha \\ M\ddot{y} &= -m\ddot{y} + m\ddot{x}\cos\alpha, \\ \text{or, } (M+m)\ddot{y} &= m\ddot{x}\cos\alpha. \end{aligned}$$

Integrating it, we get

$$(M+m)\dot{y} = m\dot{x}\cos\alpha + C_3.$$

When $\dot{x} = 0$, $\dot{y} = 0$; $\therefore$ $C_3 = 0$.

Hence $(M+m)\dot{y} = m\dot{x}\cos\alpha$.

Integrating again, we get

$$(M+m)y = mx\cos\alpha + C.$$

When $x = 0$, $y = 0$; $\therefore$ $C = 0$.

Thus

$$(M+m)y = mx\cos\alpha. \tag{4.136}$$

Equations (4.135) and (4.136) are the required equations.

Type V: Problems based on Section 4.8

Problem 10. *A thin napkin ring of radius O is projected up a plane inclined at an angle α to the horizontal with velocity V, and an initial angular velocity ω, in the sense which would cause the ring to move down the plane. If $V < 5a\omega$ and $\mu = \frac{1}{4}\tan\alpha$, show that the ring will never roll and will cease to ascend at the end of a time $\frac{4(2V-a\omega)}{9g\sin\alpha}$ and will slide back to the point of projection.*

Solution: Let C be the centre and M be the mass of the ring. The initial velocity of the point of contact is $V + a\omega$, which acts up the plane. Hence, the frictional force μR acts down the plane.

Thus, the equations of motion are

$$\begin{aligned} M\ddot{x} &= -Mg\sin\alpha - \mu R \\ M\ddot{x} &= -Mg\sin\alpha - \mu Mg\cos\alpha \quad (R = Mg\cos\alpha) \end{aligned}$$

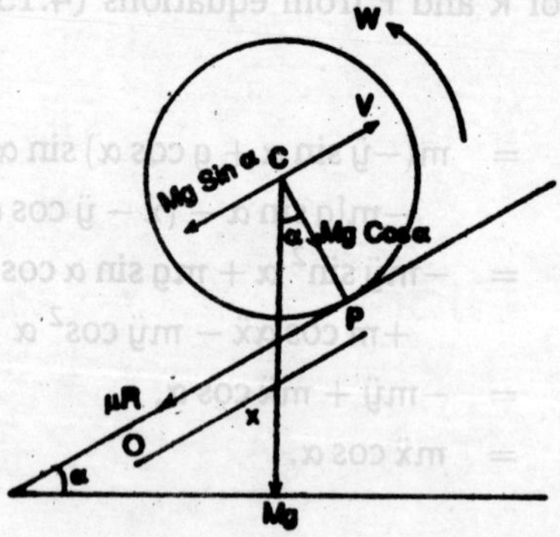

FIG 4.21

$$\text{or, } \ddot{x} = -g(\sin\alpha + \mu\cos\alpha)$$

$$= -g\left(\sin\alpha + \frac{1}{4}\tan\alpha\cos\alpha\right) \quad \left(\because \mu = \frac{1}{4}\tan\alpha\right)$$

$$= -g\left(\sin\alpha + \frac{1}{4}\sin\alpha\right) = -\frac{5g}{4}\sin\alpha \tag{4.137}$$

$$\text{and } Ma^2\ddot{\theta} = -\mu Ra = -\frac{1}{4}\tan\alpha Mg\cos\alpha\, a$$

$$Ma^2\ddot{\theta} = -\frac{1}{4}Mga\sin\alpha$$

$$\therefore a\ddot{\theta} = -\frac{g}{4}\sin\alpha. \tag{4.138}$$

Integrating equations (4.137) and (4.138), we get

$$\dot{x} = -\frac{5}{4}g\sin\alpha t + C_1 \text{ and } a\dot{\theta} = -\frac{g}{4}\sin\alpha t + C_2.$$

Initially, when $t = 0$, $\dot{x} = V$, $\dot{\theta} = \omega$.

$\therefore V = C_1$ and $C_2 = a\omega$.

Hence

$$\dot{x} = -\frac{5}{4}g\sin\alpha t + V \tag{4.139}$$

$$\text{and } a\dot{\theta} = -\frac{g}{4}\sin\alpha t + a\omega. \tag{4.140}$$

From equation (4.139), the velocity of the centre is zero after time

$$t = t_1 = \frac{4V}{5g\sin\alpha}.$$

The velocity of the point of contact at any time

$$t = \dot{x} + a\dot{\theta} = -\frac{3}{2}g\sin\alpha t + V + a\omega.$$

The point of contact will come to rest when $\dot{x} + a\dot{\theta} = 0$. If it happens at time $t = t_2$, then

$$V + a\omega - \frac{3}{2}gt_2 \sin\alpha = 0$$

$$\therefore \quad t_2 = \frac{2(V + a\omega)}{3g \sin\alpha}.$$

Clearly,

$$\frac{2(V + a\omega)}{3g\sin\alpha} < \frac{4V}{5g\sin\alpha} \quad \text{or,} \quad \frac{V + a\omega}{3} < \frac{2V}{5}$$

$$\text{or,} \quad 5V + 5a\omega < 6V$$

$$\text{or,} \quad 5a\omega < V,$$

which is given.

Thus $t_2 < t_1$.

Therefore, pure rolling may begin before the upward motion ceases if the friction is sufficient for pure rolling.

At this time t_2,

$$\begin{aligned} \dot{x} &= -\frac{5}{4}g\sin\alpha t_2 + V = -\frac{5}{4}\frac{2(V + a\omega)}{3} + V \\ &= \frac{-5V - 5a\omega + 6V}{6} = \frac{V - 5a\omega}{6}, \end{aligned}$$

which is positive and

$$\dot{\theta} = \omega - \frac{2(V + a\omega)}{3 \cdot 4a} = \frac{6a\omega - V - a\omega}{6} = \frac{5a\omega - V}{6},$$

which is negative as $V > 5a\omega$.

$$= \frac{V - 5a\omega}{6} \text{ in clockwise direction.}$$

When pure rolling commences and rotation is in clockwise direction, the friction F acts in upward direction.

The equations of motion are

$$M\ddot{y} = -Mg\sin\alpha + F \text{ and } Ma^2\ddot{\phi} = -Fa. \tag{4.141}$$

Since there is pure rolling,

$$\therefore \; y = a\phi, \text{ i.e., } \dot{y} = a\dot{\phi} \text{ and } \ddot{y} = a\ddot{\phi}.$$

Hence from equation (4.141), we get

$$-F = Ma\ddot{\phi} = M\ddot{y}.$$

Thus,

$$\begin{aligned}-F &= -Mg\sin\alpha + F\\ \text{or, } 2F &= Mg\sin\alpha\\ \therefore\ F &= \frac{1}{2}Mg\sin\alpha.\end{aligned}$$

But

$$\mu R = \frac{1}{4}\tan\alpha Mg\cos\alpha = \frac{Mg}{4}\sin\alpha < \frac{Mg}{2}\sin\alpha = F.$$

So $F > \mu R$.

Thus the friction is not sufficient for pure rolling. Hence the sliding persists and pure rolling is not possible. Therefore, the above equations of motion become,

$$\begin{aligned}M\ddot{y} &= -Mg\sin\alpha + \mu R = -Mg\sin\alpha + \frac{1}{4}\tan\alpha Mg\cos\alpha\\ &= -\frac{3}{4}Mg\sin\alpha\\ \text{or, } \ddot{y} &= -\frac{3}{4}g\sin\alpha \qquad (4.142)\\ \text{and } Ma^2\ddot{\phi} &= -\mu Ra = -\frac{1}{4}\tan\alpha Mg\cos\alpha a = -\frac{1}{4}Mga\sin\alpha\\ \text{or, } a\ddot{\phi} &= -\frac{1}{4}g\sin\alpha. \qquad (4.143)\end{aligned}$$

Integrating equations (4.142) and (4.143), we get

$$\begin{aligned}\dot{y} &= -\frac{3}{4}g\sin\alpha t + C_3\\ \text{and } a\dot{\phi} &= -\frac{1}{4}g\sin\alpha t + C_4.\end{aligned}$$

Initially, when $t = 0$,

$$\dot{y} = \frac{V - 5a\omega}{6} \text{ and } a\dot{\phi} = \frac{5a\omega - V}{6},$$

we get

$$\frac{V - 5a\omega}{6} = C_3 \text{ and } C_4 = \frac{5a\omega - V}{6}.$$

Thus,

$$\begin{aligned}\dot{y} &= -\frac{3}{4}g\sin\alpha t + \frac{V - 5a\omega}{6} \qquad (4.144)\\ \text{and } a\dot{\phi} &= -\frac{1}{4}g\sin\alpha t + \frac{5a\omega - V}{6}. \qquad (4.145)\end{aligned}$$

Clearly, $\dot{y} = 0$, after time

$$t = t_3 = \frac{2(V - 5a\omega)}{9g\sin\alpha}.$$

Putting this value of t in equation (4.145), we get

$$\begin{aligned}
a\dot{\phi} &= -\frac{1}{4}g\frac{\sin\alpha 2(V-5a\omega)}{9g\sin\alpha}+\frac{5a\omega-V}{6}\\
&= -\frac{(V-5a\omega)}{18}+\frac{5a\omega-V}{6}\\
&= \frac{-V+5a\omega+15a\omega-3V}{18}\\
&= \frac{20-4V}{18}=\frac{2(5-V)}{9}.
\end{aligned}$$

Hence total time of upward motion

$$\begin{aligned}
&= t_2+t_3=\frac{2(V+)}{3g\sin\alpha}+\frac{2(V-5)}{9g\sin\alpha}\\
&= \frac{2[3V+3+V-5]}{9g\sin\alpha}\\
&= \frac{2(4V-2)}{9g\sin\alpha}-\frac{4(2V-)}{9g\sin\alpha}.
\end{aligned}$$

Again when the upward motion ceases, we have

$$\begin{aligned}
a\dot{\phi} &= -\frac{1}{4}g\sin\alpha t_3+\frac{1}{6}(5-V)\\
&= -\frac{1}{4}g\sin\alpha\frac{2(V-5)}{9g\sin\alpha}+\frac{1}{6}(5-V)\\
&= \frac{-(V-5)}{18}+\frac{5-V}{6}\\
&= \frac{5-V+15-3V}{18}\\
&= \frac{20-4V}{18}=\frac{4(5-V)}{18}\\
&= \frac{2(5-V)}{9},
\end{aligned}$$

which is negative as $V > 5$, hence the ring returns backwards to the point of projection.

The velocity of the point of contact

$$\begin{aligned}
&= \text{velocity of the centre} + \text{velocity relative to the centre}\\
&= \dot{y}-a\dot{\phi}=0-\frac{2}{9}(5-V)\\
&= \frac{2}{9}(V-5)>0 \text{ as } V>5.
\end{aligned}$$

The velocity of the point of contact is up the plane, hence the friction μR acts downwards.

$\therefore$ the equations of motion are

$$M\ddot{z} = Mg\sin\alpha + \mu R = Mg\sin\alpha + \frac{1}{4}\tan\alpha Mg\cos\alpha$$

$$= Mg\sin\alpha + \frac{Mg}{4}\sin\alpha = \frac{5}{4}Mg\sin\alpha$$

$$\therefore \ddot{z} = \frac{5}{4}g\sin\alpha \tag{4.146}$$

$$\text{and } Ma^2\ddot{\psi} = -\mu Ra = -\frac{1}{4}\tan\alpha Mg\cos\alpha\, a,$$

$$\therefore a\ddot{\psi} = -\frac{g}{4}\sin\alpha. \tag{4.147}$$

Integrating equations (4.146) and (4.147), we get

$$\dot{z} = \frac{5}{4}gt\sin\alpha + C_5$$

$$\text{and } a\dot{\psi} = -\frac{g}{4}\sin\alpha t + C_6.$$

Initially, when $t = 0$, $\dot{z} = 0$ and $a\dot{\psi} = \frac{2}{9}(V - 15)$.

$$\therefore C_5 = 0,\ C_6 = \frac{2}{9}(V - 5)$$

$$\dot{z} = \frac{5}{4}gt\sin\alpha \text{ and } a\dot{\psi} = -\frac{gt}{4}\sin\alpha + \frac{2}{9}(V - 5).$$

$\therefore$ the velocity of the point of contact down the plane

$$= \dot{z} - a\dot{\psi} = \frac{5}{4}gt\sin\alpha + \frac{gt}{4}\sin\alpha + \frac{2}{9}(V - 5)$$

$$= \frac{6}{4}gt\sin\alpha + \frac{2}{9}(V - 5)$$

$$= \frac{3}{2}gt\sin\alpha + \frac{2}{9}(V - 5)$$

which is +ve since $V > 5$.

Hence the ring slides back to the point of contact.

Type VI: Problems based on Section 4.9 and Section 4.10

Problem 11. *A heavy uniform sphere of mass M, is resting on a perfectly rough horizontal plane, and a particle, of mass m, is gently placed on it at an angular distance α from its highest point. Show that the particle will at once slip on the sphere if $\mu < \frac{\sin\alpha\{7M+5m(1+\cos\alpha)\}}{7M\cos\alpha+5m(1+\cos\alpha)}$, where μ is the coefficient of friction between the sphere and the particle.*

Solution: Let C be the centre of the sphere. Let M be its mass and radius a. The sphere rests on a perfectly rough horizontal plane.

Let the particle of mass m remains at rest placed at P on the sphere, at an angular distance α from the highest point. At time t when the sphere rolls through a distance $OA = x$, let the particle m be at rest at P such that CP is inclined at an angle $(\theta + \alpha)$ to the vertical.

Since the sphere rolls,

$$\therefore\ x = a\theta, \text{ i.e., } \dot{x} = a\dot{\theta} \text{ and } \ddot{x} = a\ddot{\theta}.$$

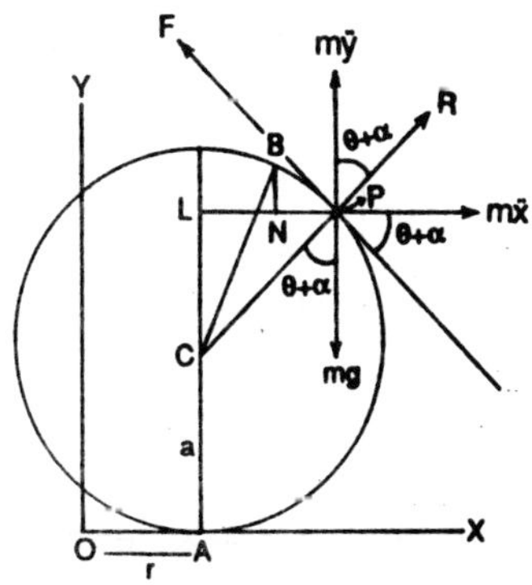

FIG 4.22

Taking O as origin and the horizontal and vertical lines through O as coordinate axes. If (x, y) be the coordinates of the particle at P, we have

$$\begin{aligned}
x &= OA + LP = x + a\sin(\theta+\alpha) = a\theta + a\sin(\alpha+\theta)\\
\text{and } y &= AL = AC + CL = a + a\cos(\theta+\alpha)\\
\therefore\ \dot{x} &= a\dot{\theta} + a\cos(\alpha+\theta)\dot{\theta}\\
\dot{x} &= a\dot{\theta}[1+\cos(\alpha+\theta)]\\
\ddot{x} &= a\left[\ddot{\theta}\{1+\cos(\alpha+\theta)\} - \sin(\alpha+\theta)\dot{\theta}^2\right]\\
&= a\ddot{\theta} + a\ddot{\theta}\cos(\alpha+\theta) - a\sin(\alpha+\theta)\dot{\theta}^2\\
\dot{y} &= -a\sin(\theta+\alpha)\dot{\theta}\\
\ddot{y} &= -a\left[\sin(\theta+\alpha)\ddot{\theta} + \cos(\theta+\alpha)\dot{\theta}^2\right]\\
\ddot{y} &= -a\sin(\alpha+\theta)\ddot{\theta} - a\cos(\alpha+\theta)\dot{\theta}^2.
\end{aligned}$$

Equations of motion of the particle m along the tangent and normal at P are

$$\begin{aligned}
mg\sin(\alpha+\theta) - F &= m\ddot{x}\cos(\alpha+\theta) - m\ddot{y}\sin(\alpha+\theta)\\
&= m\left[a\ddot{\theta}\cos(\alpha+\theta) + a\ddot{\theta}\cos^2(\alpha+\theta)\right.\\
&\quad -a\dot{\theta}^2\sin(\alpha+\theta)\cos(\alpha+\theta) + a\ddot{\theta}\sin^2(\alpha+\theta)\\
&\quad \left.+a\dot{\theta}^2\cos(\theta+\alpha)\sin(\alpha+\theta)\right]\\
&= m[a\ddot{\theta} + a\ddot{\theta}\cos(\alpha+\theta)].\\
\therefore\ mg\sin(\alpha+\theta) - F &= m[a + a\cos(\alpha+\theta)]\ddot{\theta}. \qquad (4.148)
\end{aligned}$$

$$\begin{aligned}
R - mg\cos(\alpha+\theta) &= m\ddot{x}\sin(\alpha+\theta) + m\ddot{y}\cos(\alpha+\theta)\\
&= m[a\ddot{\theta}\sin(\alpha+\theta) + a\ddot{\theta}\sin(\alpha+\theta)\cos(\alpha+\theta)\\
&\quad -a\dot{\theta}^2\sin^2(\alpha+\theta) - a\ddot{\theta}\sin(\alpha+\theta)\cos(\alpha+\theta)\\
&\quad -a\dot{\theta}^2\cos^2(\theta+\alpha)]\\
&= m[a\ddot{\theta}\sin(\alpha+\theta) - a\dot{\theta}^2].\\
\therefore\ R - mg\cos(\alpha+\theta) &= ma[\ddot{\theta}\sin(\alpha+\theta) - \dot{\theta}^2]. \qquad (4.149)
\end{aligned}$$

Energy equation gives

$$\left[\frac{1}{2}MK^2\dot{\theta}^2 + \frac{1}{2}Ma^2\dot{\theta}^2\right] + \frac{1}{2}m(\dot{x}^2+\dot{y}^2) = mgBN$$

$$\frac{1}{2}M\frac{2}{5}a^2\dot{\theta}^2 + \frac{1}{2}Ma^2\dot{\theta}^2 + \frac{1}{2}m\{a\dot{\theta} + a\cos(\theta+\alpha)\dot{\theta}\}^2$$
$$+\frac{1}{2}m\{-a\sin(\theta+\alpha)\dot{\theta}\}^2 = mg[a\cos\alpha - a\cos(\theta+\alpha)]$$

or, $$\frac{7}{10}Ma^2\dot{\theta}^2 + \frac{m}{2}[a^2\dot{\theta}^2 + a^2\dot{\theta}^2 + 2a^2\dot{\theta}^2\cos(\theta+\alpha)]$$
$$= mga[\cos\alpha - \cos(\theta+\alpha)]$$

or, $$\left[\frac{7}{10}M + m\{1+\cos(\theta+\alpha)\}\right]a^2\dot{\theta}^2 = mga[\cos\alpha - \cos(\theta+\alpha)]$$

or, $$[7M + 10m\{1+\cos(\theta+\alpha)\}]a\dot{\theta}^2 = 10mg[\cos\alpha - \cos(\theta+\alpha)].$$

Differentiating with respect to t on both the sides, we get,

$$[7M + 10m\{1+\cos(\theta+\alpha)\}]2a\dot{\theta}\ddot{\theta} - 10m\sin(\theta+\alpha)\,a\dot{\theta}^3$$
$$= 10mg\sin(\theta+\alpha)\dot{\theta}$$

or, $$[7M + 10m\{1+\cos(\theta+\alpha)\}]2a\ddot{\theta} - 10m\sin(\theta+\alpha)a\dot{\theta}^2$$
$$= 10mg\sin(\theta+\alpha)$$

or, $$[7M + 10m\{1+\cos(\theta+\alpha)\}]a\ddot{\theta} - 5m\sin(\theta+\alpha)a\dot{\theta}^2]$$
$$= 5mg\sin(\theta+\alpha). \qquad (4.150)$$

Here we want only initial motion when $\theta = 0$, $\dot{\theta} = 0$ but $\ddot{\theta}$ is not zero.

Hence, from equations (4.148), (4.149) and (4.150), we get

$$F = mg\sin\alpha - ma(1+\cos\alpha)\ddot{\theta}$$

and $$R = mg\cos\alpha + ma\ddot{\theta}\sin(\alpha+\theta)$$

and $$[7M + 10m(1+\cos\alpha)]a\ddot{\theta} = 5mg\sin\alpha$$

$$\ddot{\theta} = \frac{5mg\sin\alpha}{[7M+10m(1+\cos\alpha)]a}.$$

Putting the value of $\ddot{\theta}$ in F and R, we get

$$\begin{aligned}
F &= mg\sin\alpha - \frac{ma(1+\cos\alpha)5mg\sin\alpha}{[7M+10m(1+\cos\alpha)]a} \\
&= mg\sin\alpha - \frac{5(1+\cos\alpha)m^2 g\sin\alpha}{[7M+10m(1+\cos\alpha)]} \\
&= \frac{mg\sin\alpha[7M+10m(1+\cos\alpha)-5m(1+\cos\alpha)]}{7M+10m(1+\cos\alpha)} \\
&= \frac{mg\sin\alpha[7M+5m+5m\cos\alpha]}{7M+10m(1+\cos\alpha)} \\
&= \frac{[7M+5m(1+\cos\alpha)]mg\sin\alpha}{7M+10m(1+\cos\alpha)}. \\
R &= mg\cos\alpha + ma\frac{\sin\alpha 5mg\sin\alpha}{7Ma+10ma(1+\cos\alpha)} \\
&= mg\left[\frac{7Ma\cos\alpha+10ma(1+\cos\alpha)\cos\alpha+5ma\sin^2\alpha}{7Ma+10ma(1+\cos\alpha)}\right] \\
&= \frac{mg[7Ma\cos\alpha+5ma\{2\cos\alpha+2\cos^2\alpha+1-\cos^2\alpha\}]}{7Ma+10ma(1+\cos\alpha)} \\
&= \frac{mg[7M\cos\alpha+5m(1+\cos\alpha)^2]}{7M+10m(1+\cos\alpha)} \\
\therefore\ \frac{F}{R} &= \frac{\sin\alpha[7M+5m(1+\cos\alpha)]}{7M\cos\alpha+5m(1+\cos\alpha)^2}
\end{aligned}$$

The particle will slide on the sphere if

$$F > \mu R, \quad \text{or,} \quad \mu < \frac{F}{R}$$

$$\text{or,} \quad \mu < \frac{\sin\alpha[7M+5m(1+\cos\alpha)]}{7M\cos\alpha+5m(1+\cos\alpha)^2}.$$

Problem 12. *A sphere of radius a, whose centre of gravity G is not at its centre C, is placed on a rough horizontal table, so that C.G. is inclined at an angle α to the upward drawn vertical. Show that it will commence to slide along the table if the coefficient of friction μ be less than $\frac{c\sin\alpha(a+c\cos\alpha)}{k^2+(a+c\cos\alpha)^2}$, where C.G. = c and K is the radius of gyration about a horizontal axis through G.*

Solution: Let M be the mass, C the centre and G the C.G. of the sphere of radius a such that CG = c. Initially CG makes an angle α to the vertical. Assuming the motion to be pure rolling, let C.G. make an angle $\alpha+\theta$ to the vertical at time t. If the point of contact is shifted through a distance x, during this time, then $x = OP = a\theta$.

$$\therefore\ \dot{x} = a\dot{\theta} \text{ and } \ddot{x} = a\ddot{\theta}.$$

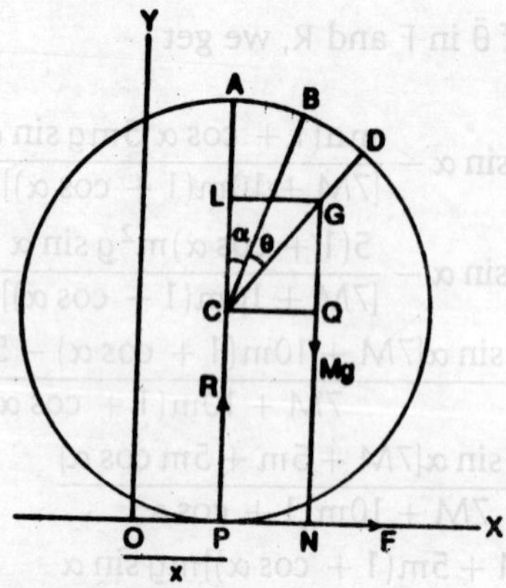

FIG 4.23

Let $(\bar{x}, \bar{y})$ be the coordinates of C.G. at G of the body referred to O as origin and horizontal and vertical through O as coordinate axes. Then

$$\bar{x} = ON = OP + PN = x + c\sin(\theta + \alpha)$$
$$\bar{x} = a\theta + c\sin(\theta + \alpha)$$
$$\bar{y} = GN = GQ + QN = GQ + CP = a + c\cos(\theta + \alpha).$$

Thus the equations of motion of the sphere are

$$F = M\frac{d^2\bar{x}}{dt^2} = M\frac{d^2}{dt^2}\{a\theta + c\sin(\theta + \alpha)\}$$
$$= M\frac{d}{dt}[a\dot{\theta} + c\cos(\theta + \alpha)\dot{\theta}]$$
$$= M\frac{d}{dt}[\{a + c\cos(\theta + \alpha)\}\dot{\theta}].$$
$$\therefore F = M\left[\{a + c\cos(\theta + \alpha)\}\ddot{\theta} - c\sin(\theta + \alpha)\dot{\theta}^2\right] \qquad (4.151)$$

$$\text{and } R - Mg = M\frac{d^2\bar{y}}{dt^2} = M\frac{d^2}{dt^2}[a + c\cos(\theta + \alpha)]$$
$$= M\frac{d}{dt}[-c\sin(\theta + \alpha)\dot{\theta}]$$
$$= -cM[\cos(\theta + \alpha)\dot{\theta}^2 + \sin(\theta + \alpha)\ddot{\theta}]. \qquad (4.152)$$

Taking moment about G, we have

$$MK^2\ddot{\theta} = R \cdot GL - F \cdot GN$$
$$= Rc\sin(\theta + \alpha) - F\{a + c\cos(\theta + \alpha)\}. \qquad (4.153)$$

Here we want only initial motion, when $\theta = 0$, $\dot{\theta} = 0$, but $\ddot{\theta}$ is not zero.
From equations (4.151), (4.152) and (4.153), we get

$$F = M[(a + c\cos\alpha)\ddot{\theta}]$$
$$R = Mg - CM[\sin\alpha\ddot{\theta}]$$
$$\text{and } MK^2\ddot{\theta} = Rc\sin\alpha - F(a + c\cos\alpha).$$

From these equations we get the initial values of F, R and $\ddot{\theta}$. Eliminating R and F from these equations, we get

$$MK^2\ddot{\theta} = (Mg - Mc \sin \alpha\dot{\theta})c \sin \alpha - M\ddot{\theta}(a + c \cos \alpha)(a + c \cos \alpha)$$

$$\ddot{\theta}[k^2 + c^2 \sin^2 \alpha + (a + c \cos \alpha)^2] = gc \sin \alpha$$

$$\therefore\ \ddot{\theta} = \frac{gc \sin \alpha}{k^2 + c^2 \sin^2 \alpha + (a + c \cos \alpha)^2}.$$

Then substituting the value of $\ddot{\theta}$, we get

$$F = \frac{M(a + c \cos \alpha)gc \sin \alpha}{K^2 + c^2 \sin^2 \alpha + (a + c \cos \alpha)^2}$$

$$\begin{aligned} \text{and } R &= Mg - Mc \sin \alpha \frac{gc \sin \alpha}{K^2 + c^2 \sin^2 \alpha + (a + c \cos \alpha)^2} \\ &= \frac{Mg[K^2 + c^2 \sin^2 \alpha + (a + c \cos \alpha)^2 - c^2 \sin^2 \alpha]}{K^2 + c^2 \sin^2 \alpha + (a + c \cos \alpha)^2} \\ &= \frac{Mg[K^2 + (a + c \cos \alpha)^2]}{K^2 + c^2 \sin^2 \alpha + (a + c \cos \alpha)^2}. \end{aligned}$$

$$\therefore\ \frac{F}{R} = \frac{(a + c \cos \alpha)c \sin \alpha}{K^2 + (a + c \cos \alpha)^2}.$$

The sphere will commence to slide if $F > \mu R$,

$$\text{i.e., if } \mu < \frac{F}{R}, \text{ or, } \mu < \frac{(a + c \cos \alpha)c \sin \alpha}{K^2 + (a + c \cos \alpha)^2}.$$

Problem 13. *A homogeneous sphere of mass M is placed on an imperfectly rough table, and a particle of mass m is attached to the end of a horizontal diameter. Show that the sphere will begin to roll or slide according as μ is greater or less than* $\frac{5(M+m)m}{7M^2+17Mm+5m^2}$.

If μ be equal to this value, show that the sphere will begin to roll if $5m^2 < M^2 + 11Mm$. **[IAS 85]**

Solution: Let M be the mass of the sphere, a be its radius and C be its centre. Let G be the common centre of gravity of two masses M at C and m at B. Let CG = c. Then

$$\begin{aligned} (m + M)\bar{x} &= M \cdot 0 + m \cdot a \\ \bar{x} &= \frac{ma}{m + M} = c \text{ (say)} \\ \therefore\ BG &= CB - CG = a - c \\ &= a - \frac{ma}{m + M} \\ &= \frac{Ma}{m + M}. \end{aligned}$$

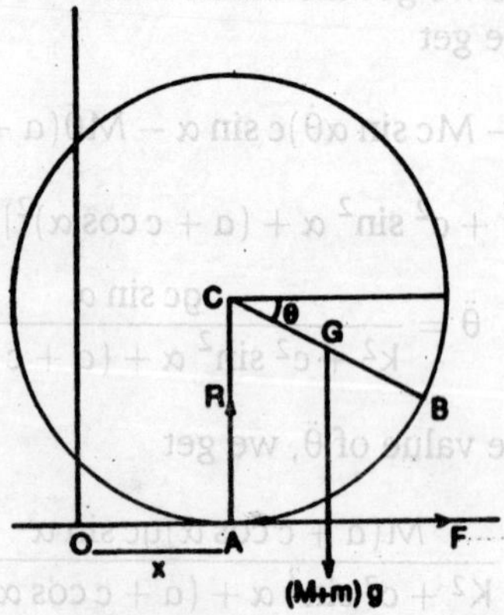

FIG 4.24

At any time t, let the radius C.G. be inclined at an angle θ to the horizontal, and now let A be the point of contact, whose distance from its initial position O is x.

We assume that the sphere rolls and let F be the force of friction sufficient for pure rolling.

Hence $x = a\theta$ and $\dot{x} = a\dot{\theta}$.

Referred to O as origin, coordinates of the centre of gravity G are $(x + c\cos\theta, a - c\sin\theta) = (\bar{x}, \bar{y})$.

The equations of motion of the sphere are

$$\begin{aligned}
F &= (M+m)\frac{d^2\bar{x}}{dt^2} = (M+m)\frac{d^2}{dt^2}(x + c\cos\theta) \\
&= (M+m)\frac{d^2}{dt^2}(a\theta + c\cos\theta)\ (\because\ x = a\theta) \\
&= (M+m)\frac{d}{dt}[a\dot{\theta} - c\sin\theta\dot{\theta}] \\
&= (M+m)[a\ddot{\theta} - c(\cos\theta\dot{\theta}^2 + \sin\theta\ddot{\theta})] \\
&= (M+m)[(a - c\sin\theta)\ddot{\theta} - c\cos\theta\dot{\theta}^2]. \qquad (4.154)
\end{aligned}$$

Also

$$\begin{aligned}
R - (M+m)g &= (M+m)\frac{d^2}{dt^2}(a - c\sin\theta) \\
&= (M+m)\frac{d}{dt}[-c(\cos\theta\dot{\theta})] \\
&= (M+m)\,[-c\{-\sin\theta\dot{\theta}^2 + \cos\theta\ddot{\theta}\}] \\
&= (M+m)[-c\cos\theta\ddot{\theta} + c\sin\theta\dot{\theta}^2] \qquad (4.155)
\end{aligned}$$

and

$$Rc\cos\theta - F(a - c\sin\theta) = (M+m)K^2\ddot{\theta}. \qquad (4.156)$$

We want to discuss only the initial motion when $\theta = 0$ and $\dot{\theta} = 0$ but $\ddot{\theta}$ is not zero.

Hence equations (4.154), (4.155) and (4.156) become

$$F = (M+m)a\ddot{\theta} \tag{4.157}$$

$$R = (M+m)g - (M+m)c\ddot{\theta} \tag{4.158}$$

$$\text{and } Rc - Fa = (M+m)K^2\ddot{\theta}. \tag{4.159}$$

These equations (4.157), (4.158) and (4.159) gives the initial values of F, R and $\ddot{\theta}$.

Putting the values of F and R from equations (4.157) and (4.158) in equation (4.159), we get

$$(M+m)gc - (M+m)c^2\ddot{\theta} - (M+m)a^2\ddot{\theta} = (M+m)K^2\ddot{\theta}$$

$$\text{or, } (M+m)\ddot{\theta}[K^2+a^2+c^2] = (M+m)gc$$

$$\text{or, } \ddot{\theta} = \frac{gc}{K^2+a^2+c^2}.$$

Hence

$$F = \frac{(M+m)agc}{K^2+a^2+c^2}$$

$$\text{and } R = (M+m)g - \frac{(M+m)cgc}{K^2+a^2+c^2}$$

$$= \frac{(M+m)g}{K^2+a^2+c^2}[K^2+a^2+c^2-c^2]$$

$$R = \frac{(M+m)g(K^2+a^2)}{K^2+a^2+c^2}.$$

The sphere will commence to slide or roll according as

$$F > \text{ or, } < \mu R, \text{ or, } \frac{F}{R} > \text{ or, } < \mu$$

$$\text{or, } \frac{ac}{K^2+a^2} > \text{ or, } < \mu. \tag{4.160}$$

Also

$$(M+m)K^2 = \left[M\frac{2a^2}{5} + Mc^2\right] + m(a-c)^2$$

$$= \frac{2Ma^2}{5} + \frac{Mm^2a^2}{(M+m)^2} + \frac{mM^2a^2}{(M+m)^2}$$

$$= \frac{2Ma^2}{5} + \frac{Mma^2}{M+m} = \frac{Ma^2[2M+2m+5m]}{5(M+m)}$$

$$\therefore K^2 = \frac{Ma^2(2M+7m)}{5(M+m)^2}.$$

Putting the value of K^2 in equation (4.160), we get

$$\frac{ac}{\frac{Ma^2(2M+7m)}{5(M+m)^2}+a^2} > \text{or}, < \mu$$

$$\Rightarrow \quad \frac{\frac{ma^2}{M+m}}{\frac{Ma^2(2M+7m)+5a^2(M+m)^2}{5(M+m)^2}} > \text{or}, < \mu$$

$$\Rightarrow \quad \frac{5ma^2(M+m)}{a^2[2M^2+7Mm+5M^2+5m^2+10Mm]} > \text{or}, < \mu$$

$$\Rightarrow \quad \frac{5m(M+m)}{[7M^2+17Mm+5m^2]} > \text{or}, < \mu.$$

When

$$\mu = \frac{5m(M+m)}{7M^2+17Mm+5m^2},$$

then sphere will roll if $K^2 > \frac{a^2}{3}$.

$$\Rightarrow \quad \frac{a^2M(2M+7m)}{5(M+m)^2} > \frac{a^2}{3}$$

$$\Rightarrow \quad 3M(2M+7m) > 5(M^2+m^2+2Mm)$$

$$\Rightarrow \quad M^2+11Mm > 5m^2$$

$$\Rightarrow \quad 5m^2 < M^2+11Mm.$$

Type VII: Problems based on Section 4.11

Problem 14. *A solid uniform sphere, resting on the top of another fixed sphere is slightly displaced and begins to roll down. Show that it will slip when the common normal makes with the vertical an angle given by* $2\sin\theta = \mu(17\cos\theta - 10\cos\alpha)$.

Solution: Let O be the centre and a be the radius of the fixed sphere and C the centre and b the radius of the moving sphere.

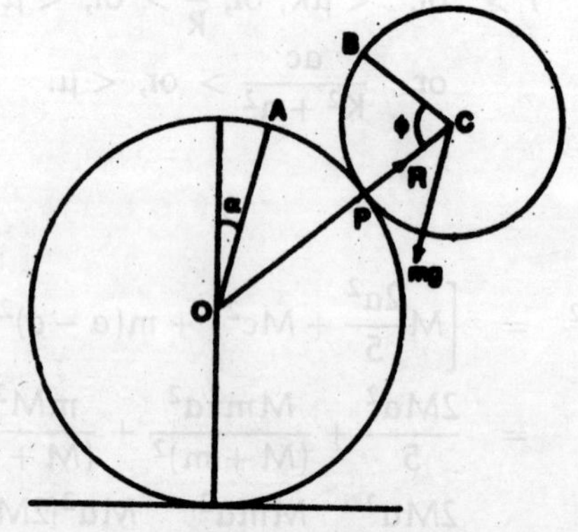

FIG 4.25

Let CB be the radius (a line fixed in the body) making an angle ϕ with the vertical.

Initially B coincided with A and the weight mg acts along a vertical line.

Since there is no slipping between the two spheres, therefore, the arc AP = arc BP,

$$\text{i.e.,} \quad a(\theta - \alpha) = b(\phi - \theta)$$
$$\text{or,} \quad b\dot{\phi} = (a + b)\dot{\theta}. \tag{4.161}$$

Taking the horizontal and vertical lines through O as coordinate axes, coordinates $(\bar{x}, \bar{y})$ of the centre C are $\{(a + b)\sin\theta, (a + b)\cos\theta\}$.

Now

$$\begin{aligned} v^2 &= (\dot{\bar{x}})^2 + (\dot{\bar{y}})^2 \\ &= [(a + b)\cos\dot{\theta}]^2 + [-(a + b)\sin\theta\dot{\theta}]^2 \\ &= (a + b)^2\dot{\theta}^2. \end{aligned}$$

Also energy equation gives

$$\frac{1}{2}m\left[\frac{2b^2}{5}\dot{\phi}^2 + (a + b)^2\dot{\theta}^2\right] = mg[(a + b)\cos\alpha - (a + b)\cos\theta]$$
$$\Rightarrow \frac{1}{2}m\left[\frac{2}{5}(a + b)^2\dot{\theta}^2 + (a + b)^2\dot{\theta}^2\right] = mg(a + b)(\cos\alpha - \cos\theta)$$
$$\Rightarrow \frac{1}{2}m\frac{7}{5}(a + b)^2\dot{\theta}^2 = mg(a + b)(\cos\alpha - \cos\theta)$$
$$\Rightarrow \frac{7}{10}(a + b)\dot{\theta}^2 = g(\cos\alpha - \cos\theta)$$
$$\Rightarrow (a + b)\dot{\theta}^2 = \frac{10g}{7}(\cos\alpha - \cos\theta). \tag{4.162}$$

Differentiating equation (4.162), we get

$$\begin{aligned} (a + b)2\dot{\theta}\ddot{\theta} &= \frac{10g}{7}\sin\theta\dot{\theta} \\ \text{or, } (a + b)\ddot{\theta} &= \frac{5g}{7}\sin\theta. \end{aligned} \tag{4.163}$$

Let F be the frictional force at P acting upwards.

Equations of motion of the sphere are

$$m(a + b)\ddot{\theta} = mg\sin\theta - F \tag{4.164}$$
$$\text{and} \quad m(a + b)\dot{\theta}^2 = mg\cos\theta - R. \tag{4.165}$$

Hence

$$
\begin{aligned}
F &= mg\sin\theta - m(a+b)\ddot{\theta} \\
&= mg\sin\theta - m\frac{5g}{7}\sin\theta \text{ [From equation (4.163)]} \\
&= \frac{2mg}{7}\sin\theta \\
\text{and } R &= mg\cos\theta - m(a+b)\dot{\theta}^2 \\
&= mg\cos\theta - m\frac{10g}{7}(\cos\alpha - \cos\theta) \text{ [From equation (4.162)]} \\
&= \frac{mg}{7}[17\cos\theta - 10\cos\alpha].
\end{aligned}
$$

$$
\therefore \frac{F}{R} = \frac{2\sin\theta}{17\cos\theta - 10\cos\alpha}.
$$

The sphere will slip when

$$
\begin{aligned}
F = \mu R, \quad \text{or,} \quad & \frac{F}{R} = \mu, \\
\text{i.e.,} \quad & \frac{2\sin\theta}{17\cos\theta - 10\cos\alpha} = \mu, \\
\text{or,} \quad & 2\sin\theta = \mu(17\cos\theta - 10\cos\alpha).
\end{aligned}
$$

Problem 15. *A homogeneous sphere rolls down an imperfectly rough fixed sphere, starting from rest at the highest point. If the spheres separate when the line joining their centers makes an angle* θ *with the vertical, prove that* $\cos\theta + 2\mu\sin\theta = Ae^{2\mu\theta}$, *where A is a function of* μ *only.*

Solution: See figure of of this chapter.

Since the fixed sphere is imperfectly rough, therefore, the moving sphere rolls as well as slides on it, so friction μR acts upwards.

Let C be the centre, M the mass and a the radius of the sphere.

Equations of motion of the sphere are

$$
M\frac{v^2}{a} = Mg\cos\theta - R \tag{4.166}
$$

$$
\text{and} \quad Mv\frac{dv}{ds} = Mg\sin\theta - \mu R. \tag{4.167}
$$

From equation (4.166), $R = Mg\cos\theta - \frac{Mv^2}{a}$.

Putting the value of R in equation (4.167), we get

$$Mv\frac{dv}{ds} = Mg\sin\theta - \mu\left[Mg\cos\theta - \frac{Mv^2}{a}\right]$$

$$\text{or,}\quad v\frac{dv}{ds} = g\sin\theta - \mu g\cos\theta + \frac{\mu v^2}{a}$$

$$\text{or,}\quad \frac{1}{2}\frac{dv^2}{ds} - \frac{\mu}{a}v^2 = g(\sin\theta - \mu\cos\theta)$$

$$\text{or,}\quad \frac{dv^2}{d\theta}\frac{d\theta}{ds} - \frac{2\mu v^2}{a} = 2g(\sin\theta - \mu\cos\theta)$$

$$\text{or,}\quad \frac{dv^2}{d\theta}\cdot\frac{1}{a} - \frac{2\mu v^2}{a} = 2g(\sin\theta - \mu\cos\theta)$$

$$\text{or,}\quad \frac{dv^2}{d\theta} - 2\mu v^2 = 2ag(\sin\theta - \mu\cos\theta), \tag{4.168}$$

which is a linear differential equation in v^2.

$$\text{I.F.} = e^{\int -2\mu d\theta} = e^{-2\mu\theta}.$$

Hence solution of equation (4.168) is

$$\begin{aligned} v^2e^{-2\mu\theta} &= C + 2ag\int e^{-2\mu\theta}(\sin\theta - \mu\cos\theta)d\theta \\ &= C + 2ag\left[\int e^{-2\mu\theta}\sin\theta d\theta - \mu\int e^{-2\mu\theta}\cos\theta d\theta\right] \\ &= C + \frac{2age^{-2\mu\theta}}{4\mu^2+1}[(-2\mu\sin\theta - \cos\theta) \\ &\quad -\mu(-2\mu\cos\theta + \sin\theta)] \\ &= C + \frac{2ag}{4\mu^2+1}e^{-2\mu\theta}[-3\mu\sin\theta - (1-2\mu^2)\cos\theta]. \end{aligned}$$

Initially, when $\theta = 0$, $v = 0$.

$$C = \frac{2ag(1-2\mu^2)}{4\mu^2+1}.$$

$$\therefore\ v^2e^{-2\mu\theta} = \frac{2ag(1-2\mu^2)}{1+4\mu^2} + \frac{2ag}{4\mu^2+1}e^{-2\mu\theta}[-3\mu\sin\theta - (1-2\mu^2)\cos\theta]. \tag{4.169}$$

The spheres will separate when $R = 0$.

Hence from equation (4.166)

$$\begin{aligned} \frac{Mv^2}{a} &= Mg\cos\theta \\ \text{or, } v^2 &= ag\cos\theta. \end{aligned}$$

Substituting in equation (4.169), we get

$$a g \cos\theta e^{-2\mu\theta} = \frac{2ag(1-2\mu^2)}{1+4\mu^2} + \frac{2ag}{1+4\mu^2} e^{-2\mu\theta} \times$$
$$[-3\mu \sin\theta - (1-2\mu^2)\cos\theta]$$
$$\Rightarrow \quad (1+4\mu^2)\cos\theta = 2(1-2\mu^2)e^{2\mu\theta} + 2[-3\mu\sin\theta - (1-2\mu^2)\cos\theta]$$
$$\Rightarrow \quad (1+4\mu^2)\cos\theta + 6\mu\sin\theta + 2(1-2\mu^2)\cos\theta = 2(1-2\mu^2)e^{2\mu\theta}$$
$$\Rightarrow \quad \cos\theta[1+2] + 6\mu\sin\theta = 2(1-2\mu^2)e^{2\mu\theta}$$
$$\Rightarrow \quad 3\cos\theta + 6\mu\sin\theta = 2(1-2\mu^2)e^{2\mu\theta}$$
$$\Rightarrow \quad \cos\theta + 2\mu\sin\theta = Ae^{2\mu\theta},$$

where $A = \frac{2}{3}(1-2\mu^2)$.

Problem 16. *A rough solid circular cylinder rolls down a second rough cylinder which is fixed with its axis horizontal. If the plane through their axis makes an angle α with the vertical when the first cylinder is at rest, show that the cylinders will separate when this angle of elevation is $\cos^{-1}\left(\frac{4}{7}\cos\alpha\right)$.*

Solution: See figure of Problem of this chapter.

Let O be the centre and a be the radius of the fixed cylinder. Let C be the centre and b the radius of the cylinder resting on the fixed one with its point B in contact to the point A of the fixed cylinder such that OA makes an angle α to the vertical. The upper cylinder rolls at time t, let P be the point of contact of the two cylinder such that the line OC joining centers make an angle θ to the vertical. Let CB makes an angle ϕ to the vertical at time t. Since there is pure rolling,

$$\therefore \quad \text{arc } AP = \text{arc } BP$$
$$\Rightarrow \quad a(\theta - \alpha) = b(\phi - \theta)$$
$$\Rightarrow \quad b\phi = (a+b)\theta - a\alpha,$$
$$\Rightarrow \quad b\dot{\phi} = c\dot{\theta}. \tag{4.170}$$

Let R be the normal reaction and F the friction acting on the upper sphere. Therefore, the equation of motion of the cylinder along CO is

$$Mc\dot{\theta}^2 = Mg\cos\theta - R, \tag{4.171}$$

where $c = a + b$.

Let $(\bar{x}, \bar{y})$ be the coordinates of C referred to the horizontal and vertical lines through O as axes and be given by

$$\bar{x} = OC\sin\theta = c\sin\theta; \; \bar{y} = OC\cos\theta = c\cos\theta.$$

So its $(\text{velocity})^2 = \dot{\bar{x}}^2 + \dot{\bar{y}}^2 = c^2\dot{\theta}^2$.

The energy equation gives

$$\frac{1}{2}MK^2\dot{\theta}^2 + \frac{1}{2}Mvc^2 = Mg(c\cos\alpha - c\cos\theta)$$

or, $$\frac{1}{2}M\frac{1}{2}b^2\dot{\phi}^2 + \frac{1}{2}Mc^2\dot{\theta}^2 = Mgc(\cos\alpha - \cos\theta)$$

$$\Rightarrow \frac{1}{4}M(c\dot{\theta})^2 + \frac{1}{2}Mc^2\dot{\theta}^2 = Mgc(\cos\alpha - \cos\theta) \text{ [From (4.170)]}$$

$$\Rightarrow \frac{3}{4}c^2\dot{\theta}^2 = gc(\cos\alpha - \cos\theta)$$

$$\Rightarrow \dot{\theta}^2 = \frac{4g}{3c}(\cos\alpha - \cos\theta). \qquad (4.171A)$$

Thus from equation (4.171A), we have

$$\begin{aligned} Mc\frac{4g}{3c}(\cos\alpha - \cos\theta) &= Mg\cos\theta - R \\ \Rightarrow R &= Mg\left[\cos\theta - \frac{4}{3}(\cos\alpha - \cos\theta)\right] \\ &= \frac{Mg}{3}[7\cos\theta - 4\cos\alpha]. \end{aligned}$$

The cylinders will separate when $R = 0$.

$$\frac{Mg}{3}(7\cos\theta - 4\cos\alpha) = 0$$

$$\Rightarrow \cos\theta = \frac{4}{7}\cos\alpha$$

$$\therefore \theta = \cos^{-1}\left(\frac{4}{7}\cos\alpha\right).$$

Type VIII: Problem based on Section 4.12

Problem 17. *A circular plate rolls down the inner circumference of a rough circle under the action of gravity, the planes of both the plate and the circle being vertical, when the line joining their centers is inclined at an angle θ to the vertical, show that the friction between the bodies is $\frac{\sin\theta}{3}$ times the weight of the plate.*

Solution: Let O be the centre of the fixed circle of radius a. Let C be the centre, m the mass and b the radius of the circular plate which rolls down the inner circumference of the fixed circle.

Let the point B of the plate be at the point A of the fixed circle initially such that $\angle AOC = \alpha$.

At time t, let the plate roll down to the point P such that $\angle COP = \theta$ and ϕ the angle that the line CB fixed in the space makes with the vertical.

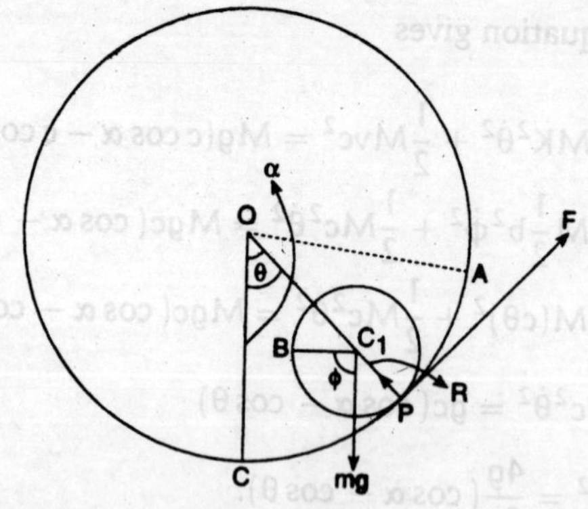

FIG 4.26

Since there is no slipping between the two bodies,

$$\therefore \quad \text{arc } AP = \text{arc } BP$$
$$\text{or,} \quad a(\alpha - \theta) = b[2\pi - (\theta + \phi)]$$
$$\text{or,} \quad -a\dot{\theta} = -b(\dot{\theta} + \dot{\phi}),$$
$$\text{i.e.,} \quad b\dot{\phi} = (a - b)\dot{\theta}$$
$$\therefore \quad b\ddot{\phi} = (a - b)\ddot{\theta}. \tag{4.172}$$

Equation of motion of the plate perpendicular to CO is

$$m(a - b)\ddot{\theta} = F - mg\sin\theta. \tag{4.173}$$

Also for the motion relative to C

$$mK^2\ddot{\phi} = -Fb$$
$$\text{or,} \quad m\frac{b^2}{2}\ddot{\phi} = -Fb$$
$$\Rightarrow \quad \frac{m}{2}b\ddot{\phi} = -F$$
$$\Rightarrow \quad \frac{m}{2}(a - b)\ddot{\theta} = -F \text{ [From equation (4.172)]}$$
$$\Rightarrow \quad m(a - b)\ddot{\theta} = -2F.$$

Hence from equation (4.173), we get

$$-2F = F - mg\sin\theta$$
$$\text{or,} \quad 3F = mg\sin\theta$$
$$\therefore \quad F = \frac{1}{3}mg\sin\theta = \left(\frac{1}{3}\sin\theta\right) \text{ times the weight of the plate.}$$

Problem 18. *A solid spherical ball rests in equilibrium at the bottom of a fixed spherical globe whose inner surface is perfectly rough. The ball is struck by a horizontal blow of such a magnitude that the initial speed of its centre is v, prove that if v lies between $\sqrt{\frac{10}{7}gd}$ and $\sqrt{\frac{27}{7}gd}$, the ball would leave the globe, d being the difference between the radii of the ball and the globe.*

Solution: Let O be the centre and a be the radius of the fixed globe. Let a solid ball of radius b, centre C and mass M rest at the lowest point A of the globe. At time t, let the ball roll to the point C such that $\angle AOP = \theta$.

At this time t, let the line CB fixed in the ball make an angle ϕ to the vertical. At time $t = 0$, the point B coincided with A.

Since there is no slipping,

$$\text{arc } AP = \text{arc } BP$$

$$\text{or,} \quad a\theta = b(\theta + \phi)$$

$$\Rightarrow \quad (a - b)\theta = b\phi$$

$$\Rightarrow \quad d\theta = b\phi \qquad (\text{given } d = a - b)$$

$$\therefore \quad d\dot{\theta} = b\dot{\phi} \tag{4.174}$$

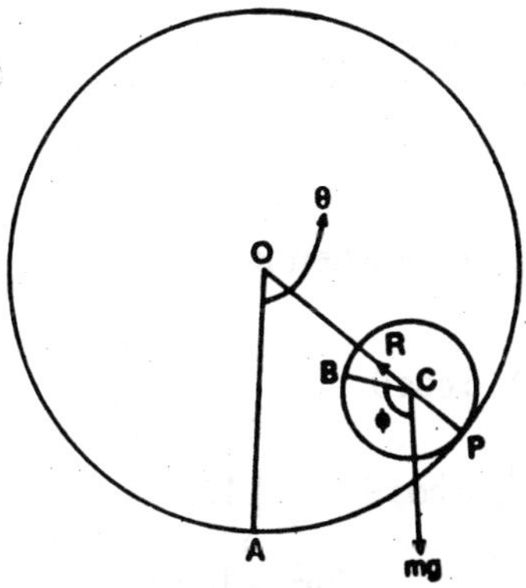

FIG 4.27

Let R be the normal reaction and F the friction at the point P. The equations of motion of the ball along and perpendicular to CO are

$$Md\dot{\theta}^2 = R \quad Mg\cos\theta \tag{4.175}$$

$$\text{and} \quad Md\ddot{\theta} = F - Mg\sin\theta. \tag{4.176}$$

Referred to O coordinates of C are $(d\sin\theta, d\cos\theta)$. So at any time t square of the velocity of the centre C

$$= (d\cos\theta\dot{\theta})^2 + (-d\sin\theta\dot{\theta})^2 = d^2\dot{\theta}^2.$$

Since initial velocity of the centre = v (given),

$$\therefore v^2 = d^2\dot{\theta}^2$$

$$\therefore v = d\dot{\theta}. \tag{4.177}$$

At time t, kinetic energy of the ball

$$= \frac{1}{2}M\left(\frac{2b^2}{5}\dot{\phi}^2 + d^2\dot{\theta}^2\right) = \frac{1}{2}M\left(\frac{2}{5}d^2\dot{\theta}^2 + d^2\dot{\theta}^2\right)$$

$$= \frac{1}{2}M\frac{7}{5}d^2\dot{\theta}^2 = \frac{7}{10}Md^2\dot{\theta}^2 = \frac{7}{10}Mv^2 \text{ [From equation (4.177)]}$$

Hence energy equation gives change in kinetic energy = Work done by the gravity.

$$\frac{1}{2}M\frac{7}{5}d^2\dot{\theta}^2 - \frac{1}{2}M\frac{7}{5}v^2 = -Mg(d - d\cos\theta)$$
$$\Rightarrow \quad \frac{7}{10}Md^2\dot{\theta}^2 = \frac{7}{10}Mv^2 - Mgd(1-\cos\theta)$$
$$\Rightarrow \quad \frac{7}{10}d^2\dot{\theta}^2 = \frac{7}{10}v^2 - gd(1-\cos\theta)$$
$$\Rightarrow \quad d\dot{\theta}^2 = \frac{v^2}{d} - \frac{10g}{7}(1-\cos\theta).$$

Thus from equation (4.175), we get

$$\begin{aligned} R &= Mg\cos\theta + Md\dot{\theta}^2 \\ &= Mg\cos\theta + \frac{Mv^2}{d} - \frac{10Mg}{7}(1-\cos\theta). \end{aligned}$$

The ball will leave the globe when $R = 0$.

$$Mg\cos\theta + \frac{Mv^2}{d} - \frac{10Mg}{7}(1-\cos\theta) = 0$$
$$\Rightarrow \quad \frac{17}{7}g\cos\theta = \frac{1}{7}\left(10g - \frac{7v^2}{d}\right)$$
$$\Rightarrow \quad \cos\theta = -\frac{7v^2 - 10gd}{17gd}. \tag{4.178}$$

We know that the ball would leave the globe when its centre rises above the horizontal through O, therefore, θ is obtuse, hence $\cos\theta$ is negative.

$$\text{or,} \quad \cos\theta < 0, \text{ i.e.,} \ \frac{7v^2 - 10gd}{17gd} > 0,$$
$$\text{i.e.,} \quad 7v^2 > 10gd, \text{ or, } v > \sqrt{\frac{10gd}{7}}.$$

Also $|\cos\theta| < 1$,

$$\text{i.e.,} \ \frac{7v^2 - 10gd}{17gd} < 1, \ \text{or,} \quad 7v^2 - 10gd < 17gd$$
$$\text{or,} \quad 7v^2 < 27gd; \text{ or, } v < \sqrt{\frac{27}{7}gd}.$$

Hence v must lie between

$$\sqrt{\frac{10gd}{7}} \text{ and } \sqrt{\frac{27gd}{7}}.$$

Type IX: Problems based on Motion of One Body on another, when the Lower Body is free to turn about its axes

Problem 19. *A rough cylinder, of mass M, is capable of motion about its axis, which is horizontal. A particle of mass m is placed on it vertically above the axis and the system is slightly disturbed. Show that the particle will slip on the cylinder when it has moved through an angle θ given by* $\mu(M+6m)\cos\theta - M\sin\theta = 4m\mu$, *where* μ *is the coefficient of friction.* [IAS 94]

Solution: Let O be the centre and M the mass of the rough cylinder, which is capable of motion about its axis which is horizontal. A particle of mass m is placed vertically above the axis. When the system is slightly displaced, let at time t, the cylinder turns through an angle θ. Thus at time t the particle is at P and $\angle AOP = \theta$.

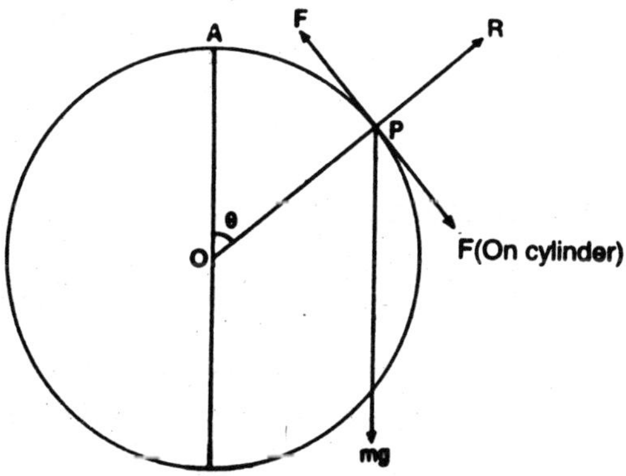

FIG 4.28

Since the particle of mass m describes a circle of radius OP = a about O, hence its acceleration along and perpendicular to PO are $ma\dot{\theta}^2$ and $ma\ddot{\theta}$.

Therefore, the equations of motion of the particle of mass m are

$$ma\dot{\theta}^2 = mg\cos\theta - R \tag{4.179}$$

$$\text{and} \quad ma\ddot{\theta} = mg\sin\theta - F. \tag{4.180}$$

The coordinates of P referred to the horizontal and vertical lines through O as axes are $(a\sin\theta, a\cos\theta)$.

So $(\text{velocity})^2$ of $P = (a\cos\theta\dot{\theta})^2 + (-a\sin\theta\dot{\theta})^2 = a^2\dot{\theta}^2$.

Thus energy equation gives

$$\frac{1}{2}MK^2\dot{\theta}^2 + \frac{1}{2}ma^2\dot{\theta}^2 = mg(a - a\cos\theta)$$

$$\Rightarrow \quad \frac{1}{2}M\frac{1}{2}a^2\dot{\theta}^2 + \frac{1}{2}ma^2\dot{\theta}^2 = mga(1-\cos\theta)$$

$$\Rightarrow \quad (M+2m)a^2\dot{\theta}^2 = 4mga(1-\cos\theta). \tag{4.181}$$

Differentiating with respect to t, we get

$$(M+2m)a^2 2\dot{\theta}\ddot{\theta} = 4mga\sin\theta\dot{\theta}$$

$$\Rightarrow \quad (M+2m)a\ddot{\theta} = 2mg\sin\theta. \tag{4.182}$$

From equations (4.179) and (4.181), we get

$$R = mg\cos\theta - ma\dot{\theta}^2 = mg\cos\theta - \frac{m4mg(1-\cos\theta)}{(M+2m)}$$

$$= \frac{mg}{M+2m}[\cos\theta(M+2m) - 4m(1-\cos\theta)]$$

$$= \frac{mg}{M+2m}[(M+6m)\cos\theta - 4m]. \quad (4.183)$$

Again from equation (4.180) and (4.182), we have

$$F = mg\sin\theta - ma\ddot{\theta} = mg\sin\theta - m\frac{2mg\sin\theta}{M+2m}$$

$$= \frac{mg\sin\theta[M+2m-2m]}{M+2m} = \frac{Mmg\sin\theta}{M+2m}. \quad (4.184)$$

Hence

$$\frac{F}{R} = \frac{M\sin\theta}{(M+6m)\cos\theta - 4m}.$$

The particle slips off from the cylinder, when

$$F = \mu R, \quad \text{or,} \quad \mu = \frac{F}{R} = \frac{M\sin\theta}{(M+6m)\cos\theta - 4m},$$

$$\Rightarrow \mu(M+6m)\cos\theta - 4m\mu = M\sin\theta,$$

$$\Rightarrow \mu(M+6m)\cos\theta - M\sin\theta = 4m\mu.$$

Problem 20. *The mass of a sphere is $\frac{1}{5}$ of that of another sphere of the same material which is free to move about its centre as a fixed point, the first sphere rolls down the second from rest at the highest point, the coefficient of friction being μ. Prove that sliding will begin when the angle θ which the line of centres makes with the vertical is given by* $\sin\theta = 2\mu(5\cos\theta - 3)$. **[IAS 48]**

Solution: Let M be the mass and a be the radius of the sphere which can move about its centre O as a fixed point. Let C be the centre, b the radius and m the mass of the sphere which rolls down the first sphere starting from rest from its highest point.

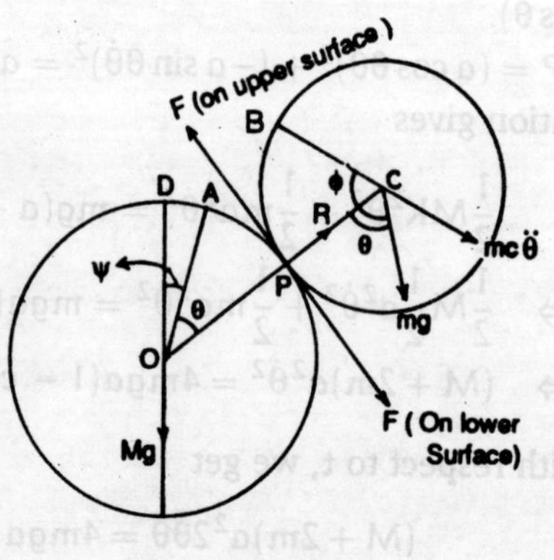

FIG 4.29

Given $M = 5m$.

Let OA be a line fixed in the lower sphere and CB a line fixed in the upper sphere. Initially, OA and CB were vertical and B coincided with A. At time t, let OA and CB make angle ψ and ϕ with the vertical when line of centres OC makes an angle θ with the vertical. Let F be the friction sufficient for pure rolling.

Since there is no slipping between the two spheres,

$$\therefore \quad \text{arc } AP = \text{arc } BP$$
$$\Rightarrow \quad a(\theta - \psi) = b(\phi - \theta)$$
$$\Rightarrow \quad a\psi + b\phi = (a + b)\theta$$
$$\Rightarrow \quad a\dot{\psi} + b\dot{\phi} = c\dot{\theta}, \text{ where } c = a + b. \tag{4.185}$$

Considering the motion of the lower and upper spheres and taking moments about their centres, we have

$$M\frac{2a^2}{5}\ddot{\psi} = Fa \text{ and } m\frac{2b^2}{5}\ddot{\phi} = Fa.$$
$$\therefore \quad \frac{Ma\ddot{\psi}}{mb\ddot{\phi}} = 1, \text{ or, } Ma\ddot{\psi} = mb\ddot{\phi}.$$

Integrating, we get

$$Ma\dot{\psi} = mb\dot{\phi} + C.$$

Initially, when $t = 0$, $\dot{\psi} = 0$, $\dot{\phi} = 0$, $C = 0$.

Hence $Ma\dot{\psi} = mb\dot{\phi}$.

$$\therefore \quad \frac{a\dot{\psi}}{m} = \frac{b\dot{\phi}}{M} = \frac{a\dot{\psi} + b\dot{\phi}}{m + M} = \frac{c\dot{\theta}}{m + M}. \tag{4.186}$$

[From equation (4.185)]

Since the centre C describes a circle of radius $OC = a+b = c$ about O, therefore, its accelerations along and perpendicular to CO are $mc\dot{\theta}^2$ and $mc\ddot{\theta}$, respectively.

Hence equations of motion of the upper sphere are

$$mc\dot{\theta}^2 = mg\cos\theta - R \tag{4.187}$$
$$\text{and } mc\ddot{\theta} = mg\sin\theta - F. \tag{4.188}$$

The coordinates of C referred to horizontal and vertical lines through O as axes are $(c\sin\theta, c\cos\theta)$.

Hence $(\text{velocity})^2$ of C $= (c\cos 0\dot{\theta})^2 + (-c\sin\theta\dot{\theta})^2 = c^2\dot{\theta}^2$.

Therefore, energy equation gives

$$\frac{1}{2}M\frac{2}{5}a^2\dot{\psi}^2 + \left(\frac{1}{2}m\frac{2}{5}b^2\dot{\phi}^2 + \frac{1}{2}mc^2\dot{\theta}^2\right) = mg(c - c\cos\theta)$$

$$\Rightarrow \quad \frac{1}{2}M\frac{2}{5}\frac{m^2c^2\dot{\theta}^2}{(m+M)^2} + \frac{1}{2}m\frac{2}{5}\left(\frac{Mc\dot{\theta}}{m+M}\right)^2 + \frac{1}{2}mc^2\dot{\theta}^2 = mgc(1-\cos\theta)$$

$$\Rightarrow \quad \frac{2}{5}M\frac{m^2c^2\dot{\theta}^2}{(m+M)^2} + \frac{2}{5}m\frac{M^2c^2\dot{\theta}^2}{(m+M)^2} + mc^2\dot{\theta}^2 = 2mgc(1-\cos\theta)$$

$$\Rightarrow \quad \left[\frac{2mM}{5}\frac{(m+M)}{(m+M)^2} + m\right]c^2\dot{\theta}^2 = 2mgc(1-\cos\theta)$$

$$\Rightarrow \quad \left[\frac{2}{5}\frac{mM}{(m+M)} + m\right]c^2\dot{\theta}^2 = 2mgc(1-\cos\theta)$$

$$\Rightarrow \quad \frac{2M+5m+5M}{5(m+M)}c\dot{\theta}^2 = 2g(1-\cos\theta)$$

$$\Rightarrow \quad \frac{7M+5m}{5(m+M)}c\dot{\theta}^2 = 2g(1-\cos\theta)$$

$$\Rightarrow \quad \frac{8M}{5\cdot 6m}c\dot{\theta}^2 = 2g(1-\cos\theta) \ [\because M = 5m]$$

$$\Rightarrow \quad \frac{c\dot{\theta}^2 8}{6} = 2g(1-\cos\theta)$$

$$\Rightarrow \quad c\dot{\theta}^2\frac{4}{3} = 2g(1-\cos\theta)$$

$$\Rightarrow \quad c\dot{\theta}^2 = \frac{3}{2}g(1-\cos\theta).$$

Differentiating it with respect to t, we get

$$2c\dot{\theta}\ddot{\theta} = \frac{3}{2}g\sin\theta\dot{\theta}, \text{ or, } c\ddot{\theta} = \frac{3}{4}g\sin\theta.$$

Hence from equations (4.187) and (4.188), we have

$$R = mg\cos\theta - m\frac{3}{2}g(1-\cos\theta)$$

$$= \frac{mg}{2}[2\cos\theta - 3 + 3\cos\theta] = \frac{mg}{2}(5\cos\theta - 3)$$

$$\text{and } F = mg\sin\theta - mc\ddot{\theta}$$

$$= mg\sin\theta - m\frac{3}{4}g\sin\theta = \frac{mg\sin\theta}{4}.$$

$$\therefore \frac{F}{R} = \frac{\sin\theta}{2(5\cos\theta - 3)}.$$

Sliding of the upper spheres begins when $\frac{F}{R} = \mu$.

$$\text{or, } \frac{\sin\theta}{2(5\cos\theta - 3)} = \mu$$

$$\text{or, } \sin\theta = 2\mu(5\cos\theta - 3).$$

Problem 21. *A uniform circular cylinder of mass M is free to rotate about its axis, which is smooth and horizontal, and about which its radius of gyration is equal to its radius. A uniform solid sphere of mass m is placed with its lowest point in contact with the highest generator of the cylinder, both sphere and cylinder being initially at rest. The sphere is then slightly disturbed and rolls down the cylinder. Show that the slipping takes place before the sphere leaves the cylinder, and begins when* $2M\sin\theta = \mu[(17M+6m)\cos\theta - (10M+4m)]$, *where* θ *is the inclination to the vertical of the plane through their axes and* μ *the coefficient of friction.*

Solution: Let O be the centre, M the mass and a the radius of the cylinder which is free to move about its axis which is fixed horizontally. Let C be the centre, m the mass and b the radius of the rolling sphere which is initially placed at rest at the highest point of the cylinder. In time t, let the sphere roll down to the point P of the cylinder such that the line CB fixed in this sphere makes an angle ψ to the vertical. Initially B coincided with A which was the highest point of the cylinder, i.e., initially CB and OA were vertical. In time t, let the cylinder turn through an angle ψ.

$$\therefore \angle DOA = \psi.$$

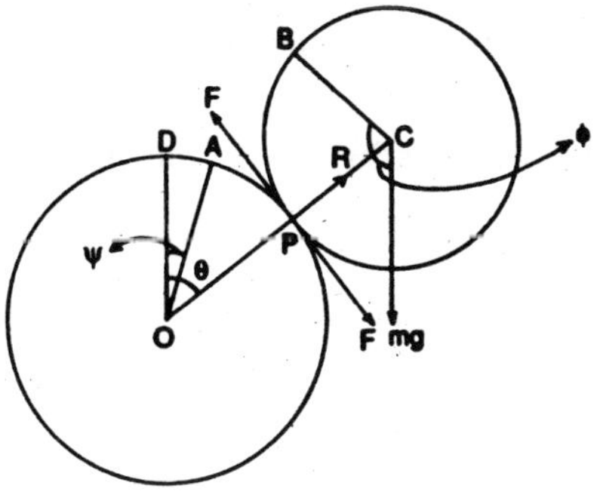

FIG 4.30

Also let the plane through the axes make an angle θ to the vertical at time t. Since there is no slipping, therefore,

$$\begin{aligned} &\text{arc AP} = \text{arc BP} \\ \text{or,}\quad & a(\theta-\psi) = b(\phi-\theta) \\ \text{or,}\quad & b\phi + a\psi = (a+b)\theta \\ \text{or,}\quad & a\dot{\psi} + b\dot{\phi} = c\dot{\theta}, \end{aligned} \tag{4.189}$$

where $c = a+b$.

Considering the motion of the cylinder and the sphere, respectively and taking moments about their centres, we have, $Ma^2\ddot{\psi} = Fa$ for the cylinder and $m\frac{2b^2}{5}\ddot{\phi} = Fb$ for the sphere.

Dividing these two, we get

$$\frac{5Ma^2\ddot{\psi}}{2mb^2\ddot{\phi}} = \frac{a}{b}, \text{ or, } 5Ma\ddot{\psi} = 2mb\ddot{\phi}.$$

Integrating, we get

$$Ma\dot{\psi} = \frac{2m}{5}b\dot{\phi} + C.$$

Initially, when $t = 0, \dot{\phi} = 0, \dot{\psi} = 0.$

Hence $C = 0.$

Thus,

$$Ma\dot{\psi} = \frac{2}{5}mb\dot{\phi},$$

$$\text{or,} \quad \frac{a\dot{\psi}}{2m} = \frac{b\dot{\phi}}{5M} = \frac{a\dot{\psi} + b\dot{\theta}}{2m + 5M} = \frac{c\dot{\theta}}{2m + 5M} \quad \text{[From equation (4.189)]}$$

Hence

$$a\dot{\psi} = \frac{2mc\dot{\theta}}{2m + 5M} \text{ and } b\dot{\phi} = \frac{5Mc\dot{\theta}}{2m + 5M}.$$

The coordinates of C referred to horizontal and vertical lines through O as axes are $(c \sin\theta, c\cos\theta)$.

Hence $(\text{velocity})^2$ of $C = (c\cos\theta\dot{\theta})^2 + (-c\sin\theta\dot{\theta})^2 = c^2\dot{\theta}^2$.

Therefore, energy equation gives

$$\frac{1}{2}Ma^2\dot{\psi}^2 + \frac{1}{2}m\left[\frac{2b^2}{5}\dot{\phi}^2 + c^2\dot{\theta}^2\right] = mg(c - c\cos\theta)$$

$$\Rightarrow \quad Ma^2\dot{\psi}^2 + m\left(\frac{2b^2\dot{\phi}^2}{5} + c^2\dot{\theta}^2\right) = 2mgc(1 - \cos\theta)$$

$$\Rightarrow \quad \frac{M4m^2c^2\dot{\theta}^2}{(2m + 5M)^2} + m\left[\frac{2}{5}\frac{25M^2c^2\dot{\theta}^2}{(2m + 5M)^2} + c^2\dot{\theta}^2\right] = 2mgc(1 - \cos\theta)$$

$$\Rightarrow \quad \left[\frac{4Mm^2}{(2m + 5M)^2} + m\left\{\frac{10M^2}{(2m + 5M)^2} + 1\right\}\right]c^2\dot{\theta}^2 = 2mgc(1 - \cos\theta)$$

$$\Rightarrow \quad \left[\frac{4Mm^2 + 10mM^2 + m(2m + 5M)^2}{(2m + 5M)^2}\right]c^2\dot{\theta}^2 = 2mgc(1 - \cos\theta)$$

$$\Rightarrow \quad \frac{[2Mm(2m + 5M) + m(2m + 5M)^2]c^2\dot{\theta}^2}{(2m + 5M)^2} = 2mgc(1 - \cos\theta)$$

$$\Rightarrow \quad \frac{(2m + 5M)m(2M + 2m + 5M)c^2\dot{\theta}^2}{(2m + 5M)^2} = 2mgc(1 - \cos\theta)$$

$$\Rightarrow \quad \frac{(2m + 7M)}{2m + 5M}c\dot{\theta}^2 = 2g(1 - \cos\theta)$$

$$\Rightarrow \quad c\dot{\theta}^2 = 2\left(\frac{2m + 5M}{2m + 7M}\right)g(1 - \cos\theta). \tag{4.190}$$

Differentiating with respect to $\dot{\theta}$, we get

$$
\begin{aligned}
2c\dot{\theta}\ddot{\theta} &= 2\left(\frac{2m+5M}{2m+7M}\right) g\sin\theta\dot{\theta} \\
\text{or, } c\ddot{\theta} &= \left(\frac{2m+5M}{2m+7M}\right) g\sin\theta. \qquad (4.191)
\end{aligned}
$$

Since the centre C describes a circle of radius $OC = a+b = c$ about O, therefore, its accelerations along and perpendicular to CO are $mc\dot{\theta}^2$ and $mc\ddot{\theta}$, respectively.

Hence equations of motion of the sphere are

$$
\begin{aligned}
mc\dot{\theta}^2 &= mg\cos\theta - R \qquad (4.192) \\
\text{and } mc\ddot{\theta} &= mg\sin\theta - F. \qquad (4.193)
\end{aligned}
$$

Now

$$
\begin{aligned}
R &= mg\cos\theta - mc\dot{\theta}^2 \\
&= mg\cos\theta - m\frac{2(2m+5M)g(1-\cos\theta)}{2m+7M} \quad \text{[From equation (4.190)]} \\
&= \frac{mg}{2m+7M}[\cos\theta(2m+7M) - 2(2m+5M)(1-\cos\theta)] \\
&= \frac{mg}{2m+7M}[\cos\theta(2m+7M+4m+10M) \\
&\qquad -2(2m+5M)] \\
&= \frac{mg}{2m+7M}[\cos\theta(6m+17M) - 2(2m+5M)] \\
\text{and } F &= mg\sin\theta - mc\ddot{\theta} \\
&= mg\sin\theta - mg\sin\theta\frac{(2m+5M)}{2m+7M} \quad \text{[From equation (4.191)]} \\
&= \frac{mg\sin\theta}{2m+7M}[2m+7M - 2m - 5M] \\
F &= \frac{mg\sin\theta 2M}{2m+7M} \\
\therefore \frac{F}{R} &= \frac{2M\sin\theta}{[(17M+6m)\cos\theta - (10M+4m)]}.
\end{aligned}
$$

Slipping of the sphere will begin, when

$$
\begin{aligned}
F = \mu R, \quad \text{or,} \quad & \frac{F}{R} = \mu \\
\text{or,} \quad & \mu = \frac{2M\sin\theta}{[(17M+6m)\cos\theta - (10M+4m)]} \\
\text{or,} \quad & 2M\sin\theta = \mu[(17M+6m)\cos\theta - (10M+4m)],
\end{aligned}
$$

which gives the value of θ.

Also when slipping begins, $\theta < \pi$.

Now

$$
R = \frac{F}{\mu} = \frac{2mMg\sin\theta}{\mu(2m+7M)}.
$$

which is positive for all values of θ between 0 and π. Hence, the slipping begins before the sphere leaves the cylinder.

Type X: Problems based on Motion of One Body on another when the bodies are free to move

Problem 22. *Two homogeneous spheres of equal radii, and masses m and m' rest on a smooth horizontal plane with m' on the highest point of m. If the system be disturbed, show that the inclination θ of their common normal to the vertical is given by $a\dot{\theta}^2(7m + 5m'\sin^2\theta) = 5g(m + m')(1 - \cos\theta)$.*

Solution: Let C_1 be the centre and m the mass of the sphere resting on a smooth horizontal plane. Let C_2 be the centre and m' the mass of another sphere of equal radius resting on the highest point of the first sphere. In time t, let the lower sphere move through a distance OA = x, on the table, while the line of centres C_1C_2 turns through an angle θ with the vertical. Initially, C_1C_2 was vertical. During this time t, let the two spheres turn through angles ϕ and ψ to the vertical. Initially C_1B, C_1C_2 and C_2D coincided vertically.

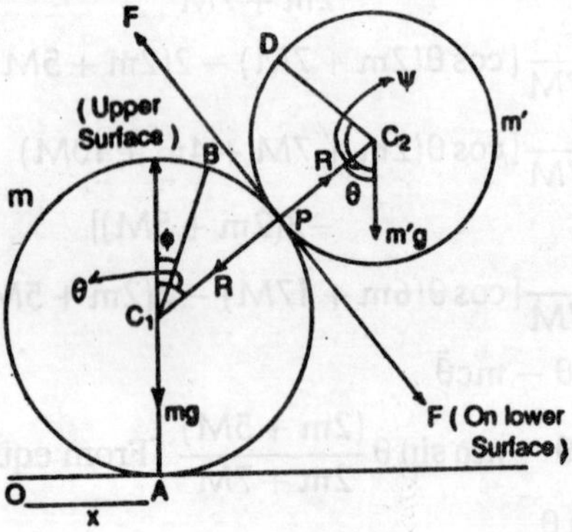

FIG 4.31

Since there is no slipping between the two spheres,

$$\therefore \quad \text{arc } BP = \text{arc } DP$$

$$\Rightarrow \quad a(\theta - \phi) = a(\psi - \theta)$$

$$\Rightarrow \quad 2\theta = \phi + \psi$$

$$\Rightarrow \quad 2\dot{\theta} = \dot{\phi} + \dot{\psi}. \tag{4.194}$$

Considering the motions of the two spheres separately and taking moments about their centres C_1 and C_2, we get

$$m\frac{2a^2}{5}\ddot{\phi} = Fa \quad \text{(For lower sphere)} \tag{4.195}$$

$$\text{and} \quad m'\frac{2a^2}{5}\ddot{\psi} = Fa \quad \text{(For upper sphere)}. \tag{4.196}$$

From equations (4.195)and (4.196), we get

$$m\ddot{\phi} = m'\ddot{\psi}.$$

Integrating, we get

$$m\dot{\phi} = m'\dot{\psi} + C.$$

Initially, when $t = 0$, $\dot{\phi} = 0$, $\dot{\psi} = 0$,

$$\therefore\ C = 0.$$

Hence, $m\dot{\phi} = m'\dot{\psi}$

$$\text{or,}\quad \frac{\dot{\phi}}{m'} = \frac{\dot{\psi}}{m} = \frac{\dot{\phi}+\dot{\psi}}{m+m'} = \frac{2\dot{\theta}}{m+m'} \quad \text{[From equation (4.194)]}$$

$$\therefore\ \dot{\phi} = \frac{2m'\dot{\theta}}{m+m'}$$

$$\text{and } \dot{\psi} = \frac{2m\dot{\theta}}{m+m'}.$$

Referred to the horizontal and vertical lines through O as axes the coordinates of centre C_1 and C_2 are (x, a) and $(x + 2a\sin\theta, a + 2a\cos\theta)$, respectively.

Since there is no horizontal force on the system, hence

$$\frac{d}{dt}[m\dot{x} + m'(\dot{x} + 2a\cos\theta\dot{\theta})] = 0.$$

Integrating, we get

$$m\dot{x} + m'(\dot{x} + 2a\cos\theta\dot{\theta}) = C_1.$$

Initially, when $t = 0$, $\dot{x} = 0$, $\dot{\theta} = 0$, $C_1 = 0$.
Hence $m\dot{x} + m'(\dot{x} + 2a\cos\theta\dot{\theta}) = 0$,

$$\text{or,}\quad \dot{x}(m+m') = -2am'\cos\theta\dot{\theta}$$

$$\therefore\quad \dot{x} = -\frac{2am'\cos\theta\dot{\theta}}{m+m'}. \tag{4.197}$$

The energy equation gives

$$\frac{1}{2}m\frac{2}{5}a^2\dot{\phi}^2 + \frac{1}{2}m(\dot{x}^2 + 0^2) + \frac{1}{2}m'\frac{2}{5}a^2\dot{\psi}^2 + \frac{1}{2}m'[(\dot{x} + 2a\cos\theta\dot{\theta})^2 + (-2a\sin\theta\dot{\theta})^2] = m'g(2a - 2a\cos\theta)$$

$$\text{or,}\quad \frac{2}{5}ma^2\dot{\phi}^2 + m\dot{x}^2 + \frac{2}{5}m'a^2\dot{\psi}^2 + m'[\dot{x}^2 + 4a^2\dot{\theta}^2 + 4a\dot{x}\cos\theta\dot{\theta}] = 4am'g(1 - \cos\theta)$$

$$\text{or,}\quad \frac{2}{5}ma^2\dot{\phi}^2 + \frac{2}{5}m'a^2\dot{\psi}^2 + \dot{x}^2(m+m') + 4a^2\dot{\theta}^2m' + 4a\dot{x}\cos\theta\dot{\theta}m' = 4am'g(1 - \cos\theta).$$

Putting the values of $\dot{\phi}$, $\dot{\psi}$ and $\dot{x}$, we get

$$\frac{2}{5}ma^2\frac{4m'^2\dot{\theta}^2}{(m+m')^2}+\frac{2}{5}\frac{m'a^24m^2\dot{\theta}^2}{(m+m')^2}+\frac{4a^2m'^2\cos^2\theta\dot{\theta}^2}{(m+m')^2}(m+m')$$

$$+4a^2\dot{\theta}^2m'-4a\cos\theta\dot{\theta}\frac{m'2am'\cos\theta\dot{\theta}}{m+m'}$$

$$=4am'g(1-\cos\theta)$$

$$\Rightarrow\quad \frac{8a^2\dot{\theta}^2mm'(m'+m)}{5(m+m')^2}-\frac{4a^2m'^2\cos^2\theta\dot{\theta}^2}{m+m'}+4a^2\dot{\theta}^2m'$$

$$=4am'g(1-\cos\theta)$$

$$\Rightarrow\quad \frac{4[2mm'-5m'^2\cos^2\theta+5m'(m+m')]a^2\dot{\theta}^2}{5(m+m')}$$

$$=4am'g(1-\cos\theta)$$

$$\Rightarrow\quad [2m-5m'\cos^2\theta+5m+5m']a\dot{\theta}^2$$

$$=5g(1-\cos\theta)(m+m')$$

$$\Rightarrow\quad (7m+5m'\sin^2\theta)a\dot{\theta}^2$$

$$=5(m+m')g(1-\cos\theta).$$

Problem 23. *A uniform solid cylinder rests on a smooth horizontal and on it there is placed a second equal cylinder touching it along its highest generator. If there is no slipping between the cylinders, and the system moves from rest, show that the cylinders separate when the plane of either axes makes an angle θ with vertical given by the equation $2\cos^3\theta+4\cos^2\theta-35\cos\theta+20=0$.*

Also show that until the cylinders separate, the same generators remain in contact.

Solution: Let C_1 be the centre, a the radius and m the mass of uniform cylinder resting initially on a smooth horizontal plane.

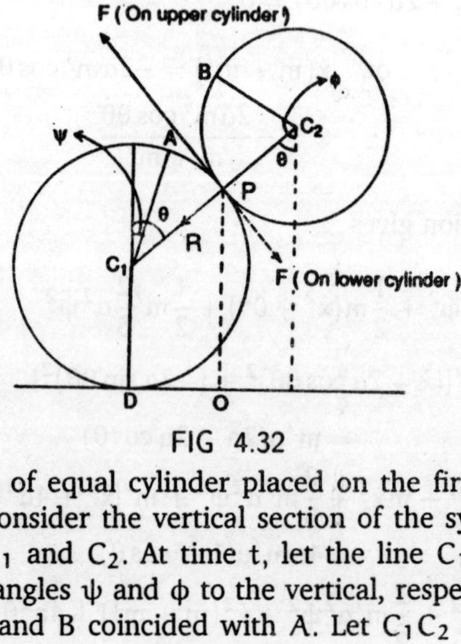

FIG 4.32

Let C_2 be the centre of equal cylinder placed on the first touching along its highest generator. We consider the vertical section of the system by the vertical plane through centres C_1 and C_2. At time t, let the line C_1A and C_2B fixed in the two cylinders make angles ψ and ϕ to the vertical, respectively while initially C_1A, C_2B were vertical and B coincided with A. Let C_1C_2 make angle θ to the vertical at time t.

Since there is no slipping and cylinders are equal,

$$\text{arc}\, C_1P = \text{arc}\, C_2P,$$

where P is the point of contact.

$$\therefore \quad \angle AC_1P = \angle BC_2P$$

$$\text{or,} \quad \theta - \psi = \phi - \theta$$

$$\text{or,} \quad 2\theta = \phi + \psi. \tag{4.198}$$

Considering motion of the two cylinders and taking moments about C_1 and C_2, we get

$$m\frac{a^2}{2}\ddot{\psi} = Fa \text{ (For lower cylinder)} \tag{4.199}$$

$$\text{and} \quad m\frac{a^2}{2}\ddot{\phi} = Fa \text{ (For upper cylinder).} \tag{4.200}$$

$\therefore \ \ddot{\psi} = \ddot{\phi}$.

Integrating, we get

$$\dot{\psi} = \dot{\phi} + C.$$

Initially, when $t = 0$, $\dot{\psi} = 0$, $\dot{\phi} = 0$, $C = 0$, $\dot{\psi} = \dot{\phi}$.

Integrating again we get $\psi = \phi + C_1$.

Initially, when $t = 0$, $\psi = 0$, $\phi = 0$.

$$\therefore \ C_1 = 0; \quad \psi - \phi.$$

Hence from equation (4.198), we get $2\theta = 2\phi$; $\theta = \phi = \psi$.

Hence A and B coincide with P, at time t. Hence the same generators remain in contact untill the cylinders separate.

Since cylinders are of same masses and $PC_1 = PC_2$, hence the point of contact of the two cylinders is the common centre of gravity of the system. Since there is no horizontal force on the system, therefore the common C.G. at P descends vertically. Let the vertical line through P cuts the horizontal plane at O, then O is the fixed point. Referred to O as origin, horizontal and vertical lines through O as coordinate axes, the coordinates of C_1 and C_2 are $(-a\sin\theta, a)$ and $(a\sin\theta, a + 2a\cos\theta)$.

The energy equation gives

$$\frac{1}{2}m\left[\frac{a^2}{2}\dot{\phi}^2 + a^2\cos^2\theta\dot{\theta}^2\right] + \frac{1}{2}m\left[\frac{a^2}{2}\dot{\psi}^2 + (a^2\cos^2\theta\dot{\theta}^2 + 4a^2\sin^2\theta\dot{\theta}^2)\right]$$
$$= mg(2a - 2a\cos\theta).$$

Since $\dot\psi = \dot\phi = \dot\theta$, hence, the energy equation becomes

$$\frac{m}{2}\left[\frac{a^2}{2}\dot\theta^2 + a^2\cos^2\theta\dot\theta^2\right] + \frac{m}{2}\left[\frac{a^2}{2}\dot\theta^2 + a^2\dot\theta^2(\cos^2\theta + 4\sin^2\theta)\right]$$
$$= 2amg(1 - \cos\theta)$$
$$\Rightarrow \quad ma^2\dot\theta^2\left[\frac{1}{4} + \frac{\cos^2\theta}{2} + \frac{1}{4} + \frac{\cos^2\theta}{2} + 2\sin^2\theta\right]$$
$$= 2amg(1 - \cos\theta)$$
$$\Rightarrow \quad a\dot\theta^2\left[\frac{1}{2} + \cos^2\theta + 2\sin^2\theta\right] = 2g(1 - \cos\theta)$$
$$\Rightarrow \quad \frac{a\dot\theta^2}{2}[1 + 2\cos^2\theta + 4(1 - \cos^2\theta)] = 2g(1 - \cos\theta)$$
$$\Rightarrow \quad a\dot\theta^2(5 - 2\cos^2\theta) = 4g(1 - \cos\theta). \tag{4.201}$$

Differentiating with respect to t, we get

$$2a\dot\theta\ddot\theta(5 - 2\cos^2\theta) + 4\cos\theta\sin\theta a\dot\theta^3 = 4g\sin\theta\dot\theta$$
$$\Rightarrow \quad a\ddot\theta(5 - 2\cos^2\theta) + 2\cos\theta\sin\theta a\dot\theta^2 = 2g\sin\theta. \tag{4.202}$$

Now equation of horizontal motion of the upper cylinder is

$$R\sin\theta - F\cos\theta = m\frac{d^2}{dt^2}(a\sin\theta)$$
$$\Rightarrow \quad R\sin\theta - F\cos\theta = ma(\cos\theta\ddot\theta - \sin\theta\dot\theta^2). \tag{4.203}$$

Eliminating F between equations (4.200) and (4.203), we get

$$R\sin\theta - \frac{ma}{2}\ddot\phi\cos\theta = ma(\cos\theta\ddot\theta - \sin\theta\dot\theta^2)$$
$$\Rightarrow \quad R\sin\theta = \frac{ma}{2}[3\cos\theta\ddot\theta - 2\sin\theta\dot\theta^2] \ (\because \ddot\phi = \ddot\theta).$$

The cylinders will separate, when $R = 0$.

$$\Rightarrow \quad \frac{ma}{2}(3\cos\theta\ddot\theta - 2\sin\theta\dot\theta^2) = 0$$
$$\Rightarrow \quad \ddot\theta = \frac{2}{3}\tan\theta\dot\theta^2. \tag{4.204}$$

From equation (4.201),

$$\dot\theta^2 = \frac{4g(1 - \cos\theta)}{a(5 - 2\cos^2\theta)}.$$

Hence from equation (4.204), we get

$$\ddot\theta = \frac{8g}{3a}\tan\theta\left(\frac{1 - \cos\theta}{5 - 2\cos^2\theta}\right).$$

Substituting the values of $\ddot{\theta}$ and $\dot{\theta}^2$ in equation (4.202), we get

$$a(5-2\cos^2\theta)\frac{8g}{3a}\tan\theta\left(\frac{1-\cos\theta}{5-2\cos^2\theta}\right)$$
$$+2a\sin\theta\cos\theta\frac{4g}{a}\left(\frac{1-\cos\theta}{5-2\cos^2\theta}\right)=2g\sin\theta$$

$$\Rightarrow\quad \frac{8g\sin\theta(1-\cos\theta)}{3\cos\theta(5-2\cos^2\theta)}[5-2\cos^2\theta+3\cos^2\theta]=2g\sin\theta$$
$$\Rightarrow\quad 4(1-\cos\theta)(5+\cos^2\theta)=3\cos\theta(5-2\cos^2\theta)$$
$$\Rightarrow\quad 4[5+\cos^2\theta-5\cos\theta-\cos^3\theta]=15\cos\theta-6\cos^3\theta$$
$$\Rightarrow\quad 20+4\cos^2\theta-20\cos\theta-4\cos^3\theta=15\cos\theta-6\cos^3\theta$$
$$\Rightarrow\quad 2\cos^3\theta+4\cos^2\theta-35\cos\theta+20=0.$$

Problem 24. *A hemisphere of mass M is free to slide with its base on a smooth horizontal table. A particle of mass m is placed on the hemisphere at an angular distance α from the vertex.*

Show that the radius to the point of contact at which the particle leaves the surface, makes with the vertical an angle θ and is given by the equation $m\cos^3\theta-(M+m)(3\cos\theta-2\cos\alpha)=0$. **[Agra 71, 83]**

Solution: In time t, let the hemisphere move through a distance on the horizontal plane. At time t let the particle be at the point P at an angular distance θ which was initially at an angular distance α from the vertex. The velocities of the particle m at P are $\dot{x}$ and $a\dot{\theta}$ along horizontal and the tangent at P and angle between them is θ.

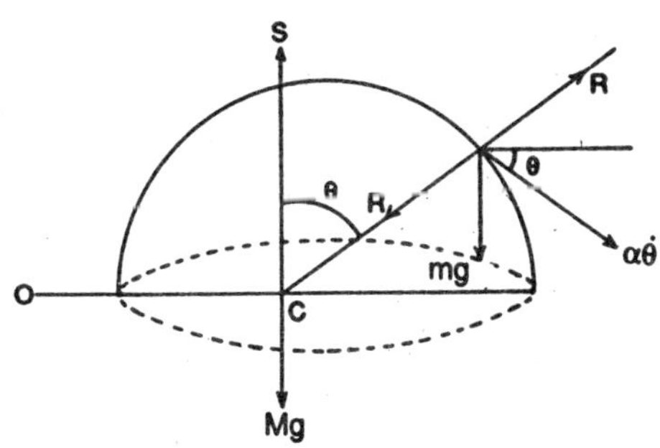

FIG 4.33

The coordinates of P referred to the horizontal and vertical lines through O as axes are $(x+a+a\sin\theta, a\cos\theta)$.

Since there is no horizontal force on the system,

$$\therefore\quad \frac{d}{dt}[M\dot{x}+\mu(\dot{x}+a\cos\theta\dot{\theta})]=0.$$

Integrating, we get $M\dot{x}+m(\dot{x}+a\cos\theta\dot{\theta})=C$.

Initially, when $\dot{x}=0$, $\dot{\theta}=0$.

$$\therefore\quad C=0.$$

$\therefore\ M\dot{x} + m(\dot{x} + a\cos\theta\dot{\theta}) = 0.$

$$\therefore\ \dot{x} = -\frac{ma\cos\theta\dot{\theta}}{M+m}. \tag{4.205}$$

Now kinetic energy of the hemisphere $= \frac{1}{2}M\dot{x}^2$ and kinetic energy of the particle

$$= \frac{1}{2}m\left[(\dot{x} + a\cos\theta\dot{\theta})^2 + (-a\sin\theta\dot{\theta})^2\right]$$
$$= \frac{1}{2}m\left[\dot{x}^2 + a^2\dot{\theta}^2 + 2a\dot{x}\dot{\theta}\cos\theta\right].$$

As there are no forces to turn the hemisphere, so there is no rotational energy. Hence the energy equation gives

$$\frac{1}{2}M\dot{x}^2 + \frac{1}{2}m(\dot{x}^2 + \dot{\theta}^2 + 2a\dot{x}\dot{\theta}\cos\theta) = mg(a\cos\alpha - a\cos\theta)$$

$$\Rightarrow (M+m)\dot{x}^2 + ma^2\dot{\theta}^2 + 2am\dot{\theta}\dot{x}\cos\theta = 2mga(\cos\alpha - \cos\theta)$$

$$\Rightarrow (M+m)\frac{m^2a^2\cos^2\theta\dot{\theta}^2}{(M+m)^2} + ma^2\dot{\theta}^2 + 2am\dot{\theta}\cos\theta\left(\frac{-ma\cos\theta\dot{\theta}}{M+m}\right)$$
$$= 2mga(\cos\alpha - \cos\theta) \text{ [From equation (4.205)]}$$

$$\Rightarrow\ a^2\dot{\theta}^2\left[1 - \frac{m\cos^2\theta}{M+m}\right] = 2ga(\cos\alpha - \cos\theta)$$

$$\Rightarrow\ a\dot{\theta}^2[M + m(1-\cos^2\theta)] = 2g(M+m)(\cos\alpha - \cos\theta)$$

$$\Rightarrow\ (M + m\sin^2\theta)a\dot{\theta}^2 = 2g(M+m)(\cos\alpha - \cos\theta). \tag{4.206}$$

The equation of horizontal motion of the hemisphere is

$$M\ddot{x} = -R\sin\theta$$
$$\Rightarrow\ M\frac{d}{dt}(\dot{x}) = -R\sin\theta.$$

The particle will leave the hemisphere if $R = 0$, i.e.,

$$\frac{d}{dt}(\dot{x}) = 0,\ \Rightarrow\ \frac{d}{dt}\left(-\frac{ma\cos\theta\dot{\theta}}{M+m}\right) = 0$$

$$\Rightarrow\ \cos\theta\ddot{\theta} - \dot{\theta}^2\sin\theta = 0$$

$$\Rightarrow\ \ddot{\theta}\cos\theta = \dot{\theta}^2\sin\theta. \tag{4.207}$$

Differentiating equation (4.206) with respect to t, we get from equation (4.206)

$$a\left[(M+m\sin^2\theta)2\dot{\theta}\ddot{\theta}+2m\sin\theta\cos\theta\dot{\theta}^3\right]$$
$$=2g(M+m)\sin\theta\dot{\theta},$$

$$\Rightarrow\quad a\left[(M+m\sin^2\theta)\ddot{\theta}+m\sin\theta\cos\theta\dot{\theta}^2\right]=g(M+m)\sin\theta$$

$$\Rightarrow\quad (M+m\sin^2\theta)\frac{\dot{\theta}^2\sin\theta}{\cos\theta}+m\sin\theta\cos\theta\dot{\theta}^2=\frac{g}{a}(M+m)\sin\theta$$

[From equation (4.207)]

$$\Rightarrow\quad [M+m\sin^2\theta+m\cos^2\theta]\dot{\theta}^2=\frac{g}{a}(M+m)\cos\theta$$

$$\Rightarrow\quad (M+m)\dot{\theta}^2=\frac{g}{a}(M+m)\cos\theta$$

$$\Rightarrow\quad \dot{\theta}^2=\frac{g}{a}\cos\theta.$$

Hence from equation (4.206), we get

$$(M+m\sin^2\theta)g\cos\theta=2g(M+m)(\cos\alpha-\cos\theta)$$
$$\text{or,}\quad [M+m-m\cos^2\theta]\cos\theta=2(M+m)(\cos\alpha-\cos\theta)$$
$$\text{or,}\quad (M+m)\cos\theta-m\cos^3\theta=2(M+m)\cos\alpha-2(M+m)\cos\theta$$
$$\text{or,}\quad m\cos^3\theta-(M+m)(3\cos\theta-2\cos\alpha)=0.$$

4.14 Exercise 4

Type I

1. One end of a thread, which is wound on a reel, is fixed, and the reel falls in a vertical line, its axis being horizontal and the unwound part of the thread being vertical. If the reel be a solid cylinder of radius a and weight w, show that the acceleration of the centre of the real is $\frac{2}{3}g$ and the tension of the thread is $\frac{w}{3}$.

2. A circular cylinder, whose centre of inertia is at a distance c from its axis, rolls on a horizontal plane. If it be just started from a position of unstable equilibrium, show that the normal reaction of the plane when the centre of mass is in its lowest position is $\left[1+\frac{4c^2}{(a-c)^2+K^2}\right]$ times its weight, where K is the radius of gyration about an axis through the centre of mass. **[Kanpur 86]**

3. Two equal cylinders, each of mass m, are bound together by an elastic string, whose tension is T, and roll with their axes horizontal down a rough plane of inclination α. Show that their acceleration is $\frac{2}{3}g\sin\alpha\left(1-\frac{2\mu T}{mg\sin\alpha}\right)$, where μ is the coefficient of friction between the cylinders. **[Kanpur 82]**

Type II

4. A uniform rod is held at an inclination 45° to the vertical with one end in contact with a horizontal table whose coefficient of friction is μ. If it is then released, show that it will commence to slide if $\mu<\frac{3}{5}$.

5. A uniform beam lies on a rough horizontal table at right angles to the edge, and is held so that one third of its length is contact with the table. Prove that after it is released it will begin to slide over the edge of the table when it has turned through an angle $\tan^{-1}\left(\frac{1}{2}\mu\right)$, μ being the coefficient of friction between the table and the beam. **[CU 97H]**

6. The lower end of a uniform rod, inclined, initially at an angle α to the horizon, is placed on a smooth horizontal table. A horizontal force is applied to its lower end of such a magnitude that the rod rotates in a vertical plane with constant angular velocity ω. Show that when the rod is inclined at an angle θ to the horizon, the magnitude of the force is $mg\cot\theta - ma\omega^2\cos\theta$, where m is the mass of the rod. **[Kanpur 83]**

7. A uniform rod is held nearly vertically with one end resting on an imperfectly by rough plane. It is released from rest and falls forwards. The inclination to the vertical at any instant is θ. Prove that:

   (a) If the coefficient of frictions is less than a certain finite amount, the lower end of the rod will slip backwards before $\sin^2\frac{\theta}{2} = \frac{1}{6}$.

   (b) However great the coefficient of friction may be, the lower end will begin to slip forwards at a value of $\sin^2\frac{\theta}{2}$ between $\frac{1}{6}$ and $\frac{1}{3}$. **[Agra 79]**

8. A uniform rod of length $2a$ is placed with one end in contact with a horizontal table, and is then at an inclination α to the horizon and is allowed to fall. When it becomes horizontal, show that its angular velocity is $\sqrt{\frac{3g}{2a}\sin\alpha}$, whether the plane be perfectly smooth or perfectly rough. Show also that the end of the rod will not leave the curve in either case. **[CU 92H, 99H]**

Type IV

9. If a sphere be projected up an inclined plane, for which $\mu = \frac{1}{7}\tan\alpha$, with velocity V and an angular velocity Ω (in the direction in which it would roll up) and if $V > a\Omega$, show that the friction acts downwards at first, and upwards afterwards, and prove that the whole time during which the sphere rises is $\frac{17V+4a\Omega}{18g\sin\alpha}$. **[VU 89H]**

10. A hoop is projected with velocity V down an inclined plane of inclination α, the coefficient of friction being $\mu(> \tan\alpha)$. It has initially such a backward spin Ω that after a time t_1 it starts moving uphill and continues to do so for a time t_2 after which it once more descends. The motion being in a vertical plane at right an angles to the given inclined plane, show that

$$(t_1 + t_2)g\sin\alpha = a\Omega - V.$$

11. A uniform sphere, of radius a, is rotating about a horizontal diameter with angular velocity Ω and is gently placed on a rough plane which is inclined at angle α to the horizontal, the sense of rotation being such as to tend to cause the sphere to move up the plane along the line of greatest slope. Show that, if the coefficient of friction be $\tan\alpha$, the centre of the sphere will remain at rest for a time $\frac{2a\Omega}{5g\sin\alpha}$ and will then move downwards with acceleration $\frac{5}{7}\sin\alpha$.

If the body be a thin circular hoop instead of sphere, show that the time is $\frac{a\Omega}{g\sin\alpha}$ and the acceleration is $\frac{1}{2}g\sin\alpha$.

12. A homogeneous sphere of radius a, rotating with an angular velocity ω about horizontal diameter is gently placed on a table whose coefficient of friction is μ. Show that there will be slipping at the point of contact for a time $\frac{2\omega a}{7\mu g}$ and then the sphere will roll with angular velocity $\frac{2\omega}{7}$

[Garhwal 82; Agra 78, CU 91H, 93H, 02H]

13. Three uniform spheres, each of radius a and of mass m, attract one another according to the l of the inverse square of the distance. Initially, they are placed on a perfectly rough horizontal plane with their centres forming a triangle whose sides are each of length $4a$. Show that the velocity of their centres when they collide is $\sqrt{\gamma\frac{5m}{14a}}$, where γ is the constant of gravitation.

Type V

14. A napkin ring, of radius a, is projected forward on a rough horizontal table with a linear velocity u and a backward spin Ω which is $> \frac{u}{a}$. Find the motion and show that the ring will return to the point of projection in time $\frac{(\mu+a\Omega)^2}{4\mu g(a\Omega-u)}$, where μ is the coefficient of friction. What happens if $u > a\Omega$. **[Agra 65]**

[*Ans.* The rolling will commence before the forward motion ceases.]

Type VI

15. A homogeneous solid hemisphere, of mass M and radius a, rests with its vertex in contact with a rough horizontal plane, and a particle of mass m, is placed on its base, which is smooth at a distance c from the centre. Show that the hemisphere will commence to roll or slide according as the coefficient of μ is greater or less then $\frac{25mac}{26(M+m)a^2+40mc^2}$.

16. If a uniform semi-circular wire be placed in a vertical plane with one extremity on a rough horizontal plane, and the diameter through that extremity vertical, show that the semi-circle will begin to roll or slide according as μ be greater than or less than $\frac{\Omega}{\Omega^2-2}$.

If μ has this value, prove that wire will roll.

Type VII

17. A uniform sphere is placed on the top of a fixed rough circular cylinder whose generators are horizontal. Show that, if slightly displaced, it will roll on the cylinder until it reaches a plane, where the inclination of the tangent plane to the horizon is given by $2\sin\theta = \mu(17\cos\theta - 10)$, μ being the coefficient of friction.

18. A solid uniform sphere, resting on the top of another fixed. Sphere is slightly displaced and begins to roll down. If the plane through their axes makes an

angle α with the vertical when first sphere is at rest, show that it will step when the common normal makes with the vertical an angle given by

$$K^2 \sin\theta = \mu[(K^2 + 3b^2)\cos\theta - 2b^2\cos\alpha],$$

where b is the radius of the moving sphere and K is the radius of gyration.

The upper sphere will leave the fixed sphere if

$$\theta = \cos^{-1}\left(\frac{2b^2\cos\alpha}{K^2+3b^2}\right).$$

19. A uniform sphere of radius a is gently placed on the top of a thin vertical pole of height $h(> a)$ and then allowed to fall over. Show that however rough the pole may be the sphere will slip on the pole before it finally falls off it.

[MU 71; Agra 64]

Type VIII

20. A disc rolls on the inside of a fixed hollow circular cylinder whose axis is horizontal, the plane of the disc being vertical and perpendicular to the axis of the cylinder. If, when in the lowest position, its centre is moving with a velocity $\sqrt{\frac{8}{3}g(a-b)}$, show that the centre of the disc will describe an angle ϕ about the centre of the cylinder in time $\sqrt{\frac{3(a-b)}{2g}}\log\tan\left(\frac{\Omega}{4}+\frac{\phi}{4}\right)$.

[Vikram 64]

21. A circular cylinder of radius a and radius of gyration K rolls without slipping inside a fixed, hollow cylinder of radius b. Show that the plane through their axes moves a circular pendulum of length $(b-a)\left(1+\frac{K^2}{a^2}\right)$. [Vikram 62]

22. A solid homogeneous sphere is rolling on the inside of a fixed hollow sphere, the two centres being always in the same vertical plane. Show that the smaller sphere will make complete revolutions if, when it is in its lowest position, the pressure on it is greater than $\frac{34}{7}$ times its own weight.

23. A cylinder of radius a, lies within a rough fixed cylinder cavity of radius $2a$. The centre of gravity of the cylinder is at a distance c from the axis, and the initial state is that of stable equilibrium at the lowest point of the cavity. Show that the smallest angular velocity with which the cylinder must be started so that it may roll right round the cavity is given by $\Omega^2(a+c) = g\left[1+\frac{4(a+c)^2}{(a-c)^2+K^2}\right]$, where K is the radius of gyration about the centre of gravity.

Find also the normal reaction between the cylinders in any position. [IAS 74]

[*Ans.* $R = mg\cos\theta + m[c\sin 2\theta\ddot{\theta} + (a + c\cos 2\theta)\dot{\theta}^2]$]

Type IX

24. A thin hollow cylinder of radius a and mass M is free to turn about its axis which is horizontal, and a similar cylinder of radius b and mass m rolls inside it without slipping, the axes of the two cylinders being parallel. Show that when the plane of the two axes is inclined at an angle θ to the vertical, the angular velocity of the larger cylinder is given by $a^2(M+m)(2M+m)\Omega^2 = 2gm^2(a-b)(\cos\theta - \cos\alpha)$, provided both the cylinders are at rest when $\theta = \alpha$.

25. A rough cylinder of mass $2nm$, capable of motion about its horizontal axes, has a particle of mass m and coefficient of friction μ placed on it vertically above the axis. The system is then slightly disturbed. Show that the particle will slip on the cylinder after it has moved through an angle θ given by $(n+3)\cos\theta - 2 = \frac{n\sin\theta}{\mu}$.

    Hence find the value of θ, when the particle will leave the curve. **[IAS 52]**

    [*Ans.* $\theta = \cos^{-1}\left(\frac{2}{n+3}\right)$]

26. A circular cylinder of radius a and of radius of gyration K rolls without slipping inside a hollow cylinder of radius b which is free to move about its axis. Show that the plane through their axes will move like a simple circular pendulum of length $(b-a)(1+n)$, where $n = \frac{K^2/a^2}{1+\frac{b^2}{a^2}} \frac{mk^2}{MK^2}$, where k and K are the radii of gyration of the inner and outer cylinder, respectively about their axes and m and M are their masses. **[Vikram 62]**

27. A uniform thin hollow cylinder of radius a and mass M is free to turn about its axis which is horizontal. A uniform solid cylinder of mass m and radius b rolls without slipping on the inner surface of the first cylinder. Show that, if the plane containing the axes of the cylinders oscillates through an angle α, its period is same as that of a simple pendulum of length $\frac{(a-b)(3M+m)}{2M+m}$ oscillating through the same angle α.

Type X

28. Two equal perfectly rough spheres are placed in unstable equilibrium, one on the top of the other, the lower resting on a perfectly smooth table. A slight disturbance being given, show that the spheres will continue to touch each other of the same points, and that if θ be the inclination to the vertical of the straight line joining the centres, than $(K^2 + a^2 + a^2\sin^2\theta)\dot{\theta}^2 = 2ga(1-\cos\theta)$, where K is the radius of gyration of each sphere about an axis through its centre.

29. Two unequal smooth spheres one placed on the top of the other are in unstable equilibrium, the lower sphere resting on a smooth table. The system is slightly disturbed, show that the spheres will separate when the line joining their centres makes an angle θ with the vertical given by the equation $m\cos^3\theta = (M+m)(3\cos\theta - 2)$, where M is the mass of the lower and, m that of the upper sphere. **[Agra 66, 67, 71, 83]**

30. A perfectly rough ball is at rest within a hollow cylindrical garden roller and the roller is then drawn along a level path with uniform velocity v. If $v^2 > \frac{27}{g}(b - a)$, show that the ball will roll completely round the inside of the roller, a and b being the radii of the ball and roller. [Garhwal 84]

Chapter 5

Motion in Two Dimensions (Under Impulsive Forces)

5.1 Impulsive Force

A force of a very great magnitude acting on a body for a very short time is called an *impulsive force.*

Impulse [CU 93H, 01H, 03H]

If an impulsive force X is acting on a body for a very short time T, the impulse X' is defined as

$$X' = \int_0^T X dt.$$

It may be noted that impulse is not a force.

5.2 Equations of Motion under Impulsive Forces

The equations of motion of a rigid body under finite forces are

$$M\frac{d^2\bar{x}}{dt^2} = \Sigma X, \text{ i.e., } \frac{d}{dt}\left(M\frac{d\bar{x}}{dt}\right) = \Sigma X \tag{5.1}$$

$$M\frac{d^2\bar{y}}{dt^2} = \Sigma Y, \text{ i.e., } \frac{d}{dt}\left(M\frac{d\bar{y}}{dt}\right) = \Sigma Y \tag{5.2}$$

$$\text{and } MK^2\ddot{\theta} = \Sigma(x'Y - y'X),$$

$$\text{i.e., } \frac{d}{dt}(MK^2\dot{\theta}) = \Sigma(x'Y - y'X), \tag{5.3}$$

where $(\bar{x}, \bar{y})$ are the coordinates of C.G. at G and X, Y the components of the force parallel to the axes, acting on the particle m at the point $P(x, y)$ whose coordinates referred to the parallel axes through G are (x', y').

Now if the forces are impulsive, then the above equations of motion do not remain in these forms, but are transformed. Thus, when the forces are impulsive, let T be the short-time during which the impulsive force acts on the body, then integrating equation (5.1), we have

$$\left[M\frac{d\bar{x}}{dt}\right]_0^T = \Sigma\int_0^T X\,dt, \text{ or, } \left[M\frac{d\bar{x}}{dt}\right]_0^T = \Sigma X', \tag{5.4}$$

where $X' = \int_0^T x\,dt$ is the impulsive force acting at the point $P(x, y)$ parallel to x-axis.

Let u and u' be the velocities of the particle parallel to x-axis of the body just before and just after the application of the impulsive forces, then from equation (5.4), we get

$$M(u' - u) = \Sigma X'. \tag{5.5}$$

Similarly, if v and v' are the velocities of the C.G. of the body parallel to y-axis, just before and just after the action of an impulsive force, then integrating equation (5.2), we get

$$\left[M\frac{d\bar{y}}{dt}\right]_v^{v'} = \Sigma\int_0^T Y\,dt$$

$$\therefore\; M(v' - v) = \Sigma Y'. \tag{5.6}$$

Equations (5.5) and (5.6) state that the change in momentum of the total mass M of the body supposed collected at the centre of gravity in any direction is equal to the sum of the impulses in that direction.

If ω and ω' are angular velocities of the body just before and just after the action of impulsive forces, then integrating equation (5.3), we get

$$[MK^2\dot{\theta}]_0^T = \Sigma\left[x'\int_0^T Y\,dt - y'\int_0^T X\,dt\right]$$

$$\text{or, } MK^2(\omega' - \omega) = \Sigma(x'Y' - y'X'). \tag{5.7}$$

The equation (5.7) states that the change in the moment of momentum (angular momentum) of the body about the centre of gravity is equal to the sum of moments of the impulses of the forces about the centre of gravity.

Equations (5.5), (5.6) and (5.7) are the equations of motion in two dimensions of a body of mass M under the action of impulsive forces.

5.3 An imperfectly elastic sphere impinging on a fixed plane

An imperfectly elastic uniform sphere, rotating with an angular velocity ω about an axis perpendicular to the plane of motion of its centre, impinges on a horizontal plane. Find the resulting changes in its motion.

Case I. *First of all we assume that the plane is rough enough to prevent any sliding.*

Let u and v be the components of the linear velocities in the two assigned directions and ω the angular velocity just before impact while u', v' be the components and ω' be the angular velocity after the impact.

Let F be the impulsive friction and R the normal impulsive reaction, then the equations of motion are

$$M(u' - u) = -F \tag{5.8}$$

$$M(v' + v) = R \tag{5.9}$$

$$\text{and } MK^2(\omega' - \omega) = Fa. \tag{5.10}$$

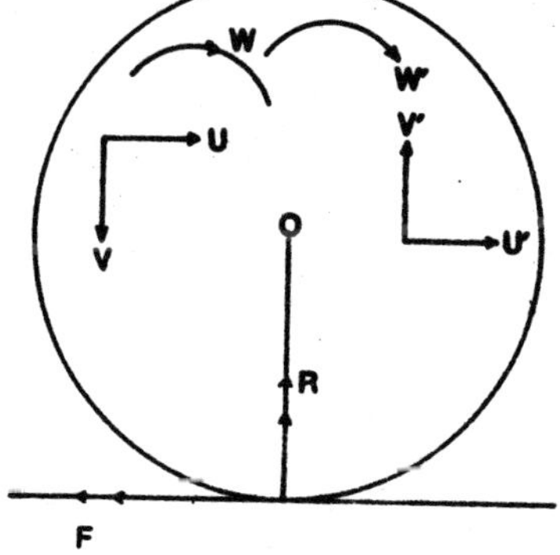

FIG 5.1

Here

$$K^2 = \frac{2}{5}a^2$$

$$M\frac{2}{5}a^2(\omega' - \omega) = Fa. \tag{5.11}$$

Since P, the point of contact, is instantaneously reduced to rest, there being no sliding, hence

$$u' - a\omega' = 0. \tag{5.12}$$

Also by Newton's law,

$$v' = ev, \tag{5.13}$$

where e is the coefficient of restitution.

From equations (5.9) and (5.11), we get

$$M(u' - u) = -\frac{2}{5}Ma(\omega' - \omega)$$

$$\Rightarrow \quad u' - u = -\frac{2}{5}a(\omega' - \omega)$$

$$\Rightarrow \quad a\omega' - u = -\frac{2}{5}a(\omega' - \omega)$$

$$\Rightarrow \quad a\omega'\left(1 + \frac{2}{5}\right) = u + \frac{2}{5}a\omega$$

$$\Rightarrow \quad a\omega'\frac{7}{5} = u + \frac{2}{5}a\omega$$

$$\therefore \quad u' = a\omega' = \frac{1}{7}(5u + 2a\omega). \tag{5.14}$$

$$F = M(u - u') = M\left[u - \frac{1}{7}(5u + 2a\omega)\right] = \frac{M}{7}(2u - 2a\omega) \tag{5.15}$$

$$\text{and } R = M(v' + v) = M(ev + v) = Mv(e + 1). \tag{5.16}$$

Case II. *If $u = a\omega$. In this case, $F = 0$, i.e., there is no frictional force.*

Hence from equation (5.9), $u' = u$.

Also from equation (5.11), we have $\omega' = \omega$.

Hence in this case u and ω remain unchanged.

Case III. *If $u < a\omega$. In this case, $F < 0$. Therefore, F acts in the direction of horizontal velocity u and $\omega' < \omega$, $u' > u$. Also velocity of A (before impact) $= u - a\omega$ is negative, i.e., velocity of A is in the direction opposite to the direction of horizontal velocity u. In this case, the angular velocity after the impact is decreased while the horizontal velocity is increased.*

Case IV. *If $u > a\omega$. In this case, $F > 0$, i.e., F acts in the direction opposite to the horizontal velocity u. Here $u' < u$ and $\omega' > \omega$. Thus when the point of contact before impact is moving in the direction of u, the angular velocity is increased while the horizontal velocity is decreased.*

Case V. *If angular velocity before impact is in anti-clockwise direction. In this case, the sign of ω is changed. So $u' = a\omega' = \frac{1}{7}(5u - 2a\omega)$ and $F = \frac{2}{7}M(u + a\omega)$.*

Now if $u = \frac{2}{5}a\omega$, then $u' = 0$, $\omega' = 0$.

Thus, the sphere rebounds from the plane vertically with no spin.

Also if $u < \frac{2}{5}a\omega$, then $u' < 0$.

Thus, the sphere rebounds towards the direction from which it came.

In both of these cases $v' = ev$ and $R = M(1 + e)v$.

Case VI. *Case of pure rolling.*

In the case of pure rolling we must have $F < \mu R$ in order that the point of contact may be instantaneously brought to rest, i.e.,

$$\frac{2}{7}(u - a\omega) < \mu(1 + e)v.$$

But if $\frac{2}{7}(u - a\omega) > \mu(1 + e)v$, the friction is not sufficient to bring the point of contact P to instantaneous rest, and hence in this case equation (5.12) will not hold good. Also instead of F, maximum frictional force μR will act. Hence the equations of motion (5.9), (5.10) and (5.11) becomes

$$M(u' - u) = -\mu R \tag{5.17}$$

$$M(v' + v) = R \tag{5.18}$$

$$\text{and} \quad MK^2(\omega' - \omega) = \mu Ra. \tag{5.19}$$

From equations (5.17) and (5.18), we have

$$\begin{aligned} M(u' - u) &= -\mu M(v' + v) \\ \text{or, } u' - u &= -\mu(v' + v) \\ \text{or, } u' &= u - \mu(v' + v) = u - \mu(ev + v) \text{ [From equation (5.13)]} \\ &= u - \mu v(e + 1). \end{aligned}$$

Again from equations (5.18) and (5.19), we get

$$
\begin{aligned}
MK^2(\omega' - \omega) &= \mu a M(v' + v) \\
\text{or, } \omega' - \omega &= \frac{\mu a}{K^2}(v' + v) \\
\text{or, } \omega' &= \omega + \frac{\mu a}{\frac{2a^2}{5}}(v' + v) \quad \left[\because K^2 = \frac{2a^2}{5}\right] \\
\text{or, } \omega' &= \omega + \frac{5\mu}{2a}(v' + v) = \omega + \frac{5\mu}{2a}(ev + v) \text{ [From equation (5.13)]} \\
&= \omega + \frac{5\mu}{2a}v(e + 1).
\end{aligned} \tag{5.20}
$$

If $F < \mu R$, then from equation (5.20), we have

$$\tfrac{2}{7}M(u + a\omega) < M\mu(ev + v) = \mu v(e + 1)M,$$

$$\text{i.e., } \tfrac{2}{7}(u + a\omega) < \mu(e + 1)v.$$

But if

$$\tfrac{2}{7}(u + a\omega) > \mu v(1 + e). \tag{5.21}$$

The friction will be insufficient and then we have similar type of equations (5.17), (5.18), (5.19) with sign of ω changed. Solving these, we get

$$
\begin{aligned}
u' &= u - \mu v(1 + e), \\
v' &= ev \\
\text{and } \omega' &= \frac{5\mu}{2a}v(1 + e) - \omega.
\end{aligned}
$$

Hence from equation (5.21) in this case, it will be possible for u to be less than $\mu v(1+e)$, if ω be large enough. Hence u' will be negative, if the ball has sufficiently large enough under cut, i.e., the ball can rebound backwards.

5.4 Work done by an Impulse

THEOREM. *A body whose mass is M, is acted upon at a given point P by a blow of impulse X. If v and v_1 be the velocities of P in the direction of X just before and just after the action of X, show that the change in the kinetic energy of the body, that is the work done on it by the impulse is $\frac{1}{2}(v + v_1)X$.*

Proof. Let G be the centre of gravity of the body of mass M, which is acted on by a blow of impulse X at the point P.

We consider the cross-section of the body through the C.G. at G and the direction of impulse X as shown in the figure.

Let a line through G parallel to X be taken as x-axis and the line perpendicular to X as y-axis. Let (x, y) be the coordinates of the point P with respect to these axes through G.

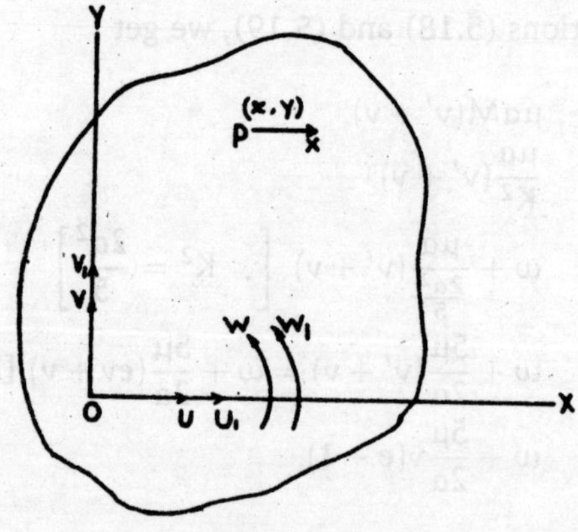

FIG 5.2

Just before the impact, let u and v be the velocities of G parallel to x-axis and y-axis, respectively and ω the angular velocity of the body. Just after the impact, let these velocities be u_1, v_1 and ω_1, respectively.

The equations of motion of the body are

$$M(u_1 - u) = X \tag{5.22}$$

$$M(v_1 - v) = 0 \tag{5.23}$$

$$\text{and} \quad MK^2(\omega_1 - \omega) = -yX. \tag{5.24}$$

Now

$$\begin{aligned} V &= \text{velocity of P parallel to GX before impulse} \\ &= \text{velocity of G parallel to GX + velocity of P relative to G} \\ &= u - y\omega. \end{aligned}$$

Similarly,

$$v_1 = u_1 - y\omega_1.$$

Now change in kinetic energy of the body, i.e., the work done by the impulse

$$\begin{aligned} &= \tfrac{1}{2}M(u_1^2 + v_1^2 + K^2\omega_1^2) - \tfrac{1}{2}M(u^2 + v^2 + K^2\omega^2) \\ &= \tfrac{1}{2}M(u_1^2 - u^2) + \tfrac{1}{2}M(v_1^2 - v^2) + \tfrac{1}{2}MK^2(\omega_1^2 - \omega^2) \\ &= \tfrac{1}{2}M(u_1 - u)(u_1 + u) + \tfrac{1}{2}M(v_1 - v)(v_1 + v) + \tfrac{1}{2}MK^2(\omega_1 - \omega)(\omega_1 + \omega) \\ &= \tfrac{1}{2}X(u_1 + u) + \tfrac{1}{2}0(v_1 + v) - \tfrac{1}{2}yX(\omega_1 + \omega) \\ &= \tfrac{1}{2}[(u - y\omega) + (u_1 - y\omega_1)]\,X \\ &= \tfrac{1}{2}(v + v_1)X. \end{aligned}$$

5.5 Worked out Examples

Type I: Problems based on Section 5.1

Problem 1. AB, BC *are two equal similar rods freely hinged at* B *and lie in a straight line on a smooth table. The end* A *is struck by a blow perpendicular to* AB, *show that the resulting velocity of* A *is* $3\frac{1}{2}$ *times of* B.

Solution: Let AB, BC be two equal rods each of mass M and length $2a$, freely hinged at B and resting in a straight line on the smooth table.

Let P be the impulsive force applied at A by a blow. After the blow, let u_1, ω_1 and u_2, ω_2 be the linear and angular velocities of the rods AB and BC, respectively. When a blow is struck, there will be an impulsive action at B. Let Q be the impulsive action at the joint B in opposite directions on the two rods.

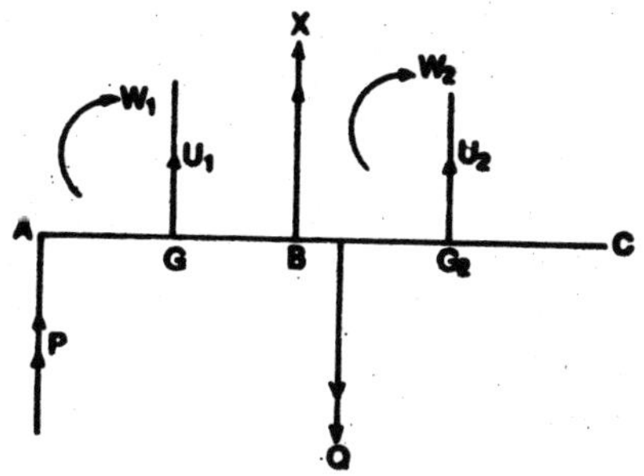

FIG 5.3

The equations of motion of the rods AB and BC are for the rod AB,

$$M(u_1 - 0) = (P+Q)a \tag{5.25}$$

$$\text{and } M\frac{a^2}{3}(\omega_1 - 0) = (P-Q)a \tag{5.26}$$

and for the rod BC,

$$M(u_2 - 0) = -Q \tag{5.27}$$

$$M\frac{a^2}{3}(\omega_2 - 0) = -Qa. \tag{5.28}$$

Since the rods are connected at B.

∴ the velocity of the point B of the rod AB = velocity of the point B of the rod BC,

$$\text{i.e., } u_1 - a\omega_1 = u_2 + a\omega_2 \tag{5.29}$$

From equations (5.25), (5.26), (5.27) and (5.28), we get

$$u_1 = \tfrac{1}{M}(P+Q), \quad a\omega_1 = \tfrac{3}{M}(P-Q),$$
$$u_2 = \tfrac{1}{M}Q, \quad a\omega_2 = -\tfrac{3}{M}Q.$$

Hence from equation (5.29), we have

$$\frac{1}{M}(P+Q) - \frac{3}{M}(P-Q) = -\frac{1}{M}Q - \frac{3}{M}Q$$
$$\Rightarrow \frac{P+Q-3P+3Q}{M} = -\frac{4Q}{M}$$
$$\Rightarrow -2P+4Q = -4Q, \text{ or, } 8Q = 2P$$
$$\Rightarrow P = 4Q. \therefore Q = \frac{1}{4}P.$$

From equation (5.25)

$$u_1 = \frac{1}{M}(P+Q) = \frac{1}{M}\left(P + \frac{P}{4}\right) = \frac{1}{M}\frac{5}{4}P = \frac{5}{4}\frac{P}{M}.$$

Also from equation (5.26)

$$a\omega_1 = \frac{3}{M}\left(P - \frac{1}{4}P\right) = \frac{3}{M}\frac{3}{4}P = \frac{9}{4}\frac{P}{M}$$

$$\therefore \quad \frac{\text{velocity of A}}{\text{velocity of B}} = \frac{u_1 + a\omega_1}{u_1 - a\omega_1} = \frac{\frac{5}{4}\frac{P}{M} + \frac{9}{4}\frac{P}{M}}{\frac{5}{4}\frac{P}{M} - \frac{9}{4}\frac{P}{M}} = \frac{14}{-4} = -\frac{7}{2} = -3\frac{1}{2}.$$

Hence velocity of A $= 3\frac{1}{2}$ times the velocity of B.

Problem 2. *Two rods AB and BC of length 2a and 2b and of masses proportional to their lengths, are freely jointed at B and are lying in a straight line. A blow is communicated to the end A. Show that the resulting kinetic energy when the system is free to the kinetic energy when C is fixed as* $\frac{(4a+3b)(3a+4b)}{12(a+b)^2}$.

Solution: Let ρ be the mass per unit length of the rod.

Let m_1 and m_2 be the masses of the rods AB and BC. Then $m_1 = 2a\rho$, $m_2 = 2b\rho$.

Let P be the impulse applied at A at right angles to the rod AB. Just after the blow, let u_1, ω_1 and u_2, ω_2 be the linear and angular velocities of the two rods AB and BC, respectively.

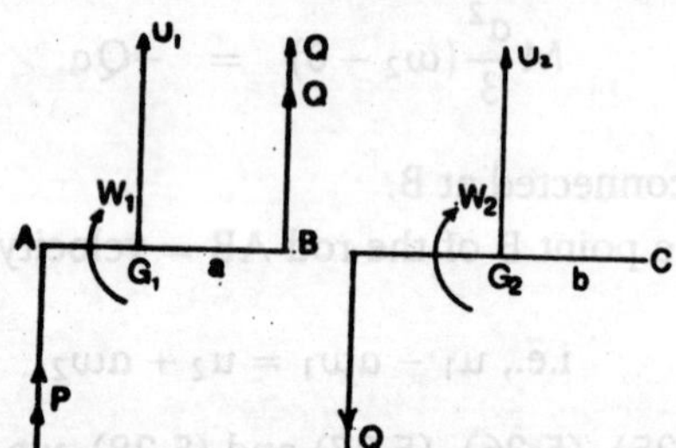

FIG 5.4

Since P is the impulse applied at A at right angles to AB, hence there will be an impulsive action between the two rods at B; let it be Q, in opposite directions on the two rods.

Case I. *When the system is free.*

The equations of motion are for the rod AB,

$$m_1(u_1 - 0) = P + Q \tag{5.30}$$

$$\text{and} \quad m_1\frac{a^2}{3}(\omega_1 - 0) = Pa - Qa \tag{5.31}$$

and for the rod BC,

$$m_2(u_2 - 0) = -Q \tag{5.32}$$

$$\text{and} \quad m_2 \tfrac{1}{3} b^2(\omega_2 - 0) = -Qb. \tag{5.33}$$

From equations (5.30), (5.31), (5.32) and (5.33), we get

$$u_1 = \frac{P+Q}{m_1}, \quad \omega_1 = \frac{3(P-Q)}{am_1}$$
$$u_2 = -\frac{Q}{m_2}, \quad \omega_2 = -\frac{Q3}{bm_2}.$$

Since the rods are connected at B,

∴ the velocity of the point B of the rod AB = velocity of the point B of the rod BC, i.e., $u_1 - a\omega_1 = u_2 + b\omega_2$

$$\frac{P+Q}{m_1} - \frac{3(P-Q)}{m_1} = -\frac{Q}{m_2} - \frac{3Q}{m_2}$$

$$\text{or,} \quad -\frac{2P+4Q}{m_1} = -\frac{4Q}{m_2}, \text{ or, } -\frac{P+2Q}{m_1} = -\frac{2Q}{m_2}$$

$$\text{or,} \quad -Pm_2 + 2Qm_2 = -2Qm_1, \text{ or, } 2Q(m_1+m_2) = Pm_2$$

$$\text{or,} \quad Q = \frac{Pm_2}{2(m_1+m_2)} = \frac{P2b\rho}{2\cdot 2\rho(a+b)} = \frac{Pb}{2(a+b)}.$$

$$u_1 = \frac{P+Q}{m_1} = \frac{P + \frac{Pb}{2(a+b)}}{m_1} = \frac{P[2a+2b+b]}{2(a+b)m_1}$$
$$= \frac{P(2a+3b)}{2(a+b)2a\rho} = \frac{P(2a+3b)}{4a\rho(a+b)}$$

$$\omega_1 = \frac{3}{am_1}(P-Q) = \frac{3}{am_1}\left[P - \frac{Pb}{2(a+b)}\right]$$
$$= \frac{3P}{am_1}\left[\frac{2a+b}{2(a+b)}\right] = \frac{3P(2a+b)}{a2a\rho2(a+b)} = \frac{3(2a+b)P}{4a^2\rho(a+b)}$$

$$u_2 = -\frac{Q}{m_2} = -\frac{bP}{2(a+b)2b\rho} = -\frac{P}{4\rho(a+b)}$$

$$\omega_2 = -\frac{3}{bm_2}\frac{Pb}{2(a+b)} = -\frac{3P}{2b\rho2(a+b)} = -\frac{3P}{4b\rho(a+b)}.$$

So in this case total kinetic energy

$$= \text{K.E. of the rod AB} + \text{K.E. of the rod BC}$$
$$= \left[\frac{1}{2}m_1\frac{a^2}{3}\omega_1^2 + \frac{1}{2}m_1u_1^2\right] + \left[\frac{1}{2}m_2\frac{b^2}{3}\omega_2^2 + \frac{1}{2}m_2u_2^2\right]$$
$$= \frac{1}{2}m_1\left[\frac{a^2}{3}\cdot\frac{9P^2(2a+b)^2}{16a^4\rho^2(a+b)^2} + \frac{P^2(2a+3b)^2}{16a^2\rho^2(a+b)^2}\right]$$
$$+ \frac{1}{2}m_2\left[\frac{b^2}{3}\cdot\frac{9P^2}{16b^2\rho^2(a+b)^2} + \frac{P^2}{16\rho^2(a+b)^2}\right]$$

$$= \frac{2a\rho}{2}\frac{P^2}{16a^2\rho^2(a+b)^2}[3(2a+b)^2+(2a+3b)^2]$$

$$+\frac{2b\rho}{2}\frac{P^2}{16P^2(a+b)^2}[3+1]$$

$$= \frac{a\rho P^2}{16a^2\rho^2(a+b)^2}[3(2a+b)^2+(2a+3b)^2]+\frac{b\rho P^2\cdot 4}{16\rho^2(a+b)^2}$$

$$= \frac{P^2}{16a\rho(a+b)^2}[3(2a+b)^2+(2a+3b)^2]+\frac{bP^2}{4\rho(a+b)^2}$$

$$= \frac{P^2}{a16\rho(a+b)^2}[3(2a+b)^2+(2a+3b)^2+4ab]$$

$$= \frac{P^2[3(4a^2+b^2+4ab)+4a^2+9b^2+12ab+4ab]}{a16\rho(a+b)^2}$$

$$= \frac{P^2[16a^2+12b^2+28ab]}{16a\rho(a+b)^2} = \frac{P^2 4(4a^2+3b^2+7ab)}{16a\rho(a+b)^2}$$

$$= \frac{P^2 4[4a^2+4ab+3ab+3b^2]}{16a\rho(a+b)^2} = \frac{P^2[4a(a+b)+3b(a+b)]}{4a\rho(a+b)^2}$$

$$= \frac{P^2(4a+3b)}{4a\rho(a+b)}.$$

Case II. *When C is fixed.*

The equations of motion are for the rod AB,

$$m_1(u_1-0) = P+Q \tag{5.34}$$

$$\text{and} \quad m_1\frac{a^2}{3}(\omega_1-0) = Pa-Qa. \tag{5.35}$$

For the rod BC, taking moment about C,

$$m_2\frac{4}{3}b^2\omega_2 = -Q\cdot 2b. \tag{5.36}$$

Since the rods are connected at B,

∴ the velocity of the point B of the rod AB = velocity of the point B of the rod BC

$$\therefore\ u_1 - a\omega_1 = 2b\omega_2 \tag{5.37}$$

From equations (5.34), (5.35) and (5.36), we get

$$u_1 = \frac{P+Q}{m_1};\ \omega_1 = \frac{3(P-Q)}{am_1};\ \omega_2 = -\frac{3Q}{2bm_2}.$$

Putting these values in equation (5.37), we get

$$\frac{P+Q}{m_1} - \frac{3(P-Q)}{m_1} = \frac{2b(-3Q)}{2bm_2}$$

$$\Rightarrow \quad \frac{-2P+4Q}{m_1} = -\frac{3Q}{m_2}, \text{ or, } \frac{-2P+4Q}{2a\rho} = -\frac{3Q}{2b\rho}$$

$$\Rightarrow \quad \frac{-2P+4Q}{a} = -\frac{3Q}{b}, \text{ or, } -2Pb+4Qb = -3Qa$$

$$\Rightarrow \quad Q(4b+3a) = 2Pb, \text{ or, } Q = \frac{2Pb}{4b+3a}$$

$$\therefore \; u_1 = \frac{1}{2a\rho}\left[P + \frac{2Pb}{4b+3a}\right] = \frac{P}{2a\rho}\left[\frac{6b+3a}{4b+3a}\right] = \frac{3(a+2b)P}{2a\rho(4b+3a)}$$

$$\omega_1 = \frac{3}{a2a\rho}\left[P - \frac{2Pb}{4b+3a}\right] = \frac{3P}{2a^2\rho}\frac{(2b+3a)}{(4b+3a)}$$

$$\omega_2 = -\frac{3}{2b2b\rho}\frac{2Pb}{4b+3a} = -\frac{3P}{2\rho b(4b+3a)}.$$

Hence total kinetic energy

$$= \text{K.E. of the rod AB + K.E. of the rod BC}$$

$$= \left[\frac{1}{2}m_1\left(\frac{a^2}{3}\right)\omega_1^2 + \frac{1}{2}m_1u_1^2\right] + \frac{1}{2}m_2\frac{4}{3}b^2\omega_2^2$$

$$= \frac{1}{2}2a\rho\left[\frac{a^2}{3}\frac{9P^2(3a+2b)^2}{4a^4\rho^2(3a+4b)^2} + \frac{9P^2(a+2b)^2}{4a^2\rho^2(3a+4b)^2}\right]$$

$$+\frac{1}{2}2b\rho\frac{4}{3}b^2\frac{9P^2}{4b^2\rho^2(3a+4b)^2}$$

$$= \frac{a\rho P^2}{4a^2\rho^2}\frac{3}{(3a+4b)^2}\left[(3a+2b)^2+3(a+2b)^2\right] + \frac{3P^2b}{\rho(3a+4b)^2}$$

$$- \frac{3P^2}{4a\rho(3a+4b)^2}\left[(3a+2b)^2+3(a+2b)^2\right] + \frac{3P^2b}{\rho(3a+4b)^2}$$

$$= \frac{3P^2}{4a\rho(3a+4b)^2}\left[(3a+2b)^2+3(a+2b)^2+4ab\right]$$

$$= \frac{3P^2\left[9a^2+4b^2+12ab+3(a^2+4b^2+4ab)+4ab\right]}{4a\rho(3a+4b)^2}$$

$$= \frac{3P^2\left[12a^2+16b^2+28ab\right]}{4a\rho(3a+4b)^2} = \frac{3P^2 4(3a^2+4b^2+7ab)}{4a\rho(3a+4b)^2}$$

$$= \frac{3P^2\left[3a^2+3ab+4ab+4b^2\right]}{a\rho(3a+4b)^2} = \frac{3P^2\left[3a(a+b)+4b(a+b)\right]}{a\rho(3a+4b)^2}$$

$$= \frac{3P^2(3a+4b)(a+b)}{a\rho(3a+4b)^2} = \frac{3P^2(a+b)}{a\rho(3a+4b)}.$$

Hence the ratio of the two kinetic energies of case I and II is

$$\frac{(4a+3b)P^2}{4a\rho(a+b)}\frac{a\rho(3a+4b)}{3P^2(a+b)} = \frac{(4a+3b)(3a+4b)}{12(a+b)^2}.$$

Problem 3. *Two uniform rods AB and BC of the same material are smoothly jointed to B and placed in a horizontal line; the rod BC is struck at G by a blow at right angles to it, find the position of G so that the angular velocities of AB and BC may be equal in magnitude.*

Solution: Let $AB = 2a$, $BC = 2b$, let ρ be the mass of each rod per unit length. Then

$$m_1 = \text{mass of the rod AB} = 2a\rho$$
$$\text{and } m_2 = \text{mass of the rod BC} = 2b\rho.$$

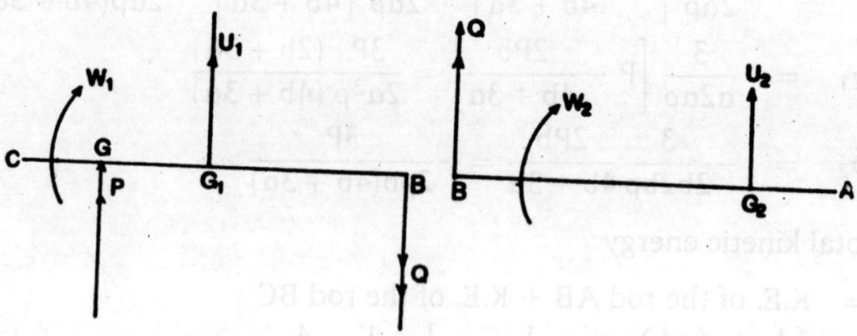

FIG 5.5

Let G_1 and G_2 be the centre of gravities of the two rods AB and BC, respectively.

Let P be the impulse applied at the point G of the rod BC, where $GG_1 = x$. Just after the blow, let u_1, ω_1 and u_2, ω_2 be the linear and angular velocities of the rods; AB and BC, respectively.

When the blow is struck, there will be an impulsive action at B between the two rods let its impulse be Q, in opposite directions on the two rods.

The equations of motion of the rods AB and BC are

For the rod AB, $\quad m_1(u_1 - 0) = Q \quad (5.38)$

and $\quad m_1\frac{1}{3}a^2(\omega_1 - 0) = Q \cdot a \quad (5.39)$

For the rod BC, $\quad m_2(u_2 - 0) = P - Q \quad (5.40)$

and $\quad m_2\frac{1}{3}b^2(\omega_2 - 0) = P \cdot GG_1 + Q \cdot GG_1 = Px + Qb. \quad (5.41)$

Also, since the rods are connected at B, therefore, velocity of the point B of the rod AB = velocity of the point B of the rod BC,

$$\text{i.e., } u_1 + a\omega_1 = u_2 - b\omega_2. \quad (5.42)$$

From equations (5.38), (5.39), (5.40) and (5.41), we get

$$u_1 = \frac{Q}{m_1}, \quad a\omega_1 = \frac{3Q}{m_1}$$
$$u_2 = \frac{P-Q}{m_2}; \quad b\omega_2 = \frac{3(Px+Qb)}{bm_2}.$$

Substituting these values in equation (5.42), we get

$$\frac{Q}{m_1}+\frac{3Q}{m_1}=\frac{P-Q}{m_2}-\frac{3(Px+Qb)}{bm_2}$$

$$\Rightarrow \quad \frac{4Q}{m_1}=\frac{1}{bm_2}[bP-bQ-3Px-3Qb]$$

$$\Rightarrow \quad \frac{4Q}{m_1}=\frac{1}{bm_2}[b(P-Q)-3Px-3Qb]$$

$$\Rightarrow \quad 4Qbm_2=bm_1(P-Q)-3Pm_1x-3Qbm_1$$

$$\Rightarrow \quad Q[4bm_2+bm_1+3bm_1]=P(bm_1-3m_1x)$$

$$\Rightarrow \quad Q4b(m_2+m_1)=Pm_1(b-3x).$$

$$\therefore\ Q = \frac{Pm_1(b-3x)}{4b(m_1+m_2)}=\frac{P2a\rho(b-3x)}{4b(2a\rho+2b\rho)}$$

$$= \frac{P2a\rho(b-3x)}{4b2\rho(a+b)}=\frac{Pa(b-3x)}{4b(a+b)}$$

$$\omega_1 = \frac{3}{am_1}Q=\frac{3}{a2a\rho}\frac{Pa(b-3x)}{4b(a+b)}=\frac{3P(b-3x)}{8ab\rho(a+b)} \tag{5.43}$$

$$\text{and } \omega_2 = \frac{3}{b^2m_2}(Px+Qb)=\frac{3}{b^22b\rho}\left[Px+\frac{Pa(b-3x)}{4b(a+b)}b\right]$$

$$= \frac{3P}{b^22b\rho}\left[x+\frac{a(b-3x)}{4(a+b)}\right]$$

$$= \frac{3P}{b^22b\rho}\left[\frac{4ax+4bx+ab-3ax}{4(a+b)}\right]$$

$$= \frac{3P}{b^22b\rho}\frac{(ax+4bx+ab)}{4(a+b)}=\frac{3P[x(a+4b)+ab]}{b^22b\rho4(a+b)}$$

$$= \frac{3P[x(a+4b)+ab]}{8b^3\rho(a+b)}. \tag{5.44}$$

Case I. *If* $\omega_1 = \omega_2$ *(equal in magnitude and in the same sense), then from* (5.43) *and* (5.44),

$$\frac{3P(b-3x)}{8ab\rho(a+b)}=\frac{3P[x(a+4b)+ab]}{8b^3\rho(a+b)}$$

$$\Rightarrow \quad \frac{b-3x}{a}=\frac{x(a+4b)+ab}{b^2}$$

$$\Rightarrow \quad b^3-3xb^2=ax(a+4b)+a^2b$$

$$\Rightarrow \quad b^3-a^2b=x[a^2+4ab+3b^2]$$

$$\Rightarrow \quad b(b-a)(b+a)=x[a^2+3ab+ab+3b^2]$$

$$\Rightarrow \quad b(b-a)(b+a)=x[a(a+3b)+b(a+3b)]$$

$$\Rightarrow \quad b(b-a)(b+a)=x(a+b)(a+3b)$$

$$\therefore \quad x=\frac{b(b-a)}{a+3b}.$$

Case II. *If* $\omega_1 = -\omega_2$ *(equal in magnitude but opposite in sense), then from* (5.43) *and* (5.44), *we get*

$$\frac{3(b-3x)P}{8ab\rho(a+b)} = -\frac{3[x(a+4b)+ab]P}{8b^3\rho(a+b)}$$

$$\Rightarrow \quad \frac{b-3x}{a} = -\frac{[x(a+4b)+ab]}{b^2}$$

$$\Rightarrow \quad b^3 - 3b^2x = -ax(a+4b) - a^2b$$

$$\Rightarrow \quad x[a^2+4ab-3b^2] = -a^2b - b^3$$

$$\Rightarrow \quad -x[3b^2-4ab-a^2] = -b(a^2+b^2)$$

$$\Rightarrow \quad x = \frac{a^2+b^2}{3b^2-4ab-a^2}.$$

Problem 4. *Three particles of equal masses are attached to the ends A and C and the middle point B of a light rigid rod ABC, and the system is at rest on a smooth table. The particle C is struck a blow at right angles to the rod. Show that the energy communicated to the system when A is fixed, is to the energy communicated when the system is free as 24 : 25.*

Solution: Let m be the mass of each particle and 2a be the length of the rod ABC. Let P be the impulse of the blow imparted at C.

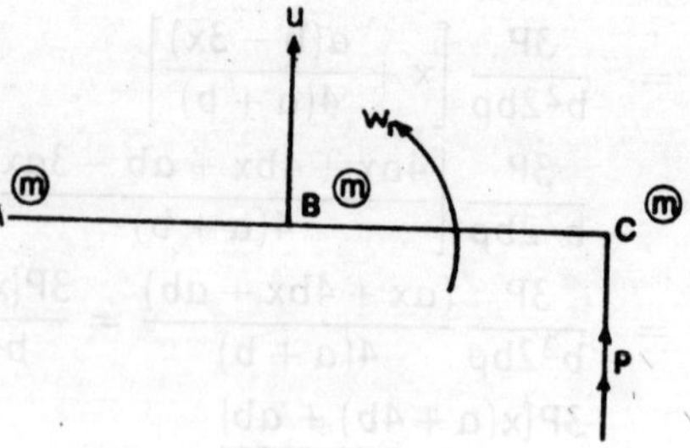

FIG 5.6

Case I. *When A is not fixed.*

The centre of gravity A of the three particles is at B and the system is at rest before the impulse. Let u be the velocity of the point B after the impulse. Let ω be the angular velocity of the rod after the impulse.

Equations of motion are

$$3mu = P, \tag{5.45}$$

i.e., motion of the C.G. which is at B and

$$3mK^2\omega = Pa. \tag{5.46}$$

On taking moments about B, where $3mK^2 = ma^2 + ma^2$ so that

$$K^2 = \frac{2a^2}{3}.$$

$$\begin{aligned}
\text{K.E.} &= \frac{1}{2}3mu^2 + \frac{1}{2}3mK^2\omega^2 \\
&= \frac{3m}{2}\left[\frac{P}{9m^2} + \frac{2a^2}{3}\frac{P^2a^2}{9m^2K^4}\right] \\
&= \frac{3m}{2}\left[\frac{P^2}{9m^2} + \frac{2a^4P^2.9}{27m^24a^4}\right] \quad \text{[From equations (5.45) and (5.46)]} \\
&= \frac{3m}{2}\left[\frac{P^2}{9m^2} + \frac{P^2}{6m^2}\right] = \frac{3mP^2}{2}\frac{(2+3)}{18m^2} \\
&= \frac{3mP^2}{2}\frac{5}{18m^2} = \frac{5P^2}{12m}.
\end{aligned}$$

Case II. *When A is fixed.*

In this case, the mass m attached at A does not move, the rod turns about A, say with angular velocity ω_1. Hence the particle m at B describes a circle of radius a about the fixed point A, and the particle m at C describes a circle of radius $2a$ about the point A, so that their linear velocities are $a\omega_1$, $2a\omega_1$, respectively.

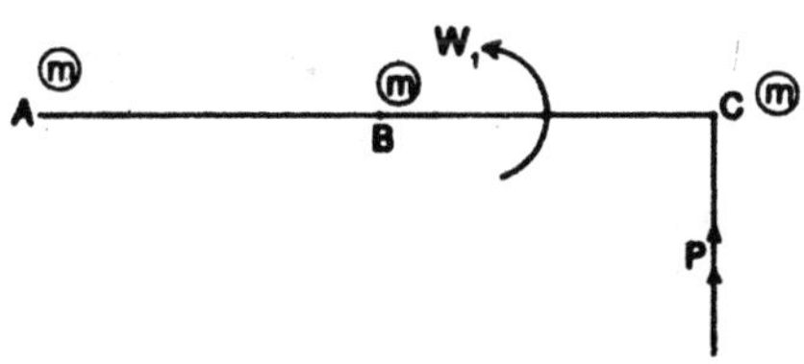

FIG 5.7

Taking moments about A, we get

$$ma^2\omega_1 + m\cdot 4a^2\omega_1 = P2a, \text{ or, } 5a^2m\omega_1 = P2a$$

$$\therefore \quad \omega_1 = \frac{2P}{5ma}.$$

Now

$$\begin{aligned}
\text{K.E.} &= \frac{1}{2}ma^2\omega_1^2 + \frac{1}{2}m4a^2\omega_1^2 = \frac{5}{2}ma^2\omega_1^2 \\
&= \frac{5}{2}ma^2\frac{4P^2}{25m^2a^2} = \frac{2P^2}{5m}.
\end{aligned}$$

Hence

$$\frac{\text{K.E. when A is fixed}}{\text{K.E. when A is free}} = \frac{\frac{2P^2}{5m}}{\frac{5P^2}{12m}} = \frac{2}{5}\cdot\frac{12}{5} = \frac{24}{25}.$$

Problem 5. *Three equal uniform rods AB, BC, CD are freely jointed and placed in a straight line on a smooth table. The rod AB is struck at its ends by a blow which is perpendicular to its length. Find the resulting motion and show that the velocity of the centre of AB is 19 times that of CD and its angular velocity 11 times that of CD.*

[CU 90H, 92H, 99H, 03H]

Solution: Let M be the mass and $2a$ be the length of each of the rods AB, BC and CD, joined freely at the ends B and C. Let P be the impulse of the blow at the end A perpendicular to AB. There will be impulsive actions at B and C. Let Q and R be the impulsive actions at B and C, respectively in the opposite directions at the ends B and C of the two rods as shown in the figure.

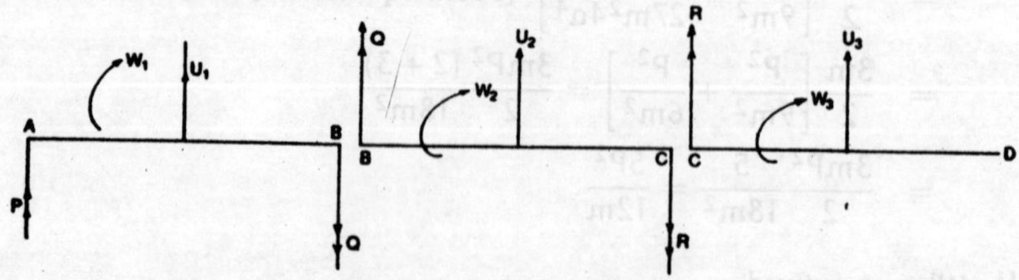

FIG 5.8

After the blow, let $u_1, \omega_1, u_2, \omega_2$ and u_3, ω_3 be the linear and angular velocities of the rods AB, BC and CD, respectively.

The equations of motion are:

For the rod AB

$$M(u_1 - 0) = P - Q \tag{5.47}$$

$$\text{and} \quad M\frac{a^2}{3}(\omega_1 - 0) = Pa + Qa. \tag{5.48}$$

For the rod BC

$$M(u_2 - 0) = Q - R \tag{5.49}$$

$$\text{and} \quad M\frac{a^2}{3}(\omega_2 - 0) = Qa + Ra. \tag{5.50}$$

For the rod CD

$$M(u_3 - 0) = R \tag{5.51}$$

$$\text{and} \quad M\frac{a^2}{3}(\omega_3 - 0) = Ra. \tag{5.52}$$

Also the rods are connected at B and C,

$\therefore$ the velocity of the point B, the end of the rod AB = velocity of the point B of the rod BC,

$$\text{i.e., } u_1 - a\omega_1 = u_2 + a\omega_2. \tag{5.53}$$

The velocity of C, the end of the rod BC = velocity of the point C of the rod CD,

$$\text{i.e., } u_2 - a\omega_2 = u_3 + a\omega_3. \tag{5.54}$$

From equations (5.47) to (5.52), we get

$$u_1 = \frac{P-Q}{M}, \quad \omega_1 = \frac{3(P+Q)}{aM}$$
$$u_2 = \frac{Q-R}{M}, \quad \omega_2 = \frac{3(Q+R)}{aM}$$
and $$u_3 = \frac{R}{M}, \quad \omega_3 = \frac{3R}{aM}.$$

Substituting these values of $u_1, \omega_1, u_2, \omega_2, u_3, \omega_3$ in equations (5.53) and (5.54) we get

$$\frac{P-Q}{M} - \frac{3(P+Q)}{M} = \frac{Q-R}{M} + \frac{3(Q+R)}{M}$$
or, $$\frac{-2P-4Q}{M} = \frac{4Q+2R}{M}$$
or, $$-P-2Q = 2Q+R, \text{ or, } -P = 4Q+R. \tag{5.55}$$

From equation (5.54), we get

$$\frac{Q-R}{M} - \frac{3(Q+R)}{M} = \frac{R}{M} + \frac{3R}{M}$$
$$\Rightarrow \frac{-2Q-4R}{M} = \frac{4R}{M}$$
$$\Rightarrow -Q-2R = 2R, \text{ or, } -Q = 4R$$
$$\therefore R = -\frac{Q}{4}.$$

Hence from equation (5.55), we get

$$-P = -15R$$
$$\therefore R = P/15.$$

Thus

$$Q = -\frac{4P}{15}.$$

Substituting these value of Q and R in equations (5.47) to (5.52), we have

$$u_1 = \frac{P + \frac{4P}{15}}{M} = \frac{19P}{15M},$$
$$\omega_1 = \frac{3}{aM}\left(P - \frac{4P}{15}\right) = \frac{3 \cdot 11P}{15aM} = \frac{11P}{5aM},$$
$$u_2 = \frac{-\frac{4P}{15} - \frac{P}{15}}{M} = -\frac{5P}{15M} = -\frac{P}{3M},$$
$$\omega_2 = \frac{3}{aM}\frac{P}{15} = \frac{P}{5aM},$$
$$u_3 = \frac{P}{15M},$$
$$\omega_3 = \frac{3}{aM}\frac{P}{15} = \frac{P}{5aM}.$$

These values of $u_1, \omega_1, u_2, \omega_2, u_3, \omega_3$ determine the resulting motion of the system.

Here
$$\frac{u_1}{u_3} = \frac{19P}{15M}\frac{15M}{P} = 19, \text{ or, } u_1 = 19u_3.$$

Hence the velocity of centre G_1 of AB is 19 times that of CD.

Also
$$\frac{\omega_1}{\omega_3} = \frac{11P}{5aM}\frac{15aM}{3P} = 11.$$

Hence the angular velocity of AB is 11 times that of CD.

Problem 6. *Three uniform rods AB, BC, CD are hinged freely at their ends, B and C, so as to form three sides of a square and are laid on a smooth table. The end A is struck by a horizontal blow P at right angles to AB. Show that the initial velocity of A is nineteen times that of D and that the impulsive actions at B and C are $\frac{5}{12}$P and $\frac{1}{12}$P respectively.* **[CU 93H]**

Solution: Let M be the mass and $2a$ the length of each rod. Let P be the impulse at A perpendicular to AB, therefore the initial velocity of the end B will be along BC. Also the impulsive action at C will be along CB. Let Q and R be the impulsive actions at the ends B and C, respectively in the two rods at these ends in opposite directions as shown in figure.

After the impulse, let u_1, u_2, u_3 be the linear velocities of the rods AB, BC and CD and ω_1, ω_2 be the angular velocities of rods AB and CD, respectively.

The equations of motion are:

For the rod AB,
$$M(u_1 - 0) = P + Q \tag{5.56}$$
$$\text{and} \quad M\frac{a^2}{3}(\omega_1 - 0) = Pa - Qa. \tag{5.57}$$

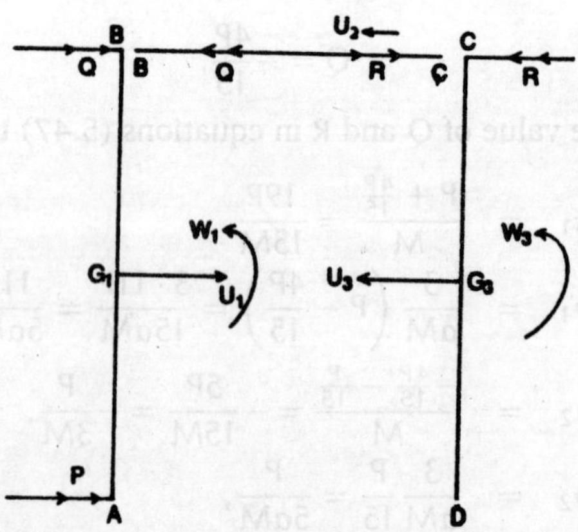

FIG 5.9

For the rod BC,
$$M(u_2 - 0) = Q - R. \tag{5.58}$$

For the rod CD,

$$M(u_3 - 0) = R \tag{5.59}$$

$$\text{and} \quad M\tfrac{1}{3}a^2(\omega_2 - 0) = Ra. \tag{5.60}$$

Since the rods are connected at B and C,
velocity of the point B of the rod AB = velocity of the point B of the rod BC,

$$\text{i.e., } u_1 - a\omega_1 = -u_2. \tag{5.61}$$

Also, velocity of the point C of the rod BC = velocity of the point C of the rod CD,

$$\text{i.e., } u_2 = u_3 + a\omega_2. \tag{5.62}$$

From equations (5.56), (5.57), (5.58), (5.59) and (5.60), we get

$$u_1 = \frac{P+Q}{M}, \quad a\omega_1 = \frac{3(P-Q)}{M},$$
$$u_2 = \frac{Q-R}{M}, \quad u_3 = \frac{R}{M}, \quad a \cdot \omega_2 = \frac{3R}{M}.$$

With these values, equations (5.61) and (5.62) becomes

$$\frac{P+Q}{M} - \frac{3(P-Q)}{M} = -u_2$$
$$\Rightarrow \quad \frac{P+Q-3P+3Q}{M} = -u_2$$
$$\Rightarrow \quad \frac{4Q-2P}{M} = -u_2 = \frac{R-Q}{M}$$
$$\Rightarrow \quad 4Q - 2P = R - Q$$
$$\Rightarrow \quad 5Q - R = 2P \tag{5.63}$$
$$\text{and} \quad \frac{Q-R}{M} = \frac{R}{M} + \frac{3R}{M} = \frac{4R}{M}$$
$$\Rightarrow \quad Q = 5R. \tag{5.64}$$

Thus from equation (5.63), we get

$$5Q - \frac{Q}{5} = 2P, \text{ or, } \frac{24Q}{5} = 2P, \text{ or, } P = \frac{12Q}{5}.$$

$$\therefore \quad Q = \frac{5}{12}P \text{ and } R = \frac{Q}{5} = \frac{P}{12}.$$

Substituting the values of Q and R in equations (5.56), (5.57), (5.59) and (5.60), we get

$$u_1 = \frac{P+Q}{M} = \frac{1}{M}\left(P + \frac{5}{12}P\right) = \frac{1}{M}\frac{17}{12}P$$
$$a\omega_1 = \frac{3}{M}(P-Q) = \frac{3}{M}\left(P - \frac{5}{12}P\right) = \frac{3}{M}\frac{7}{12}P = \frac{1}{M}\frac{7}{4}P$$
$$u_3 = \frac{R}{M} = \frac{P}{12M}$$
$$a\omega_2 = \frac{3}{M} \cdot \frac{P}{12} = \frac{3P}{12M} = \frac{P}{4M}.$$

Hence

$$\frac{\text{velocity of A}}{\text{velocity of B}} = \frac{u_1 + a\omega_1}{-u_3 + a\omega_2} = \frac{\frac{17P}{12M} + \frac{7P}{4M}}{-\frac{P}{12M} + \frac{P}{4M}}$$

$$= \frac{38P}{12M} \times \frac{12M}{2P} = 19.$$

Thus velocity of A is nineteen times the velocity of D.

Problem 7. AB, BC, CD *are three equal uniform rods hinged freely at* B *and* C. *They are laid on a smooth horizontal table so that* ABC *and* BCD *are at right angles on opposite side of* BC. *A blow is given to* A *in the direction* AC. *Prove that* D *begins to move in a direction* $\tan^{-1}\left(\frac{7}{4}\right)$ *with* DC.

Solution: Let AB = BC = CD = 2a.

Let P be the impulse applied at A in the direction AC.

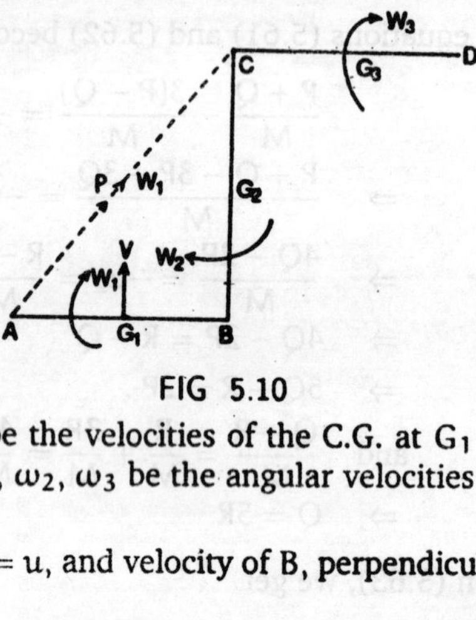

FIG 5.10

After impulse, let u, v be the velocities of the C.G. at G_1 of rod AB along and perpendicular to it. Let $\omega_1, \omega_2, \omega_3$ be the angular velocities of the three rods AB, BC, CD, respectively.

Velocity of B along AB = u, and velocity of B, perpendicular to AB

= velocity of G perpendicular to AB + velocity of B relative to G_1

= $v - a\omega_1$.

Velocity of the C.G. at G_2 of the rod BC parallel to AB

= velocity of B+ velocity of G_2 relative to B parallel to AB

= $u + a\omega_2$,

and perpendicular to AB

= velocity of B perpendicular to AB = $v - a\omega_1$.

Velocity of C parallel to AB

= velocity of B+ velocity of C relative to B parallel to AB

= $u + 2a\omega_2$.

and perpendicular to AB

$$= \text{velocity of } G_2 \text{ perpendicular to AB} = v - a\omega_1.$$

Velocity of G_2 (C.G. at G of rod CD) parallel to AB

$$= \text{velocity of } G_2 \text{ parallel to AB} = u + 2a\omega_2.$$

and perpendicular to AB

$$= \text{velocity of C+ velocity of } G_2 \text{ relative to C, perpendicular to AB}$$
$$= v - a\omega_1 - a\omega_2.$$

Velocity of D, parallel to AB

$$= \text{velocity of C, parallel to AB} = u + 2a\omega_2$$

and perpendicular to AB

$$= \text{velocity of C+ velocity of D relative to C perpendicular to AB}$$
$$= v - a\omega_1 \quad 2a\omega_3.$$

Considering motion of the system of all the three rods as one body and taking moment about A, we have

$$M\left(\frac{a^2}{3}\omega_1 - va\right) + M\left[\frac{a^2}{3}\omega_2 + a(u + a\omega_2) - (v - a\omega_1)2a\right]$$
$$+M\left[\frac{a^2}{3}\omega_2 + (u + 2a\omega_2)2a - (v - a\omega_1 - a\omega_3)3a\right] = 0$$

$$\Rightarrow \quad \frac{a^2}{3}\omega_1 - av + \frac{a^2}{3}\omega_2 + au + a^2\omega_2 - 2av + 2a^2\omega_1$$
$$+\frac{a^2}{3}\omega_3 + 2au + 4a^2\omega_2 - 3av + 3a^2\omega_1 + 3a^2\omega_3 = 0$$

$$\frac{16}{3}a^2\omega_1 + \frac{16}{3}a^2\omega_2 + \frac{10}{3}a^2\omega_3 + 3au - 6av = 0$$

$$\Rightarrow \quad a(16\omega_1 + 16\omega_2 + 10\omega_3) + 9u - 18v = 0. \tag{5.65}$$

Considering motion of the rods BC and CD together and taking moments about B, we have

$$M\left[\frac{a^2}{3}\omega_2 + (u + a\omega_2)a\right] + M\left[\frac{a^2}{3}\omega_3 - a(v - a\omega_1 - a\omega_3)\right.$$
$$\left.+2a(u + 2a\omega_2)\right] = 0$$

$$\Rightarrow \quad \frac{a^2}{3}\omega_2 + au + a^2\omega_2 + \frac{a^2}{3}\omega_3 - av + a^2\omega_1 + a^2\omega_3$$
$$+2au + 4a^2\omega_2 = 0$$

$$\Rightarrow \quad a^2\omega_1 + \frac{16}{3}a^2\omega_2 + \frac{4}{3}a^2\omega_3 + 3au - av = 0$$

$$\Rightarrow \quad a(3\omega_1 + 16\omega_2 + 4\omega_3) + 9u - 3v = 0. \tag{5.66}$$

Considering, motion of the rod CD and taking moments about C, we have

$$M\left[\frac{a^2}{3}\omega_3 - (v - a\omega_1 - a\omega_3)a\right] = 0$$

$$\Rightarrow \quad \frac{a\omega_3}{3} - v + a\omega_1 + a\omega_3 = 0$$

$$\Rightarrow \quad \frac{4}{3}a\omega_3 + a\omega_1 - v = 0$$

$$\Rightarrow a(4\omega_3 + 3\omega_1) - 3v = 0. \tag{5.67}$$

Considering motion of all the three rods and resolving at right angles to AC, we have

$$(u\cos 45° - v\cos 45°) + [(u + a\omega_2)\cos 45° - (v - a\omega_1)\cos 45°]$$
$$+ [(u + 2a\omega_2)\cos 45° - (v - a\omega_1 - a\omega_3)\cos 45°] = 0$$
$$\Rightarrow \quad u - v + u + a\omega_2 - v + a\omega_1 + u + 2a\omega_2 - v + a\omega_1 + a\omega_3 = 0$$
$$\Rightarrow \quad 3u - 3v + 2a\omega_1 + 3a\omega_2 + a\omega_3 = 0$$
$$\Rightarrow \quad 3(u - v) + a(2\omega_1 + 3\omega_2 + \omega_3) = 0. \tag{5.68}$$

If the direction of motion of D makes an angle θ to CD, (i.e., to AB), then

$$\tan\theta = \frac{\text{velocity of D at right angles to DC}}{\text{velocity of D along DC}}$$
$$= \frac{v - a\omega_1 - 2a\omega_3}{u + 2a\omega_2}. \tag{5.69}$$

Subtracting equation (5.67) from equation (5.66), we get

$$a16\omega_2 + 9u = 0 \tag{5.70}$$
$$\Rightarrow \quad 9u = -16a\omega_2$$
$$u = \frac{16}{9}a\omega_2$$
$$\Rightarrow \quad u + 2a\omega_2 = \frac{2}{3}a\omega_2. \tag{5.71}$$

Again from equation (5.67), $3v = a(3\omega_1 + 4\omega_3)$. Also $9u = -16a\omega_2$.
Putting these values in equation (5.65), we get

$$a(16\omega_1 + 16\omega_2 + 10\omega_3) - 16a\omega_2 - 6a(3\omega_1 + 4\omega_3) = 0$$
$$\Rightarrow \quad 16\omega_1 + 16\omega_2 + 10\omega_3 - 16\omega_2 - 18\omega_1 - 24\omega_3 = 0$$
$$\Rightarrow \quad -2\omega_1 - 14\omega_3 = 0$$
$$\therefore \quad \omega_3 = -\frac{\omega_1}{7}.$$

Subtracting equation (5.67) from equation (5.68), we get

$$3u + a[2\omega_1 + 3\omega_2 + \omega_3 - 4\omega_3 - 3\omega_1] = 0$$

$$\Rightarrow \quad 3u + a(-\omega_1 - 3\omega_3 + 3\omega_2) = 0$$

$$\Rightarrow \quad -\frac{16}{3}a\omega_2 + a(-\omega_1 - 3\omega_3 + 3\omega_2) = 0$$

$$\Rightarrow \quad -\frac{16}{3}\omega_2 - \omega_1 - 3\omega_3 + 3\omega_2 = 0$$

$$\Rightarrow \quad -16\omega_2 - 3\omega_1 - 9\omega_3 + 9\omega_2 = 0$$

$$\Rightarrow \quad -7\omega_2 - 3\omega_1 - 9\omega_3 = 0$$

$$\Rightarrow \quad -7\omega_2 + 3 \cdot 7\omega_3 - 9\omega_3 = 0$$

$$\Rightarrow \quad -7\omega_2 = -12\omega_3$$

$$\Rightarrow \quad \omega_2 = \frac{12}{7}\omega_3 = -\frac{12}{49}\omega_1.$$

From equation (5.67) $v - a\omega_1 = \frac{4}{3}a\omega_3$.
Hence from equation (5.69), we have

$$\tan\theta = \frac{\frac{4}{3}a\omega_3 - 2a\omega_3}{-\frac{16}{9}a\omega_2 + 2a\omega_2} = -\frac{2a\omega_3}{3} \times \frac{9}{2a\omega_2}$$

$$= -\frac{3\omega_3}{\omega_2} = \frac{(-)3(-\omega_1)}{7\omega_2} = \frac{3}{7}\frac{\omega_1}{\omega_2} = \frac{3}{7}\frac{\omega_1}{-\frac{12}{4a}\omega_1} = -\frac{7}{4}.$$

$$\therefore \theta = \tan^{-1}\left(-\frac{7}{4}\right).$$

Problem 8. *A lamina in the form of an equilateral triangle ABC lies on a smooth horizontal plane. Suddenly it receives a blow at A in a direction parallel to BC, which causes A to move with velocity v. Determine the instantaneous velocities of B and C and describe the subsequent motion of the lamina.*

Solution: Let M be the mass and $2a$ the length of each side of an equilateral triangular lamina ABC. Let P be the impulse of the blow imparted at A, parallel to BC, which causes A to move with velocity v parallel to BC. Since the impulse is parallel to BC, therefore, just after the impulse the velocity of the C.G. at G of the lamina will be parallel to BC. Let the velocity be u and let ω be the angular velocity of the lamina.

Length of each median $= a\sqrt{3}$

$$AG = \frac{2}{3}a\sqrt{3} \text{ and } GD = \frac{a}{\sqrt{3}}.$$

The equations of motion for the lamina are

$$M(u - 0) = P \tag{5.72}$$

$$\text{and} \quad \frac{M}{3}\left(\frac{a^2}{3} + \frac{a^2}{3} + \frac{a^2}{3}\right)\omega = P \cdot \frac{2}{3}a\sqrt{3}$$

$$\text{and} \quad \frac{M}{3}a^2\omega = P\frac{2}{3}a\sqrt{3}$$

$$\therefore \quad Ma\omega = 2\sqrt{3}P. \tag{5.73}$$

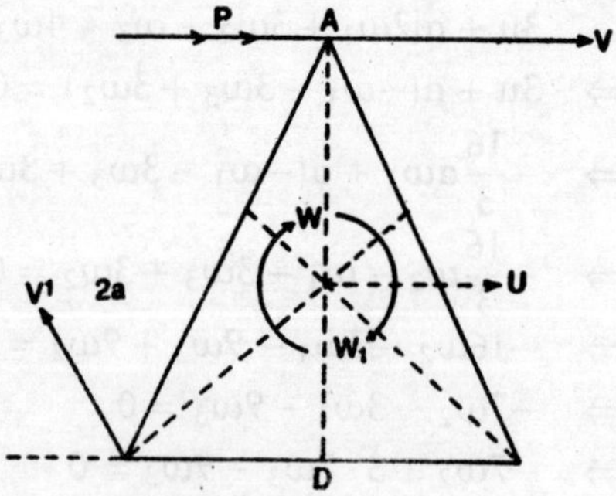

FIG 5.11

From equations (5.72) and (5.73), we have

$$Mu = \frac{Ma\omega}{2\sqrt{3}}; \text{ or, } u = \frac{a\omega}{2\sqrt{3}}.$$

Now, velocity of the point A,

$$v = \text{velocity of G} + \text{velocity of A relative to G}$$

$$v = u + GA\omega = u + \frac{2}{3}a\sqrt{3}\omega = \frac{a\omega}{2\sqrt{3}} + \frac{2a}{\sqrt{3}}\omega$$

$$v = \frac{5a\omega}{2\sqrt{3}}.$$

$$\therefore\ a\omega = \frac{2\sqrt{3}}{5}v \text{ and } u = \frac{a\omega}{2\sqrt{3}} = \frac{v}{5}.$$

Velocity of the point B parallel to BC

= velocity of G parallel to BC + velocity of B parallel to BC, relative to G

$$= u - v'\cos 60° = u - GB\omega\frac{1}{2}$$

$$= u - \frac{2}{3}a\sqrt{3}\omega\frac{1}{2} \quad \left(GA = GB = \frac{2\sqrt{3}a}{3}\right)$$

$$= \frac{v}{5} - \frac{1}{\sqrt{3}}\frac{2\sqrt{3}}{5}v = \frac{v}{5} - \frac{2v}{5} = -\frac{v}{5}$$

and velocity of the point B perpendicular to BC

= velocity of G perpendicular to BC + velocity of B perpendicular to BC, relative to G

$$= 0 + v'\sin 60° = GB\omega\frac{\sqrt{3}}{2}$$

$$= \frac{2\sqrt{3}a}{3}\omega\frac{\sqrt{3}}{2} = a\omega = \frac{2\sqrt{3}v}{5}.$$

Resultant velocity of B

$$= \sqrt{\left(-\frac{v}{5}\right)^2 + \left(\frac{2\sqrt{3}v}{5}\right)^2} = \frac{v}{5}\sqrt{1+12} = \frac{v}{5}\sqrt{13}.$$

Similarly, velocities of C are $-\frac{v}{2}$ and $\frac{2}{5}\sqrt{3}v$ along and perpendicular to BC.

Problem 9. *A rectangular lamin,; whose sides are of length 2a and 2b, is at rest when one corner is caught and suddenly made to move with prescribed speed V in the plane of the lamina; show that the greatest angular velocity which can thus be imparted to the lamina is* $\frac{3V}{4\sqrt{a^2+b^2}}$.

Solution: Let M be the mass and G the centre of the rectangle ABCD whose sides are of lengths 2a and 2b. The corner A is suddenly made to move with prescribed velocity V making an angle θ with the side AB in the plane of the rectangle.

The impulse is applied at the corner A. Let X and Y be the components of the impulse along and perpendicular to AB.

Just before the impulse, the lamina is at rest. Just after the impulse, let u, v be the velocities of the centre of gravity G parallel and perpendicular to AB and ω the angular velocity of the lamina.

We have $AB = 2a$, $BC = 2b$.

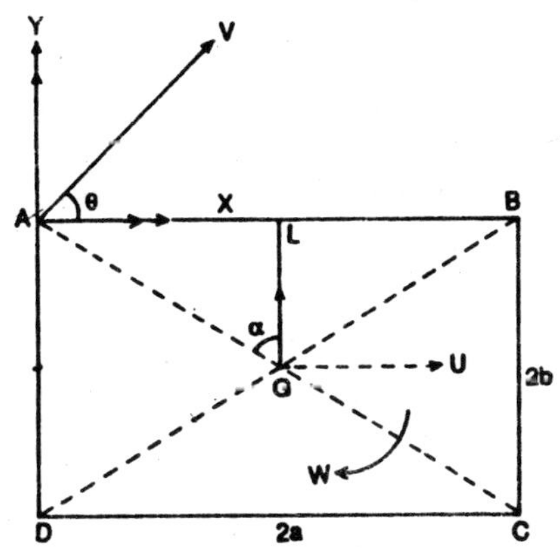

FIG 5.12

Thus, $AC = \sqrt{AB^2 + BC^2} = 2\sqrt{a^2+b^2}$

$$\therefore AG = \sqrt{a^2+b^2}, \qquad AL - LB = a.$$

Let $\angle AGL = \alpha$.

$$\therefore \sin\alpha = \frac{AL}{AG} = \frac{a}{\sqrt{a^2+b^2}}, \qquad \cos\alpha = \frac{GL}{AG} = \frac{b}{\sqrt{a^2+b^2}}.$$

The equations of motion are

$$Mu = X \tag{5.74}$$

$$Mv = Y \tag{5.75}$$

$$\text{and } M\frac{a^2+b^2}{3}\omega = Xb + Ya \text{ (Moment about G)}. \tag{5.76}$$

Now, velocity of the point A

$$= \text{velocity of } G + \text{velocity of } A, \text{ parallel to } AB \text{ relative to } G$$

$$\therefore\ v\cos\theta = u + \sqrt{a^2+b^2}\,\omega\cos\alpha,$$

$$\text{i.e., } v\cos\theta = u + b\omega. \tag{5.77}$$

Also velocity of A, perpendicular to AB

$$= \text{velocity of } G, \text{ perpendicular to } AB + \text{velocity of } A, \text{ perpendicular to } AB \text{ relative to } G$$

$$\begin{aligned}\therefore\ v\sin\theta &= v + \sqrt{a^2+b^2}\,\omega\sin\alpha\\ &= v + \sqrt{a^2+b^2}\,\omega\frac{a}{\sqrt{a^2+b^2}},\end{aligned}$$

$$\text{i.e., } v\sin\theta = v + \omega a. \tag{5.78}$$

Now, substituting the values of X and Y from equations (5.74) and (5.75) in equation (5.76), we get

$$\begin{aligned}\frac{M}{3}(a^2+b^2)\omega &= Mva + Mub\\ \Rightarrow\quad \frac{a^2+b^2}{3}\omega &= va + ub\\ \Rightarrow\quad (a^2+b^2)\omega &= 3(va+ub).\end{aligned} \tag{5.79}$$

From equation (5.77) and equation (5.78), we get

$$u = v\cos\theta - b\omega;\ v = v\sin\theta - a\omega.$$

With these values equation (5.79), becomes

$$\begin{aligned}(a^2+b^2)\omega &= 3\left[(v\sin\theta - a\omega)a + (v\cos\theta - b\omega)b\right]\\ &= 3\left[v(a\sin\theta + b\cos\theta) - \omega(a^2+b^2)\right]\\ \Rightarrow (a^2+b^2)\omega &= 3v(a\sin\theta + b\cos\theta) - 3(a^2+b^2)\omega\\ \Rightarrow 4(a^2+b^2)\omega &= 3v(a\sin\theta + b\cos\theta).\end{aligned} \tag{5.80}$$

Differentiating with respect to θ, we get

$$4(a^2+b^2)\frac{d\omega}{d\theta} = 3v(a\cos\theta - b\sin\theta). \tag{5.81}$$

But ω will be maximum when $\frac{d\omega}{d\theta} = 0$.

$$\begin{aligned}&3v(a\cos\theta - b\sin\theta) = 0\\ \Rightarrow\ &a\cos\theta = b\sin\theta\\ \Rightarrow\ &\frac{\sin\theta}{a} = \frac{\cos\theta}{b} = \frac{1}{\sqrt{a^2+b^2}}\\ \therefore\ &\sin\theta = \frac{a}{(\sqrt{a^2+b^2})},\ \cos\theta = \frac{b}{\sqrt{a^2+b^2}}.\end{aligned}$$

Differentiating equation (5.81), again we get

$$4(a^2+b^2)\frac{d^2\omega}{d\theta^2} = 3v(-a\sin\theta - b\cos\theta) < 0.$$

Thus, ω is maximum when

$$\sin\theta = \frac{a}{\sqrt{a^2+b^2}}, \quad \cos\theta = \frac{b}{\sqrt{a^2+b^2}}.$$

Thus, maximum value of ω from equation (5.80),

$$\begin{aligned} 4(a^2+b^2)\omega &= 3v\left[\frac{a^2+b^2}{\sqrt{a^2+b^2}}\right] = 3v\sqrt{a^2+b^2} \\ \Rightarrow \omega &= \frac{3v}{4\sqrt{a^2+b^2}}. \end{aligned}$$

Problem 10. *A light string is wound round the circumference of a uniform reel, of radius a and radius of gyration K about its axis. The free end of the string being tied to a fixed point, the reel is lifted up and let fall so that at the moment when the string becomes tight, the velocity of the centre of the reel is u and the string is vertical. Find the change in the motion and show that the impulsive tension is* $mu\frac{K^2}{a^2+K^2}$.

Solution: Let O be the fixed point to which the free end is attached and G the centre of gravity of the reel.

When the string becomes tight, a jerk is experienced by the string. Let T be the impulsive tension in the string.

Just before the jerk, u is the vertical velocity of the centre of gravity G of the system and just after the jerk, let v be the vertical velocity of the C.G. at G and ω the angular velocity of the reel. Before the jerk there is no angular velocity of the reel.

Since there is no horizontal impulse, therefore, there will be no velocity in horizontal direction.

Equations of motion of the reel are

$$M(v-u) = -T \tag{5.82}$$

$$\text{and } MK^2\omega = Ta. \tag{5.83}$$

Also, just after the impulse the velocity of the point of contact A is zero.

$$\therefore\ v - a\omega = 0. \tag{5.84}$$

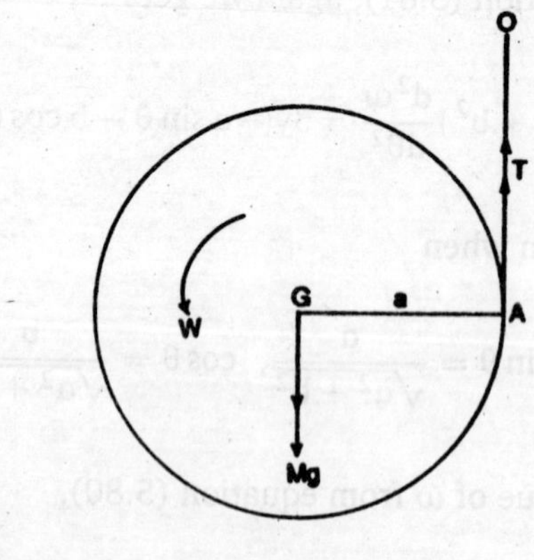

FIG 5.13

From equation (5.84), $\omega = \frac{v}{a}$.

Thus, from equation (5.83), we get

$$MK^2 \frac{v}{a} = Ta; \text{ or, } v = \frac{Ta^2}{MK^2}.$$

Thus, from equation (5.82), we get

$$M\left(\frac{Ta^2}{MK^2} - u\right) = -T \tag{5.85}$$

$$\text{or, } T\left(\frac{a^2}{K^2} + 1\right) = Mu \tag{5.86}$$

$$\text{or, } T = \frac{MuK^2}{a^2 + K^2} \tag{5.87}$$

Problem 11. *A rod of length 2a is held in a position inclined at an angle α to the vertical, and is then let fall on to a smooth inelastic horizontal plane. Show that the end which hits the plane will leave it immediately after the impact if the height through which the rod falls is greater than $\frac{1}{15} a \sec\alpha \operatorname{cosec}^2\alpha(1 + 3\sin^2\alpha)^2$.*

Solution: Let the end A of the rod AB of mass M and length $2a$ hits the smooth inelastic horizontal plane. The rod AB is inclined at an angle α to the vertical at the time of striking the plane. Let u be the velocity of the C.G. at G just before the impact, and just after the impact, let v be the velocity of G and ω the angular velocity of the rod.

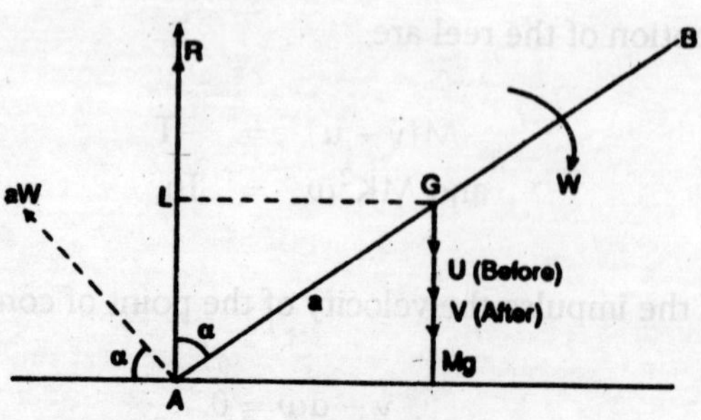

FIG 5.14

Let R be the impulsive reaction of the plane at A.

Just after the impact the inclination of the rod is also α to the vertical.

Hence the equations of motion are

$$M(v-u) = -R \tag{5.88}$$

$$\text{and} \quad M\frac{a^2}{3}\omega = R.GL = Ra\sin\alpha. \tag{5.89}$$

As the horizontal plane is inelastic, the end A does not rebound after impact, hence the vertical velocity of A just after impact $= 0$,

$$\text{i.e., } v - a\omega\sin\alpha = 0. \tag{5.90}$$

From equations (5.88) and (5.90), we get

$$R = M(u - a\omega\sin\alpha).$$

Hence from equation (5.89),

$$\frac{1}{3}Ma^2\omega = M(u - a\omega\sin\alpha)a\sin\alpha$$

$$\Rightarrow \quad a^2\omega = 3a\sin\alpha(u - a\omega\sin\alpha)$$

$$\Rightarrow \quad a^2\omega(1+3\sin^2\alpha) = 3a\sin\alpha u$$

$$\therefore \quad \omega = \frac{3u\sin\alpha}{a(1+3\sin^2\alpha)}. \tag{5.91}$$

Assuming that after impact the end A remains in contact with the plane and the rod is moving about A. Let at time t, the rod make an angle θ to the vertical. Let S be the normal reaction at this instant. Since the motion is under finite forces S and Mg, the vertical height of G above the plane is $a\cos\theta$.

The equations of motion of the rod are

$$M\frac{d^2}{dt^2}(a\cos\theta) = S - Mg \tag{5.92}$$

$$\text{and} \quad M\frac{a^2}{3}\ddot{\theta} = Sa\sin\theta. \tag{5.93}$$

From equation (5.92),

$$-aM\left[\cos\theta\dot{\theta}^2 + \ddot{\theta}\sin\theta\right] = S - Mg$$

$$\Rightarrow \quad S = Mg + Ma(-\sin\theta\ddot{\theta} - \cos\theta\dot{\theta}^2). \tag{5.94}$$

From equation (5.93),

$$Ma\ddot{\theta} = 3S\sin\theta.$$

Thus, from equation (5.94), we get

$$S = Mg - 3S\sin^2\theta - Ma\cos\theta\dot{\theta}^2$$

$$\Rightarrow \quad S(1+3\sin^2\theta) = Mg - Ma\cos\theta\dot{\theta}^2.$$

But initially, when $\theta = \alpha$, $\dot{\theta} = \omega$ and $S = R$.

$$\therefore \; R(1 + 3\sin^2\alpha) = Mg - Ma\cos\alpha\omega^2.$$

At this time (i.e., initially just after the impact) the end A will leave the plane, if R is negative.

i.e., if $Mg - Ma\cos\alpha\omega^2 < 0 \; [\because \; 1 + 3\sin^2\alpha > 0]$

$$\Rightarrow g < a\omega^2\cos\alpha$$

$$\Rightarrow \quad g < \frac{a\cos\alpha 9u^2\sin^2\alpha}{a^2(1 + 3\sin^2\alpha)^2} \quad \text{[From equation (5.91)]}$$

$$\Rightarrow \quad \frac{a(1 + 3\sin^2\alpha)^2}{9\cos\alpha\sin^2\alpha} < \frac{u^2}{g}$$

$$\Rightarrow \quad \frac{u^2}{2g} > \frac{a\sec\alpha\,\text{cosec}^2\alpha}{18}(1 + 3\sin^2\alpha)^2,$$

i.e., the height through which the rod falls is greater than

$$\frac{1}{18}a\sec\alpha\,\text{cosec}^2\alpha(1 + 3\sin^2\alpha)^2.$$

Problem 12. *An equilateral triangle formed by uniform rods freely hinged at their ends, is falling freely with one side horizontal and uppermost. If the middle point of this side be suddenly stopped, show that the impulsive actions at the upper and lower hinges are in the ratio* $\sqrt{13} : 1$.

Solution: Let ABC be the equilateral triangle, formed by three rods AB, BC, CA each of length $2a$, which is falling freely with the side BC horizontal and upper most. Let the middle point D of this side BC be stopped suddenly. Thus, there is an impulsive action at D.

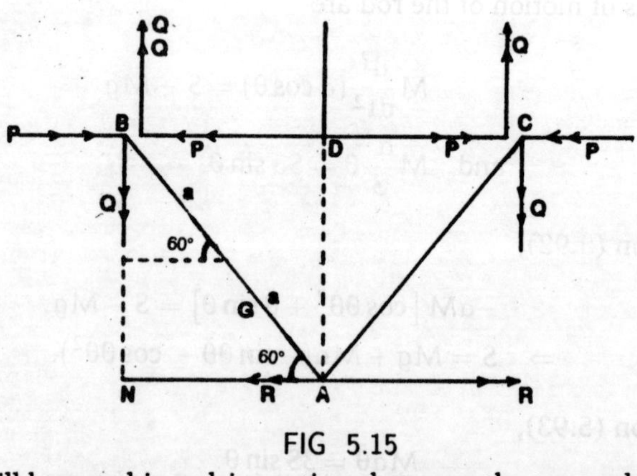

FIG 5.15

Also there will be equal impulsive actions between the two rods at B and C. Let these impulses at B and C be P horizontally and Q vertically in opposite directions on the two rods at B and C. The impulsive action at A will be horizontally only. Let it be R on the two rods at A in opposite directions.

Let G be the middle point of the rod AB.

Before the system was suddenly stopped, there was only the vertical velocity of each rod, i.e., there was no horizontal and no angular velocities of rods before impact. As the system is stopped, there is neither linear velocity nor angular velocity of any rod after the impact.

Considering the motion of the rod AB, its equations of motion are

1. $$P - R = 0 \tag{5.95}$$

   as there is no horizontal velocity of the rod AB.

2. Also there is no angular velocity of the rod AB before or after the impact. Thus, taking moment about G, we get

$$P \cdot LB - Q \cdot GL + R \cdot LN = 0$$

$$\text{or,}\quad Pa\sin 60^\circ - Qa\cos 60^\circ + Ra\sin 60^\circ = 0$$

$$\text{or,}\quad \sqrt{3}(P + R) = Q. \tag{5.96}$$

From equation (5.95) and equation (5.96), we get

$$2\sqrt{3}P = Q$$

$$\text{and}\quad P = R, \;\therefore\; Q = 2\sqrt{3}R.$$

Now,

$$\frac{\text{Impulsive action at the upper end B}}{\text{Impulsive action at the lower end A}} = \frac{\sqrt{P^2 + Q^2}}{R} = \frac{\sqrt{R^2 + 12R^2}}{R} = \frac{\sqrt{13}}{1}.$$

Problem 13. *Of two inelastic circular discs with milled edges each of mass* m *and radius* a, *one is rotating with angular velocity* ω *round its centre O which is fixed on a smooth plane, and the other is moving without spin in the plane, with velocity* v *directed towards O. Find the motion immediately afterwards, and show that the energy lost by the impact is* $\frac{1}{2}m\left(v^2 + \frac{a^2\omega^2}{5}\right)$.

Solution: Let A and B be the two discs, A is rotating about the fixed centre O with angular velocity ω, while B is moving with linear velocity v towards O.

Let F be the impulsive friction at the point of contact P.

The discs are inelastic, so they will not rebound, and hence after the impact, the velocity of the disc B will be zero along OO_1. Let u be the velocity of B at right angles to common normal. Also let ω_1 and ω_2 be the angular velocities of the disc A and B, respectively after the impact.

Hence the equations of motion are:

For the disc A,

$$\frac{ma^2}{2}(\omega_1 - \omega) = -Fa. \tag{5.97}$$

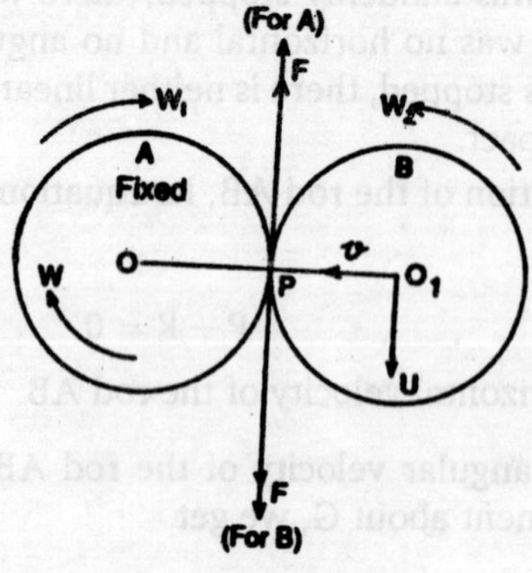

FIG 5.16

For the disc B

$$m(u-0)=F \tag{5.98}$$

$$\text{and} \quad m\cdot\frac{1}{2}a^2(\omega_2-0)=Fa. \tag{5.99}$$

Also, since the two discs touch at P, therefore, the velocity of the point of contact A as deduced from each disc, must be equal,

$$\text{i.e., } a\omega_1 = u + a\omega_2. \tag{5.100}$$

From equations (5.97), (5.98) and (5.99), we get

$$\tfrac{a}{2}(\omega_1-\omega)=-\tfrac{F}{m};\quad \omega_1-\omega=-\tfrac{2F}{am}$$

$$\omega_1=\omega-\tfrac{2F}{am};\qquad u=\tfrac{F}{m}$$

$$\text{and } \tfrac{ma}{2}\omega_2=F;\qquad \omega_2=\tfrac{2F}{am}.$$

Hence from equation (5.100), we get

$$a\left(\omega-\frac{2F}{am}\right)=u+\frac{2F}{m}=\frac{F}{m}+\frac{2F}{m}$$

$$\text{or,}\quad a\omega-\frac{2F}{m}=\frac{3F}{m},\ \text{or, } a\omega=\frac{5F}{m}$$

$$\therefore\quad F=\frac{ma\omega}{5}.$$

Now total kinetic energy before the impact

$$=\tfrac{1}{2}m\tfrac{a^2}{2}\omega^2+\tfrac{1}{2}mv^2.$$

And kinetic energy after the impact

$$= \frac{1}{2}m\frac{a^2}{2}\omega_1^2 + \frac{1}{2}m\left(u^2 + \frac{a^2}{2}\omega_2^2\right)$$
$$= \frac{1}{2}m\frac{a^2}{2}\frac{9}{25}\omega^2 + \frac{1}{2}m\left(\frac{1}{25}a^2\omega^2 + \frac{a^2}{2}\frac{4}{25}\omega^2\right)$$
$$= \frac{9a^2m\omega^2}{100} + \frac{a^2\omega^2m}{50}(1+2) = a^2m\omega^2\left(\frac{9}{100} + \frac{3}{50}\right)$$
$$= a^2m\omega^2\left(\frac{15}{100}\right) = \frac{3ma^2\omega^2}{20}.$$

Hence loss in kinetic energy

$$= \frac{1}{2}m\frac{a^2}{2}\omega^2 + \frac{1}{2}mv^2 - \frac{3ma^2\omega^2}{20} = \frac{m}{2}\left(v^2 + \frac{a^2\omega^2}{5}\right).$$

Problem 14. *A sphere of mass m falls with velocity v on a perfectly rough inclined plane of mass M and angle α, which rests on a smooth horizontal plane. Show that the vertical velocity of the centre of the sphere immediately after the impact is $\frac{5(M+m)v\sin^2\alpha}{7M+2m+5m\sin^2\alpha}$, the bodies being all supposed perfectly inelastic.*

Solution: Let G be the centre and a be the radius of the sphere.

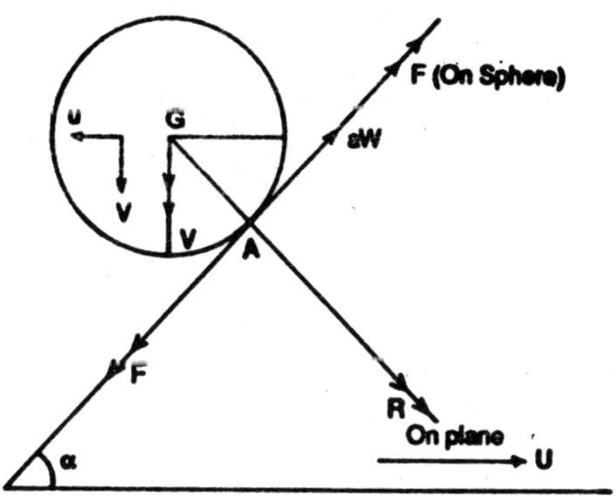

FIG 5.17

Just before the impact the velocity of the sphere is v in vertically downward direction. Therefore, there is no horizontal velocity and no vertical velocity then.

Just after the impact, let u and v be the horizontal and vertical velocities and ω be the angular velocity of the sphere. Just after the impact, the inclined plane also begins to move in the horizontal direction. Let its velocity be u.

Let R be the impulsive normal reaction and F the impulsive friction between the sphere and the inclined plane.

Now equations of motion of the sphere are

$$m(u-0) = R\sin\alpha - F\cos\alpha \tag{5.101}$$
$$m(v-V)) = -R\cos\alpha - F\sin\alpha \tag{5.102}$$
and $$m\cdot\frac{2}{5}a^2(\omega-0) = aF. \tag{5.103}$$

For the motion of the inclined plane

$$MU = R\sin\alpha - F\cos\alpha. \tag{5.104}$$

Also, the horizontal velocity of the point A of the sphere is the same as that of the same point A of the inclined plane,

$$\text{i.e., } u - a\omega\cos\alpha = -U \tag{5.105}$$

and vertical velocity of the point A of the sphere is the same as that of the same point A of the inclined plane,

$$\text{i.e., } v - a\omega\sin\alpha = 0. \tag{5.106}$$

From equations (5.101) and (5.104), we see that

$$MU = mu;\ U = \frac{mu}{M}.$$

From equation (5.105),

$$u - a\omega\cos\alpha = -\frac{mu}{M}$$

$$u = a\omega\cos\alpha - \frac{mu}{M}$$

$$\Rightarrow\quad u\left(1 + \frac{m}{M}\right) = a\omega\cos\alpha$$

$$\Rightarrow\quad u = \frac{Ma\omega\cos\alpha}{M+m}. \tag{5.107}$$

Now equation (5.101) $\times \cos\alpha$ and equation (5.102) $\times \sin\alpha$ gives

$$mu\cos\alpha = R\sin\alpha\cos\alpha - F\cos^2\alpha$$

$$\text{and}\quad m(v - V)\sin\alpha = -R\cos\alpha\sin\alpha - F\sin^2\alpha.$$

Adding these two, we get

$$mu\cos\alpha + m(v - V)\sin\alpha = -F = -\frac{2am w}{5}\ \text{[From equation (5.103)]}$$

$$\Rightarrow\quad u\cos\alpha + (v - V)\sin\alpha = -\frac{2a\omega}{5}.$$

From equation (5.107), it becomes

$$\Rightarrow\quad \frac{Ma\omega\cos^2\alpha}{M+m} + (v - V)\sin\alpha + \frac{2a\omega}{5} = 0.$$

$$\Rightarrow\quad \left[\frac{M\cos^2\alpha}{M+m} + \frac{2}{5}\right] a\omega + (v - V)\sin\alpha = 0$$

$$\Rightarrow\quad \left[\frac{M\cos^2\alpha}{M+m} + \frac{2}{5}\right] \frac{v}{\sin\alpha} + (v - V)\sin\alpha = 0.$$

From equation (5.106)

$$\Rightarrow \left[\frac{M\cos^2\alpha}{M+m}+\frac{2}{5}+\sin^2\alpha\right]v = V\sin^2\alpha$$

$$\Rightarrow [5M\cos^2\alpha + 2(M+m) + 5(M+m)\sin^2\alpha]v$$
$$= V\sin^2\alpha 5(M+m).$$

$$\Rightarrow [5M+2M+2m+5m\sin^2\alpha]v = V\sin^2\alpha 5(M+m)$$

$$\Rightarrow [7M+2m+5m\sin^2\alpha]v = 5(M+m)V\sin^2\alpha$$

$$\Rightarrow v - \frac{5(M+m)V\sin^2\alpha}{7M+2m+5m\sin^2\alpha}.$$

Type II: Problems based on Section 5.2

Problem 15. *If a hollow lawn tennis ball of elasticity e has on striking the ground supposed perfectly rough, a vertical velocity v and an angular velocity ω about a horizontal axis, find its angular velocity after impact, and prove that the range of the rebound will be $\left(\frac{4a\omega}{5g}\right)$ ev.*

Solution: Let O be the centre, M the mass of ball and A the point of contact with the ground. Just before the impact, v is the vertical velocity and ω the angular velocity of the ball.

Just after the impact, let u', v' be the horizontal and vertical velocities and ω' the angular velocity.

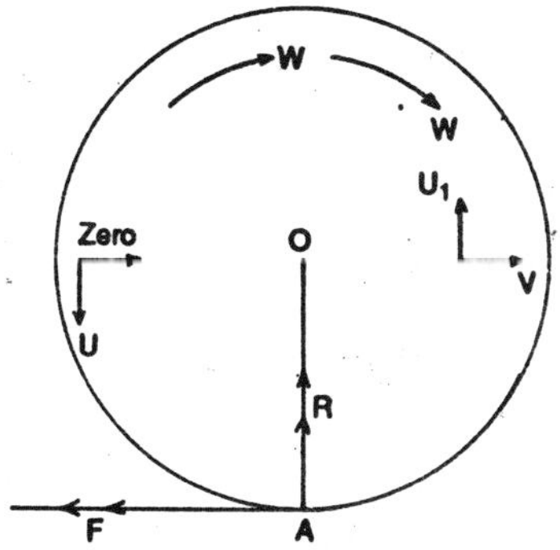

FIG 5.18

Let R be the normal impulsive reaction and F the impulsive friction. The equations of motion are

$$M(v'-0) = -F \tag{5.108}$$

$$M(u'+u) = R \tag{5.109}$$

$$\text{and}\quad M\frac{2}{3}a^2(\omega'-\omega) = Fa. \tag{5.110}$$

Also, since the point of contact A is reduced to rest instantaneously, there being no sliding.

$$v' - a\omega' = 0. \tag{5.111}$$

Since e is the coefficient of restitution,

$$\therefore\ u' = eu. \tag{5.112}$$

From equations (5.108) and (5.110), we get

$$M\frac{2}{3}a^2(\omega' - \omega) = -Mv'a$$

$$\Rightarrow\ \frac{2}{3}a(\omega' - \omega) = -v'$$

$$\Rightarrow\ \frac{2}{3}(v' - a\omega) = -v' \text{ [From equation (5.111)]}$$

$$\Rightarrow\ \frac{5}{3}v' = \frac{2}{3}a\omega, \text{ or, } v' = \frac{2}{5}a\omega. \tag{5.113}$$

Now the range after rebound

$$= \frac{2}{g}\text{(horizontal velocity) (vertical velocity)}$$

$$= \frac{2}{g}v'u' = \frac{2}{g}\frac{2}{5}a\omega \cdot eu = \frac{4a\omega}{5g}eu.$$

Problem 16. *A tennis ball of hollow spherical shape is given an 'underground cut' and hits the ground at the other side of the net at a distance c from it. If u and v be its horizontal and vertical velocities and ω its angular velocity when it hits the perfectly rough ground, show that the ball will return back towards the net of 2aω > 3u.*

Further show that it will rebound over the net if

$$c < \frac{2ev(2a\omega - 3u)}{5g}$$

and will touch the net overhead if

$$h < \frac{5c}{2}\frac{2ev(2a\omega - 3u) - 5gc}{(2a\omega - 3u)^2},$$

where e is the coefficient of restitution and h is the height of the net.

Solution: Let G be the centre, a the radius and M the mass of the ball striking the ground at the point A, at a distance OA = c, from the net ON.

Let u_1, v_1 and ω_1 be the horizontal, vertical and angular velocities of the ball just after the impact while these were u, v, ω, respectively just before the impact as shown in the figure.

Let R be the impulsive normal reaction and F the impulsive friction. Then the equations of motion of the ball are

$$M(u_1 + u) = F \tag{5.114}$$

$$M(v_1 + v) = R \tag{5.115}$$

$$\text{and}\quad M\frac{2a^2}{3}(\omega_1 - \omega) = -Fa. \tag{5.116}$$

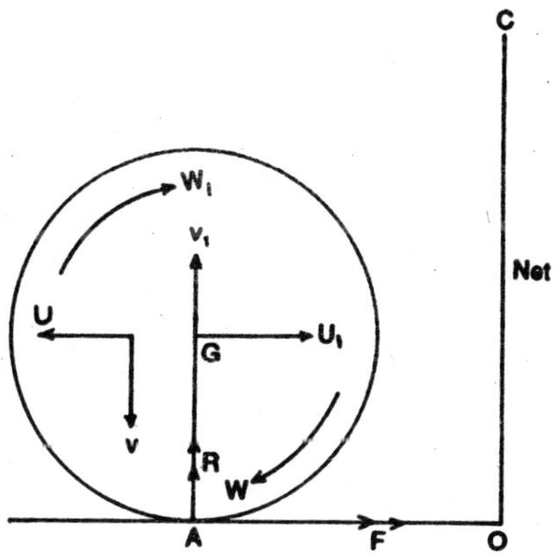

FIG 5.19

Also by Newton's law

$$v_1 = ev. \tag{5.117}$$

Also, since there is no horizontal velocity of the point of contact just after the impact

$$u_1 - a\omega_1 = 0. \tag{5.118}$$

From equations (5.114) and (5.116), we get

$$M(u_1 + u) = -\frac{2aM}{3}(\omega_1 - \omega)$$

$$\Rightarrow \quad 3(u_1 + u) = -2a(\omega_1 - \omega) = -2(a\omega_1 - a\omega)$$

$$\Rightarrow \quad 3u_1 + 3u = -2(u_1 - a\omega) \text{ [From equation (5.118)]}$$

$$\Rightarrow \quad 5u_1 = 2a\omega - 3u$$

$$\Rightarrow \quad u_1 = \frac{1}{5}(2a\omega - 3u). \tag{5.119}$$

The ball will return back towards the net if u_1 is positive,

$$\text{i.e., } 2a\omega - 3u > 0; \text{ or, } 2a\omega > 3u.$$

When the ball rebounds from the ground, it moves in a trajectory.
Its initial horizontal velocity is u_1 and its initial vertical velocity is v_1, i.e., ev.
Considering the horizontal motion of the ball, we have time t to reach the net

$$= \frac{c}{u_1} = \frac{5c}{2a\omega - 3u} \text{ [From equation (5.119)]} \tag{5.120}$$

This time must be less than the time of reaching the ground,

$$\text{i.e., } \quad \frac{5c}{2a\omega - 3u} < \frac{2ev}{g},$$

$$\text{i.e., } \quad c < \frac{2ev(2a\omega - 3u)}{5g}.$$

If h_1 is the vertical height of the ball risen in time t_2, then

$$\begin{aligned} h_1 &= v_1 t_2 - \frac{1}{2} g t_2^2 = e v t_2 - \frac{1}{2} g t_2^2 \text{ [From equation (5.117)]} \\ &= \frac{t_2}{2}(2ev - g t_2) \\ &= \frac{5c}{2a\omega - 3u}\left(2ev - \frac{g5c}{2a\omega - 3u}\right) \text{ [From equation (5.120)]} \\ &= \frac{5c}{(2a\omega - 3u)^2}[2ev(2a\omega - 3u) - 5gc]. \end{aligned}$$

The ball will touch the net over head of $h < h_1$

$$\therefore\ h < \frac{5c}{2}\frac{2ev(2a\omega - 3u) - 5gc}{(2a\omega - 3u)^2}.$$

Type III: Problems based on Section 5.3

Problem 17. *A light rigid rod ABC has three particles each of mass m attached to its A, B, C. The rod is struck by a blow P at right angles to it at a point, distance from A equal to BC. Prove that the kinetic energy set-up is* $\frac{1}{2}\frac{P^2(a^2-ab+b^2)}{m(a^2+ab+b^2)}$, *where* $AB = a$, $BC = b$.

Solution: Let D be the point in AB, where the blow is imparted.

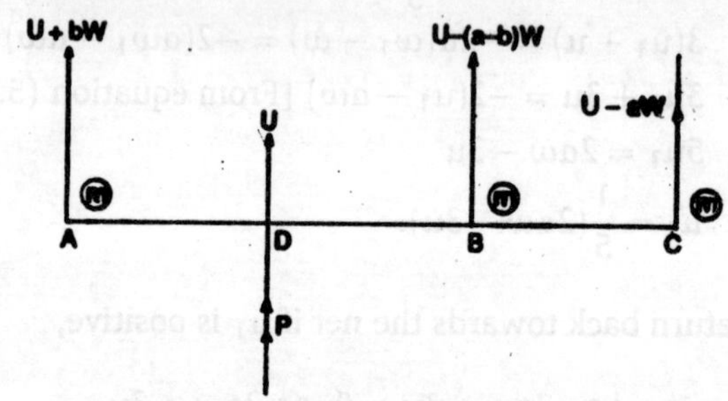

FIG 5.20

Given that

$$AD = BC = b. \therefore\ BD = a - b.$$

Before the impulse the system is at rest. Just after the impulse, let u be the velocity of the point D and ω the angular velocity of the rod. Therefore, the velocities of masses at A, B and C will be $u + b\omega$, $u - (a-b)\omega$ and $u - a\omega$, respectively.

Now, since the change in momentum = impulsive forces,

$$\begin{aligned} &\therefore\ m(u + b\omega) + m[u - (a-b)\omega] + m(u - a\omega) = P \\ &\Rightarrow\ m[u + b\omega + u - a\omega + b\omega + u - a\omega] = P \\ &\Rightarrow\ m[3u + 2b\omega - 2a\omega] = P \\ &\Rightarrow\ 3u - 2(a-b)\omega = \frac{P}{m}. \qquad (5.121) \end{aligned}$$

Also taking moment about D, the moment equation is

$$-m(u+b\omega)b + m\{u-(a-b)\omega\}(a-b) + m(u-a\omega)a = 0$$
$$\Rightarrow \quad -bu - b^2\omega + (a-b)u - (a-b)^2\omega + ua - a^2\omega = 0$$
$$\Rightarrow \quad 2u(a-b) = \omega[a^2+b^2-2ab+a^2+b^2]$$
$$\Rightarrow \quad u(a-b) = \omega(a^2-ab+b^2).$$

Thus from equation (5.121)

$$3u - \frac{2(a-b)u(a-b)}{a^2-ab+b^2} = \frac{P}{m}$$
$$\Rightarrow \quad u[3a^2-3ab+3b^2-2(a^2-2ab+b^2)] = \frac{P}{m}(a^2-ab+b^2)$$
$$\Rightarrow \quad u[a^2+ab+b^2] = \frac{P}{m}(a^2-ab+b^2)$$
$$\Rightarrow \quad u = \frac{P}{m}\frac{(a^2-ab+b^2)}{(a^2+ab+b^2)}.$$

$$\text{K.E. set up} = \tfrac{1}{2}P(V+V_1) = \tfrac{1}{2}P(O+u) = \frac{P^2(a^2-ab+b^2)}{2m(a^2+ab+b^2)}.$$

Problem 18. *Four freely-jointed rods of the same material and thickness, form a rectangle of sides 2a and 2b and of mass M′. When lying in this form on a horizontal plane an inelastic particle of mass M moving with velocity V in a direction perpendicular to the rod of length 2a impinges on it at a distance from its centre. Show that the kinetic energy lost in the impact is* $\frac{1}{2}V^2 \div \left[\frac{1}{M} + \frac{1}{M'}\left(1 + \frac{3a+3b}{a+3b}\cdot\frac{c^2}{a^2}\right)\right]$.

Solution: Let AB, BC, CD, DA be four rods forming a rectangle of sides $2a$ and $2b$ and mass M' such that $BC = AD = 2b$. If m_1 is the mass of rod AB or CD and m_2 the mass of BC or AD, then

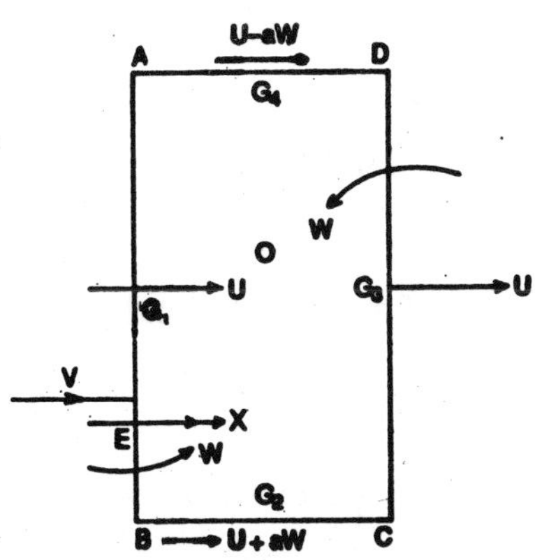

FIG 5.21

$$\frac{m_1}{2a} = \frac{m_2}{2b}$$

or, $$\frac{2m_1}{4a} = \frac{2m_2}{4b} = \frac{2m_1 + 2m_2}{4a + 4b} = \frac{M'}{4a + 4b}$$

$$m_1 = \frac{2aM'}{4(a+b)} = \frac{aM'}{2(a+b)},$$

$$m_2 = \frac{bM'}{2(a+b)}.$$

The particle of mass M strikes AB at the point E with a velocity V at right angles to AB, where $EG_1 = c$. Let X be the impulse of the blow at E.

After the blow, let u be the velocity of C.G. at G_1 of the rod AB and ω its angular velocity. Since CD is parallel to AB, so that velocity of G_3 will also be u, perpendicular to CD and ω its angular velocity.

Now velocity of E

= velocity of G_1 + velocity of E relative to G_1

$= u + c\omega$.

Velocity of G_2

= same as velocity of B

= velocity of G_1 + velocity of B relative to G_1

$= u + a\omega$.

Velocity of G_3

= velocity of C + velocity of G_3 relative to C

$= u + a\omega - a\omega = u$.

Velocity of G_4

= same as velocity of A

= velocity of G_1 + velocity of A relative to G_1

$= u - a\omega$.

The equation of motion for the system (i.e., four rods and the particle) is

$$M'u + M(u + c\omega) = MV \tag{5.122}$$

(By principle of conservation of momentum) and by taking the moments about the centre O of the rectangle.

$$\Big[m_1 \cdot \frac{1}{3}a^2\omega + m_1 \cdot \frac{1}{3}a^2\omega + m_2(u + a\omega)a - m_2(u - a\omega)a + M(u + c\omega)c\Big] - [MVc] = 0. \tag{5.123}$$

Also considering motion of the particle M only, we have

$$M[(u + c\omega) - V] = -X. \tag{5.124}$$

From equation (5.123), we have

$$\frac{2}{3}m_1 a^2\omega + 2m_2 a^2\omega + MC(u + c\omega) - McV = 0$$

$$\text{or,} \quad 2a^2\omega\left[\frac{m_1}{3} + m_2\right] + Mc(u + c\omega) - McV = 0.$$

From equation (5.121)

$$2a^2\omega\left[\frac{1}{3}\frac{aM'}{2(a+b)} + \frac{bM'}{2(a+b)}\right] + Mc(u + c\omega) - McV = 0$$

$$\Rightarrow \frac{2a^2\omega M'}{2(a+b)}\left(\frac{a}{3} + b\right) + Mc(u + c\omega) - McV = 0$$

$$\Rightarrow \frac{a^2\omega M'(a+3b)}{3(a+b)} + Mc(u + c\omega) - McV = 0$$

$$\Rightarrow \frac{M'}{K}c^2\omega + Mc(u + c\omega) - McV = 0,$$

where

$$K = \frac{3(a+b)c^2}{a^2(a+3b)}$$

$$\Rightarrow M'c\omega + K[M(u + c\omega) - Mv] = 0. \tag{5.125}$$

Adding equations (5.122) and (5.125), we get

$$M'(u + c\omega) + M(u + c\omega)(K + 1) = MVK + MV$$

$$\Rightarrow (u + c\omega)[M' + M(1 + K)] - MV(K + 1) = 0. \tag{5.126}$$

From equation (5.124)

$$u + c\omega - V = -\frac{X}{M}$$

$$u + c\omega = V - \frac{X}{M}$$

so from equation (5.126), we have

$$\left(V - \frac{X}{M}\right)[M' + M(1 + K)] = MV(1 + K)$$

$$\Rightarrow VM' = \frac{X}{M}[M' + M(1 + K)]$$

$$\Rightarrow VM' = \frac{XM'}{M} + X(1 + K)$$

$$\Rightarrow VM' = \frac{X}{M}[M' + M(1 + K)]$$

$$\Rightarrow X = \frac{MM'V}{M' + M(1 + K)}. \tag{5.127}$$

But the loss of energy $= \frac{1}{2}X(V+V_1)$.
Hence loss of energy

$$= \text{loss in energy of the particle} + \text{loss in energy of the four rods}$$
$$= \frac{1}{2}X[V+(u+c\omega)] - \frac{1}{2}X[0+(u+c\omega)]$$
$$= \frac{1}{2}XV = \frac{1}{2}\frac{MM'V^2}{M'+M(1+K)} \quad \text{[From equation (5.127)]}$$
$$= \frac{1}{2}V^2 \div \frac{[M'+M(1+K)]}{MM'}$$
$$= \frac{1}{2}V^2 \div \left[\frac{1}{M}+\frac{1}{M'}(1+K)\right]$$
$$= \frac{1}{2}V^2 \div \left[\frac{1}{M}+\frac{1}{M}\left(1+\frac{3a+3b}{a+3b}\frac{c^2}{a^2}\right)\right].$$

5.6 Exercise 5

Type I

1. Two equal uniform rods, AB and AC, are freely hinged at A and rest in a straight line on a smooth table. A blow is struck at B, perpendicular to the rods. Show that the kinetic energy generated is $\frac{7}{4}$ times which it would be if the rods were rigidly fastened together at A. **[CU 88H; VU 88H]**

2. Two uniform rods AB, BC are freely jointed at B and laid on a horizontal table AB is struck by a horizontal blow of impulse P in a direction perpendicular to AB at a distance c from its centre, the lengths of AB, BC being $2a$ and $2b$ and their masses M and M′. Find the motion immediately after the blow. **[Vikram 86]**

   [*Ans.* The four quantities which determine motion are

$$u_1 = \frac{P}{M}\left[1-\frac{1}{4}\frac{M'}{M+M'}\left(1+\frac{3c}{a}\right)\right]$$
$$\text{and } \omega_1 = \frac{3P}{Ma}\left[\frac{c}{a}-\frac{1}{4}\frac{M'}{M+M'}\left(1+\frac{3c}{a}\right)\right]$$
$$u_2 = \frac{1}{4}\frac{P}{M+M'}\left(1+\frac{3c}{a}\right)$$
$$\text{and } \omega_2 = -\frac{3}{4}\frac{P}{(M+M')b}\left(1+\frac{3c}{b}\right).]$$

3. Two equal uniform rods, AB and BC, are freely jointed at B and turn about a smooth joint at A. When the rods are in a straight line, ω being the angular velocity of AB and u the velocity of the centre of mass of BC, BC impinges on a fixed inelastic obstacle at a point D. Show that the rods are instantaneously brought to rest if $BD = 2a\frac{2u-a\omega}{3u+2a\omega}$, where $2a$ is the length of either rod.

4. Two equal uniform rods, AB and AC, are freely jointed at A, and are placed on a smooth table so as to be at right angles. The rod AC is struck by a blow at C in a direction perpendicular to itself. Show that the resulting velocities of the middle points of AB and AC are in the ratio 2 : 7. [CU 89H, 01H; VU 93H]

5. AB and CD are two equal and similar rods connected by a string BC. AB, BC and CD form three sides of a square. The point A of the rod AB is struck by a blow in a direction perpendicular to the rod. Show that the initial velocity of A is seven times that of D.

6. Two uniform rods, AB and AC, are freely jointed at A and laid on a smooth horizontal table so that the angle BAC is a right angle. The rod AB is struck by a blow P at B in a direction perpendicular to AB. Show that the initial velocity of A is $\frac{2P}{4m'+m}$, where m, m' are the masses of AB, AC, respectively.

[VU 87H, CU 96H, 04H]

7. Four equal uniform rods AB, BC, CD and DE are freely jointed at B, C and D and lie on a smooth table in the form of a square. The rod AB is struck by a blow at A at right angles to AB from the inside of the square. Show that the initial velocity of A is 79 times that of E.

8. A square lamina ABCD rests on a smooth horizontal plane. If the corner A is made to move with velocity u along the line BA produced, then determine the initial angular velocity of the lamina. [*Ans.* $-\frac{3u}{8a}$, where $AB = a$]

9. A billiard ball is at rest on a horizontal table and is struck by a horizontal blow in a vertical plane passing through the centre of the ball. If the initial motion is one of pure rolling, find the height of the point struck above the table.

[*Ans.* $\frac{7a}{5}$, where a is the radius of the ball]

10. A square plate of side $2a$ is falling with velocity u. A diagonal being vertical, when an elastic string attached to the middle point of an upper edge becomes tight in a vertical position. Show that the impulsive tension of the string is $\frac{4}{7}Mu$, where M is the mass of the plate.

11. A uniform inelastic rod falls without rotation, being inclined at any angle to the horizon and hits a smooth fixed peg at a distance from its upper end equal to one third of its length. Show that the lower end begins to descend vertically.

12. Four equal rods, each of mass m and length $2a$, are freely jointed at their ends so as to form a rhombus. The rhombus falls with a diagonal, vertical and is moving with velocity v when it hits a fixed horizontal inelastic plane. Find the motion of the rods immediately after the impact, and show that their angular velocities are each equal to $\frac{3V\sin\alpha}{2a(1+3\sin^2\alpha)}$, where α is the angle, each rod makes with the vertical.

Show also that the impact destroys a friction $\frac{3\sin^2\alpha}{1+3\sin^2\alpha}$ of the kinetic energy just before the impact.

13. An inelastic sphere of radius a, rolls down a flight of perfectly rough steps; show that if the velocity of the centre on the first step exceeds $\sqrt{ga}$ its velocity will be the same on every step, the step being such that, in its flight, the sphere never impinges on an edge.

14. AB, BC and CD are three equal uniform rods lying in a straight line on a smooth plane, and they are freely jointed at B and C. A blow is applied at the centre of BC in a direction perpendicular to BC. If ω be the initial angular velocity of AB or CD, and θ the angle they make with BC at any time, show that the angular velocity is $\frac{w}{\sqrt{1+\sin^2\theta}}$.

15. A uniform flat rod of length $2a$ rests on a rough horizontal plane with its weight uniformly distributed. A horizontal force P large enough to produce motion is applied suddenly at the end, perpendicular to the length of the rod. Show that initially the rod begins to turn about a point, distance x from the middle point, where x is given by the positive roots of the equation

$$x^2 - \left(\frac{1}{3} - \frac{2P}{\mu w}\right)a^2x - \frac{2}{3}\frac{P}{\mu w}a^3 = 0$$

ω being weight of the rod and μ the coefficient of friction.

Type II

16. An imperfectly elastic sphere descending vertically comes in contact with a fixed rough point, the impact taking place at a point, distant α from the lowest point, and the coefficient of elasticity being e. Find the motion, and show that the sphere will start moving horizontally after the impact if $\alpha = \tan^{-1}\sqrt{\frac{7e}{5}}$.

[Nagpur 81; Agra 79]

17. A rough imperfectly elastic ball is dropped vertically, and when its velocity is V, a man suddenly moves his racket forward in its own plane with velocity u, and thus subjects the ball to pure cut in a downward direction making an angle α with the horizon. Show that on striking the rough ground, the ball will not proceed beyond the point of impact, provided

$$(u - V\sin\alpha)(1 - \cos\alpha) > (1+e)\left(1 + \frac{a^2}{K^2}\right)V\sin\alpha\cos\alpha.$$

Type III

18. An elliptic disc of mass m is dropped in a vertical plane with velocity V on a perfectly rough horizontal planes. Show that the loss of kinetic energy by the impact is

$$\frac{1}{2}(1-e^2)mV^2\frac{K^2+p^2}{K^2+r^2},$$

where r is the distance of the centre of the disc from the point of contact, p is the central perpendicular on the tangent and e is the coefficient of elasticity.

19. A uniform square plate ABCD of mass M and side $2a$, lies on a smooth horizontal plane. It is struck at A by a particle of mass m moving with velocity

v in the direction of AB, the particle remaining attached to the plate. Determine the subsequent motion of the system and show that its angular velocity is $\frac{m}{M+4m}\frac{3V}{2a}$. **[MU 79; Agra 71]**

20. A rod OA can turn freely in a horizontal plane about the end O and lies at rest. An insect whose mass is one third that of the rod alights on the end A and commences crawling along the rod with uniform velocity V. At the same instant the rod is set in rotation about O in such a way that the initial velocity of A is V. When the insect reaches O, prove that the rod has rotated through a right angle that the angular velocity of the rod is then twice the initial angular velocity.

21. A disc of any form moving in its plane without rotation with velocity V at right angles to a fixed plane strikes the plane, so that the distances of its centre of gravity from the point of impact and from the plane are r and p.

    Assuming the plane to be elastic and sufficiently rough to prevent sliding, show that the impulsive pressure and friction are respectively

$$\frac{m(1+e)V(p^2+K^2)}{r^2+K^2} \text{ and } \frac{m(1+e)Vp(r^2-p^2)^{1/2}}{r^2+K^2},$$

    where K is the radius of gyration.

22. A uniform rod AB of mass m and length $2a$ is at rest. It is struck by a blow of impulse P at right angles to its length at a distance c from its middle point. Find the point about which it will begin to turn.

    Find also the kinetic energy. [CU 94H]

    [*Ans.* At the point $x = \frac{P}{mw}$, w is the angular velocity. Kinetic energy $= \frac{3P^2}{8m}\frac{(a+c)^2}{a^2}$]

    [*Hints.* Type I]

Chapter 6

Conservation of Momentum and Energy

6.1 Principle of Conservation of Linear Momentum (under Finite Forces)

[TMBU 97H, 02H, BBABU 02H]

Statement. *If a body be moving under the action of some external forces whose sum of resolved parts parallel to a certain line is zero throughout the motion, the momentum of the body parallel to that line remains constant throughout the motion.*

[MU 76, 77, 80–84]

Proof. Let the fixed straight line be taken as the axis of x. Let X be the resolved part of the external force parallel to x-axis, acting on a particle of mass m whose coordinates are (x, y, z) at time t. Then equation of motion is

$$\Sigma m \frac{d^2x}{dt^2} = \Sigma X, \text{ or, } \frac{d}{dt}\left(\Sigma m \frac{dx}{dt}\right) = \Sigma X.$$

If $\Sigma X = 0$, then

$$\frac{d}{dt}\left(\Sigma m \frac{dx}{dt}\right) = 0.$$

Integrating, we get

$$\Sigma m \frac{dx}{dt} = \text{constant.} \qquad (6.1)$$

If $\bar{x}$ be the x-coordinate of the centre of gravity of the body of mass M, then

$$\begin{aligned} \bar{x} = \frac{\Sigma mx}{\Sigma m} &= \frac{\Sigma mx}{M} \\ \text{or, } \Sigma mx &= M\bar{x} \\ \therefore \Sigma m \frac{dx}{dt} &= M \frac{d\bar{x}}{dt}. \end{aligned}$$

Thus from equation (6.1), we have $M\frac{d\bar{x}}{dt}$ = constant which shows that the total momentum of the body measured parallel to the axis of x remains constant throughout the motion.

6.2 Principle of the Conservation of Moment or Angular Momentum (under Finite Forces) [CU 91H, 94H, 96H, 00H, 02H, MU 00H, TMBU 97H, 02H, BBABU 99H, 03H]

Statement. *If the algebraic sum of the moments of all the external forces, acting on a rigid body about a given line be zero throughout the motion, the moment of momentum (or angular momentum) of the body about the line remains unaltered throughout the motion.* [MU 76, 77, 80, 83; Raj 82, CU 93H]

Proof. Let X, Y, Z be the resolved parts of the external force parallel to the axes, acting on a particle of mass m whose coordinates are (x, y, z) at time t. Taking the given line as the axis of x, the sum of the moments of the external forces about the given line is $\Sigma(yZ - zY)$.

By D' Alembert's principle, we have

$$\Sigma m \left(y\frac{d^2z}{dt^2} - z\frac{d^2y}{dt^2} \right) = \Sigma(yZ - zY). \tag{6.2}$$

Given $\Sigma(yZ - zY) = 0$, throughout the motion, then from equation (6.2), we have

$$\Sigma m \left(y\frac{d^2z}{dt^2} - z\frac{d^2y}{dt^2} \right) = 0, \text{ or, } \frac{d}{dt}\left\{ \Sigma m \left(y\frac{dz}{dt} - z\frac{dy}{dt} \right) \right\} = 0.$$

Integrating, we have

$$\Sigma m \left(y\frac{dz}{dt} - z\frac{dy}{dt} \right) = \text{constant}. \tag{6.3}$$

But

$$y\frac{dz}{dt} - z\frac{dy}{dt}$$

is the moment about the axis of x of the velocity of the particle of mass m at (x, y, z).

The equation (6.3) states that the total moment of momentum of the body about x-axis is constant throughout the motion.

6.3 Principle of Conservation of Linear Momentum (under Impulsive Forces)

Statement. *If the sum of the impulsive forces parallel to a certain fixed line vanishes, the total momentum in that direction remains the same just before and after the application of the impulses.*

[MU 83, 84]

Proof. Let us take the fixed line as the axis of x. Then

$$\Sigma m\frac{d^2x}{dt^2} = \Sigma X, \text{ or, } \frac{d}{dt}\left[\Sigma m\frac{dx}{dt} \right] = \Sigma X. \tag{6.4}$$

If τ be the time (very small) during which the impulsive forces act, then integrating equation (6.4) with respect to t, we get

$$\left[\Sigma m\frac{dx}{dt} \right]_0^\tau = \Sigma \int_0^\tau X\,dt, \text{ or, } \Sigma m(v - u) = \Sigma X', \tag{6.5}$$

where

$$X' = \int_0^{\tau} X\,dt = \text{impulse of the forces parallel to x-axis,}$$

where u and v are the velocities before and after the impulse.

But if $\Sigma X' = 0$, then from equation (6.5), we have

$$\Sigma m(v - u) = 0; \ \Sigma mv = \Sigma mu,$$

i.e., *total momentum parallel to the axis after impulse = the total momentum parallel to the axis of x before impulse.*

6.4 Principle of Conservation of Angular Momentum (under Impulsive Forces)

[MU 78; MU 94H, CU 98H]

Statement. *If the sum of the moments of the impulses about a certain fixed line vanishes, the angular momentum about that line remains the same before and after the application of the impulses.*

Proof. Let the fixed line be taken as the axis of x, and τ is the time (very small) during which the impulsive forces act on the body. Then by D' Alembert's principle we have

$$\Sigma m\left(y\frac{d^2z}{dt^2} - z\frac{d^2y}{dt^2}\right) = \Sigma(yZ - zY). \tag{6.6}$$

Equation (6.6) can be written as

$$\frac{d}{dt}\left\{\Sigma m\left(y\frac{dz}{dt} - z\frac{dy}{dt}\right)\right\} = \Sigma(yZ - zY).$$

Integrating it with respect to t, we get

$$\left[\Sigma m\left(y\frac{dz}{dt} - z\frac{dy}{dt}\right)\right]_0^{\tau} = \Sigma\int_0^{\tau}(yZ - zY)dt = \Sigma(yZ_1 - zY_1), \tag{6.7}$$

where $Y_1 = \int_0^{\tau} Y\,dt$ and $Z_1 = \int_0^{\tau} Z\,dt$.

L.H.S. of equation (6.7) is the angular momentum after impulse minus angular momentum before impulse.

R.H.S. of equation (6.7) is the sum of the moments of the impulses about x-axis.

If $\Sigma(yZ_1 - zY_1) = 0$, i.e., the sum of the moments of impulses about x-axis is zero, we have, *total Angular momentum after impulse = total Angular momentum before impulse.*

6.5 Principle of Conservation of Energy **[MU 83, 85; LNMU 93H, BBABU 00H, CU 00H]**

Statement. *If a system moves under the action of finite forces, and if the geometrical equations of the system do not contain the time explicitly, the change in the kinetic energy of the system in passing from one configuration to any other is equal to the corresponding work done by the forces.*

Proof. Let X, Y, Z be the resolved parts parallel to the axes of the external forces acting on a particle m whose coordinates are (x, y, z) at the time t.

Then by D' Alembert's principle, the forces

$$X - m\frac{d^2x}{dt^2},\ Y - m\frac{d^2y}{dt^2},\ Z - m\frac{d^2z}{dt^2},$$

acting at the point (x, y, z) and the similar forces acting on the other particles of the system form a system of forces in equilibrium.

Let δx, δy, δz be the small virtual displacements of the particle m at (x, y, z), consistent with the geometrical conditions of the system at time t, then by the principle of virtual work

$$\Sigma\left[\left(X - m\frac{d^2x}{dt^2}\right)\delta x + \left(Y - m\frac{d^2y}{dt^2}\right)\delta y + \left(Z - m\frac{d^2z}{dt^2}\right)\delta z\right] = 0. \tag{6.8}$$

If the geometrical conditions are expressed by the equations which do not contain time explicitly, then we may replace $\delta x, \delta y, \delta z$ by the actual displacements

$$\frac{dx}{dt}\delta t,\ \frac{dy}{dt}\delta t,\ \frac{dz}{dt}\delta t.$$

Hence the equation (6.8) becomes

$$\Sigma\left[\left(X - m\frac{d^2x}{dt^2}\right)\frac{dx}{dt} + \left(Y - m\frac{d^2y}{dt^2}\right)\frac{dy}{dt} + \left(Z - m\frac{d^2z}{dt^2}\right)\frac{dz}{dt}\right]\delta t = 0,$$

$$\text{or,}\quad \Sigma m\left[\frac{d^2x}{dt^2}\frac{dx}{dt} + \frac{d^2y}{dt^2}\frac{dy}{dt} + \frac{d^2z}{dt^2}\frac{dz}{dt}\right] = \Sigma\left(X\frac{dx}{dt} + Y\frac{dy}{dt} + Z\frac{dz}{dt}\right).$$

Integrating with respect to t, we have

$$\frac{1}{2}\Sigma m\left[\left(\frac{dx}{dt}\right)^2 + \left(\frac{dy}{dt}\right)^2 + \left(\frac{dz}{dt}\right)^2\right]_{t_1}^{t_2}$$

$$= \Sigma\int_{t_1}^{t_2}(X\,dx + Y\,dy + Z\,dz). \tag{6.9}$$

That is the change in the kinetic energy of the system from time t_1 to time t_2 is equal to the work done by the external forces on the body from the configuration of the body at time t_1 to the configuration at time t_2.

6.6 Vis-Viva

Vis-Viva of a particle is the product of its mass and the square of its velocity.

So if v be the velocity of a particle of mass m, then its Vis-Viva $= mv^2$.

Hence Vis-Viva of a particle is equal to twice its kinetic energy.

6.7 Principle of Vis-Viva

If a system moves under the action of finite forces and if the geometrical relations of the parts of the system be expressed by equations which do not contain the time explicitly, the change in the vis-viva of the system in passing from any position to any other is equal to twice the corresponding work done by the forces.

The proof is same as that of the principle of conservation of energy in 6.5.

6.8 Conservative Forces

If the forces acting on the body are such that $\Sigma(X\,dx + Y\,dy + Z\,dz)$ is the complete differential of some quantity V, then the forces are said to be conservative.

Thus when forces are conservative, we have

$$\Sigma(X\,dx + Y\,dy + Z\,dz) = dV$$

$$\therefore \quad \Sigma\int(X\,dx + Y\,dy + Z\,dz) = \int dV. \tag{6.10}$$

Hence the quantity $\Sigma\int(X\,dx + Y\,dy + Z\,dz)$ is independent of the path followed by the body in going from one position to the other and depends only on the configuration of the body at times t_1 and t_2.

From equation (6.10), we have

$$\Sigma\int_{t_1}^{t_2}(X\,dx + Y\,dy + Z\,dz) = \int_{t_1}^{t_2} dV. \tag{6.11}$$

Also, we have

$$\frac{1}{2}\Sigma m\left[\left(\frac{dx}{dt}\right)^2 + \left(\frac{dy}{dt}\right)^2 + \left(\frac{dz}{dt}\right)^2\right]_{t_1}^{t_2}$$

$$= \Sigma\int_{t_1}^{t_2}(X\,dx + Y\,dy + Z\,dz). \tag{6.12}$$

From equations (6.11) and (6.12), we have

$$\frac{1}{2}\Sigma m\left[\left(\frac{dx}{dt}\right)^2 + \left(\frac{dy}{dt}\right)^2 + \left(\frac{dz}{dt}\right)^2\right]_{t_1}^{t_2} = \int_{t_1}^{t_2} dV = V_{t_1} - V_{t_2}.$$

$\therefore$ kinetic energy at time t_2 − kinetic energy at time t_1

$$= V_{t_1} - V_{t_2} = V_B - V_A,$$

where A and B are the configuration of the body at times t_1 and t_2.

6.9 Conservation of Energy

When a body moves under the action of a system of conservative forces, the sum of its kinetic and potential energies is constant throughout the motion.

We have just shown that if A and B are the configurations of the body at times t_1 and t_2, then kinetic energy at the time t_2- kinetic energy at the time t_1

$$= V_B - V_A. \tag{6.13}$$

The potential energy of the body in any position A is the work the forces do in moving the body from this position A to a *standard* position C (say).

Then the potential energy at the time t_1

$$= \int_A^C (X\,dx + Y\,dy + Z\,dz) = \int_C^A dV = V_C - V_A \tag{6.14}$$

and the potential energy at the time t_2

$$= \int_B^C (X\,dx + Y\,dy + Z\,dz) = \int_B^C dV = V_C - V_B. \tag{6.15}$$

Subtracting equation (6.15) from equation (6.14), we have, potential energy at the time t_1- potential energy at the time t_2

$$= (V_C - V_A) - (V_C - V_B) = V_B - V_A. \tag{6.16}$$

Comparing equations (6.13) and (6.16), we see that

Kinetic energy at the time t_2- kinetic energy at the time t_1 = P.E. at the time t_1- P.E. at the time t_2

Kinetic energy at the time t_2+ P.E. at the time t_2 = kinetic energy at the time t_1+ P.E. at the time t_1, i.e., the sum of the kinetic and potential energies is the same at all times.

6.10 The Kinetic Energy of a rigid body moving in any manner

[MU 91H, 97H, 99H, 03H, 05H, BBABU 92H, 94H, 96H, 98H, 01H, TMBU 92H, 97H, 00H, 01H, 03H, 04H]

Theorem 6.10.1 *The kinetic energy of a rigid body, moving in any manner, is at any instant equal to the kinetic energy of the whole mass, supposed to be collected at its centre of inertia and moving with it, together with the kinetic energy of the whole mass relative to its centre of inertia.*

Proof. Let (x, y, z) be the coordinates of any particle of mass m at the point P of the body, at time t, referred to the axes OX, OY, OZ through a fixed point O. Let $(\bar{x}, \bar{y}, \bar{z})$ be the coordinates of G, the centre of gravity of the body.

We consider the axes GX′, GY′, GZ′, the axes through C, parallel to the axes OX, OY, OZ, respectively. If (x', y', z') are the coordinates of the point P referred to the axes through G, then

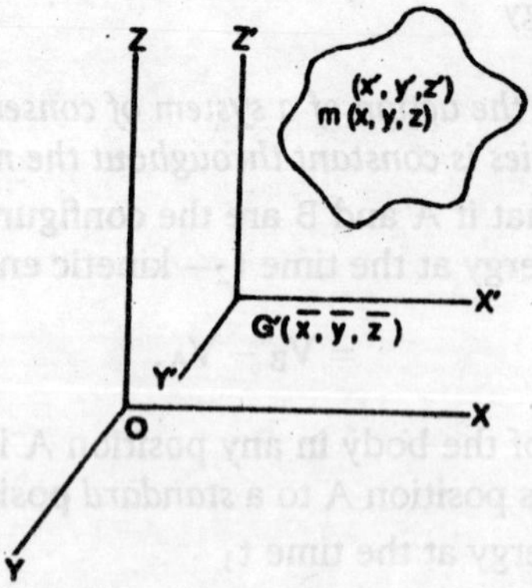

FIG 6.1

$$x = \bar{x} + x',\ y = \bar{y} + y',\ z = \bar{z} + z'$$

$$\therefore\ \frac{dx}{dt} = \frac{d\bar{x}}{dt} + \frac{dx'}{dt},\ \frac{dy}{dt} = \frac{d\bar{y}}{dt} + \frac{dy'}{dt},\ \frac{dz}{dt} = \frac{d\bar{z}}{dt} + \frac{dz'}{dt}.$$

The total kinetic energy of the body

$$= \frac{1}{2}\Sigma m\left[\left(\frac{dx}{dt}\right)^2 + \left(\frac{dy}{dt}\right)^2 + \left(\frac{dz}{dt}\right)^2\right]$$

$$= \frac{1}{2}\Sigma m\left[\left(\frac{d\bar{x}}{dt} + \frac{dx'}{dt}\right)^2 + \left(\frac{d\bar{y}}{dt} + \frac{dy'}{dt}\right)^2 + \left(\frac{d\bar{z}}{dt} + \frac{dz'}{dt}\right)^2\right]$$

$$= \frac{1}{2}\Sigma m\left[\left(\frac{d\bar{x}}{dt}\right)^2 + \left(\frac{d\bar{y}}{dt}\right)^2 + \left(\frac{d\bar{z}}{dt}\right)^2\right]$$

$$+ \frac{1}{2}\Sigma m\left[\left(\frac{dx'}{dt}\right)^2 + \left(\frac{dy'}{dt}\right)^2 + \left(\frac{dz'}{dt}\right)^2\right]$$

$$+ \Sigma m\frac{dx'}{dt}\frac{d\bar{x}}{dt} + \Sigma m\frac{dy'}{dt}\frac{d\bar{y}}{dt} + \Sigma m\frac{dz'}{dt}\frac{d\bar{z}}{dt}. \qquad (6.17)$$

Now referred to GX′, GY′, GZ′ as the axes, the coordinates of G are (0, 0, 0).

$$\therefore\ \frac{\Sigma m x'}{\Sigma m} = 0$$

$$\therefore\ \Sigma m x' = 0,\ \Sigma m y' = 0,\ \Sigma m z' = 0$$

$$\therefore\ \Sigma m\frac{dx'}{dt} = \frac{d}{dt}\Sigma m x' = 0$$

$$\therefore\ \Sigma m\frac{d\bar{x}}{dt}\frac{dx'}{dt} = \frac{d\bar{x}}{dt}\Sigma m\frac{dx'}{dt} = 0.$$

Similarly,

$$\Sigma m\frac{dy'}{dt} = 0,\quad \Sigma m\frac{dz'}{dt} = 0,$$

$$\frac{d\bar{y}}{dt}\Sigma m\frac{dy'}{dt} = 0,\quad \frac{d\bar{z}}{dt}\Sigma m\frac{dz'}{dt} = 0.$$

Thus from equation (6.17), we have:
Total kinetic energy of the body

$$= \frac{1}{2}\Sigma m\left[\left(\frac{d\bar{x}}{dt}\right)^2+\left(\frac{d\bar{y}}{dt}\right)^2+\left(\frac{d\bar{z}}{dt}\right)^2\right]$$
$$+\frac{1}{2}\Sigma m\left[\left(\frac{dx'}{dt}\right)^2+\left(\frac{dy'}{dt}\right)^2+\left(\frac{dz'}{dt}\right)^2\right]$$
$$= \frac{1}{2}\left[\left(\frac{d\bar{x}}{dt}\right)^2+\left(\frac{d\bar{y}}{dt}\right)^2+\left(\frac{d\bar{z}}{dt}\right)^2\right]\Sigma m$$
$$+\frac{1}{2}\Sigma m\left[\left(\frac{dx'}{dt}\right)^2+\left(\frac{dy'}{dt}\right)^2+\left(\frac{dz'}{dt}\right)^2\right]$$
$$= \frac{M}{2}\left[\left(\frac{d\bar{x}}{dt}\right)^2+\left(\frac{d\bar{y}}{dt}\right)^2+\left(\frac{d\bar{z}}{dt}\right)^2\right]$$
$$+\frac{1}{2}\Sigma m\,(\text{velocity of } m \text{ relative to } G)^2,$$

where M is the total mass of the body

$$= \frac{1}{2}MV^2+\frac{1}{2}\Sigma m\,(\text{velocity of } m \text{ relative to } G)^2,$$

where V is the velocity of the centre of gravity G.

Hence the total kinetic energy of the body = the kinetic energy of the total mass M supposed to be collected at G and moving with it + the kinetic energy of the body relative to G.

6.11 Worked out Examples

Type I: Problems based on Section 6.1 and 6.2

Problem 1. *A uniform circular wire of radius a lies on a smooth horizontal table and is movable about a fixed point O in its circumference. An insect of mass equal to that of the wire, starts from the other end of the diameter through O and crawls along the wire. Show that at the end of the time* t*, the wire has turned through an angle* $\frac{vt}{2a}-\frac{1}{\sqrt{3}}\tan^{-1}\left(\frac{1}{\sqrt{3}}\tan\frac{vt}{2a}\right)$.

Solution: The insect is at P after time t starting from A and consequently

$$\angle AOP = \angle OPC = \frac{\theta}{2}.$$

In ΔOAP,

$$\cos\frac{\theta}{2} = \frac{OP}{OA}=\frac{OP}{2a}$$
$$\therefore\ OP = 2a\cos\frac{\theta}{2}. \qquad (6.18)$$

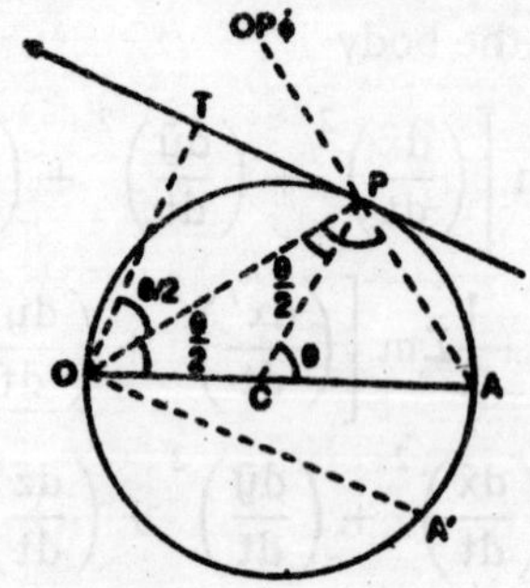

FIG 6.2

Again as the insect is moving with uniform velocity V,

$$\therefore\ \text{arc } AP = vt$$
$$\text{or, } a\theta = vt$$
$$\therefore\ \theta = \frac{vt}{a}. \tag{6.19}$$

From O we draw OT perpendicular to tangent at P, so that in ΔOPT,

$$\cos\frac{\theta}{2} = \frac{OT}{OP} = \frac{OT}{2a\cos\frac{\theta}{2}} \quad \text{[From equation (6.18)]}$$

$$OT = 2a\cos^2\frac{\theta}{2} \tag{6.20}$$

Let the diameter OA turn through an angle ϕ during this time t.

The forces acting on the system are

1. reaction at the point O acting vertically upwards;
2. weight of the insect acting vertically downwards; and
3. weight of the wire acting vertically downwards.

Since moment of all the forces about the vertical axis through O vanishes throughout the motion, the moment of momentum of the system about that axis remains unchanged throughout the motion.

Now the insect has a velocity v along the tangent PT at P and a velocity $OP\cdot\dot{\phi}$ perpendicular to OP due to rotation of the wire.

Hence the moment of momentum of the insect about a vertical through O

$$= m(v\cdot OT + OP\dot{\phi}OP)$$

$$= m\left[2av\cos^2\frac{\phi}{2} + \left(2a\cos\frac{\theta}{2}\right)^2\dot{\phi}\right] \quad \text{[By equations (6.18) and (6.19)]}$$

$$= m\left[2av\cos^2\frac{\phi}{2} + 4a^2\cos^2\frac{\theta}{2}\dot{\phi}\right]. \tag{6.21}$$

Also the moment of momentum of the circular wire about the vertical axis through O (C.G. describing a circle of radius a with angular velocity ϕ)

$$= \text{the moment of momentum of mass } m \text{ moving with the C.G.} + \text{moment of momentum about the C.G.}$$

$$= m[a\dot{\phi}\cdot a + a^2\dot{\phi}] \text{ [By the formula } m\dot{v}p + mK^2\ddot{\theta}]$$

$$= 2ma^2\dot{\phi}. \tag{6.22}$$

Hence the total moment of momentum after time t

$$= m\left[2av\cos^2\frac{\theta}{2} + 4a^2\cos^2\frac{\theta}{2}\dot{\phi}\right] + 2a^2m\dot{\phi} \text{ [By equations (6.21) and (6.22)]}$$

Initially, the body was at rest and we know that the moment of momentum of the system remains unchanged.

$$\therefore \quad m\left[2av\cos^2\frac{\theta}{2} + 4a^2\cos^2\frac{\theta}{2}\dot{\phi}\right] + 2ma^2\dot{\phi} = 0$$

$$\Rightarrow \quad 2a^2\left(1 + 2\cos^2\frac{\theta}{2}\right)\dot{\phi} = -2av\cos^2\frac{\theta}{2}$$

$$\Rightarrow \frac{d\phi}{dt} = -\frac{v\cos^2\frac{\theta}{2}}{a\left(1+2\cos^2\frac{\theta}{2}\right)} = \frac{-v\cos^2\frac{vt}{2a}}{a\left[1+2\cos^2\frac{vt}{2}\right]} \text{ [By equation (6.19)]}$$

$$\Rightarrow d\phi = \frac{-v\cos^2\frac{vt}{2a}}{a\left(1+2\cos^2\frac{vt}{2}\right)}dt = -\frac{v}{2a}\left[1 - \frac{1}{1+2\cos^2\frac{vt}{2}}\right]dt$$

$$= -\frac{v}{2a}\left[1 - \frac{\sec^2\frac{vt}{2}}{\sec^2\frac{vt}{2}+2}\right]dt$$

$$\int_0^{\phi} d\phi = -\frac{V}{2a}\int_0^t \left[1 - \frac{\sec^2\frac{vt}{2}}{\tan^2\frac{vt}{2}+3}\right]dt$$

$$\phi = -\frac{v}{2a}[t - I_1].$$

For I_1, let $y = \tan\frac{vt}{a^2}$.

$$\frac{2a}{V}dy = \sec^2\frac{vt}{2}dt$$

$$I_1 = \int \frac{2a\,dy}{v(y^2+3)} = \frac{2a}{v}\frac{1}{\sqrt{3}}\tan^{-1}\frac{y}{\sqrt{3}}$$

$$\phi = -\frac{v}{2a}\left[t - \frac{2a}{v}\frac{1}{\sqrt{3}}\tan^{-1}\left(\frac{1}{\sqrt{3}}\tan\frac{vt}{2a}\right)\right]$$

$$= -\left[\frac{vt}{2a} - \frac{1}{3}\tan^{-1}\left(\frac{1}{\sqrt{3}}\tan\frac{vt}{2a}\right)\right].$$

The negative sign shows that the wire moves in a direction opposite to the insect.

Problem 2. *A circular disc is moving with an angular velocity Ω about an axis through its centre perpendicular to its plane. An insect alights on its edge and crawls* *along a curve drawn on the disc in the form of a lemniscate with uniform relative angular velocity $\frac{1}{8}\Omega$, the curve touching the edge of the disc. The mass of the insect being $\frac{1}{16}$th of that of the disc, show that the angle turned through by the disc when the insect gets to the centre is $\frac{24}{\sqrt{7}}\tan^{-1}\frac{\sqrt{7}}{3}-\frac{\pi}{4}$.*

Solution: Let the disc turn through an angle ϕ in time t, i.e., $\angle AOA' = \phi$.

Initially, A′ coincides with A, where the insect alights the edge of the disc and starts crawling along the arc of the lemniscate.

The equation of the lemniscate referred to the centre O of the disc as pole and OA′ the initial line is

$$r^2 = a^2\cos 2\theta. \tag{6.23}$$

At time t, when the disc has turned through an angle ϕ, let the insect crawl to the point $P(r,\theta)$, where $\angle POA' = \theta$ and $OP = r$.

Since all the forces acting on the system are vertical or pass through O, so the sum of their moments about the vertical axis through O is zero and hence the moment of momentum of the system remains unaltered, i.e., about the vertical axis through O.

The moment of momentum of the system at time t

$$= \text{Moment of momentum of the system at } t = 0$$

$$\Rightarrow (MK^2\dot{\phi}+0)+\frac{1}{16}M\left(r\dot{\phi}r + r^2\cdot\frac{1}{8}\omega\right) = (MK^2\omega+0)+0$$

$$\Rightarrow \frac{1}{2}Ma^2\dot{\phi}+\frac{1}{16}Mr^2\dot{\phi}+\frac{Mr^2}{128}\omega = \frac{1}{2}Ma^2\omega$$

$$\Rightarrow \frac{(8a^2+r^2)}{16}\dot{\phi}M = \frac{M\omega(64a^2-r^2)}{128}$$

$$\Rightarrow (8a^2+r^2)\dot{\phi} = \frac{1}{8}(64a^2-r^2)\omega$$

$$\Rightarrow \dot{\phi} = \frac{d\phi}{dt} = \frac{1}{8}\left(\frac{64a^2-a^2\cos^2\theta}{8a^2+a^2\cos^2\theta}\right)\omega \text{ [From equation (6.23)]}$$

$$\Rightarrow \frac{d\phi}{d\theta}\frac{d\theta}{dt} = \frac{1}{8}\left(\frac{64-\cos 2\theta}{8+\cos 2\theta}\right)\omega$$

$$\Rightarrow \frac{d\phi}{d\theta}\frac{1}{8}\omega = \frac{1}{8}\left(\frac{64-\cos 2\theta}{8+\cos 2\theta}\right)\omega \left(\because \frac{d\theta}{dt}=\frac{1}{8}\omega\right)$$

$$\Rightarrow d\phi = \left(\frac{64-\cos 2\theta}{8+\cos 2\theta}\right)d\theta. \tag{6.24}$$

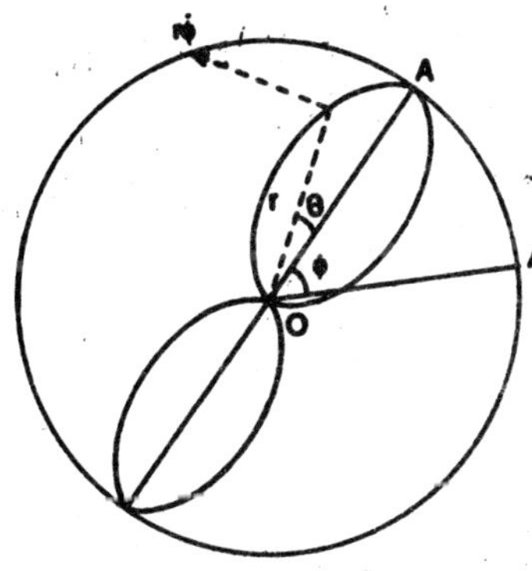

FIG 6.3

When the insect reaches the centre of the disc, then $\theta = \frac{\pi}{4}$. If during this time the disc turns through an angle ϕ, then integrating equation (6.24), we get

$$\int_0^{\phi} d\phi = \int_0^{\pi/4} \frac{64 - \cos 2\theta}{8 + \cos 2\theta} d\theta.$$

$$\begin{aligned}\phi &= \int_0^{\pi/4} \left(-1 + \frac{72}{8 + \cos 2\theta}\right) d\theta = -[\theta]_0^{\pi/4} + \int_0^{\pi/4} \frac{72}{8 + \frac{1-\tan^2\theta}{1+\tan^2\theta}} d\theta \\ &= -\frac{\pi}{4} + 72\int_0^{\pi/4} \frac{\sec^2\theta d\theta}{9 + 7\tan^2\theta} = -\frac{\pi}{4} + \frac{72}{7}\int_0^{\pi/4} \frac{\sec^2\theta d\theta}{\tan^2\theta + \frac{9}{7}}.\end{aligned}$$

Let $y = \tan\theta$.

$$\therefore dy = \sec^2\theta d\theta$$

$$\begin{aligned}\therefore \phi &= -\frac{\pi}{4} + \frac{72}{7}\int_0^{\pi/4} \frac{dy}{y^2 + \frac{9}{7}} = -\frac{\pi}{4} + \frac{72}{7}\frac{\sqrt{7}}{3}\left[\tan^{-1}\left(\frac{y\sqrt{7}}{3}\right)\right]_0^{\pi/4} \\ &= -\frac{\pi}{4} + \frac{24}{\sqrt{7}}\left[\tan^{-1}\left\{\tan\theta\frac{\sqrt{7}}{3}\right\}\right]_0^{\pi/4} = -\frac{\pi}{4} + \frac{24}{\sqrt{4}}\left[\tan^{-1}\left(\frac{\sqrt{7}}{3}\right)\right].\end{aligned}$$

Problem 3. *A uniform circular board of mass M and radius a, is placed on a perfectly smooth and horizontal plane, and free to rotate about a vertical axis through its centre. A man of mass M′, walks round the edge of the board whose upper surface is rough enough to prevent his slipping, when he has walked completely round the board to his starting point, show that the board has turned through an angle* $\frac{M'}{M+2M'}4\pi$.

Solution: At any time t let the board has turned through an angle θ and radius to the man makes an angle ϕ with the fixed line OA′.

The forces acting on the system are all vertical; therefore their moments about the vertical through O are zero throughout the motion; hence moment of momentum about this axis through O remains unaltered, throughout the motion,

$$\text{i.e., } M\frac{a^2}{2}\dot{\theta} + M'a^2\dot{\phi} = 0.$$

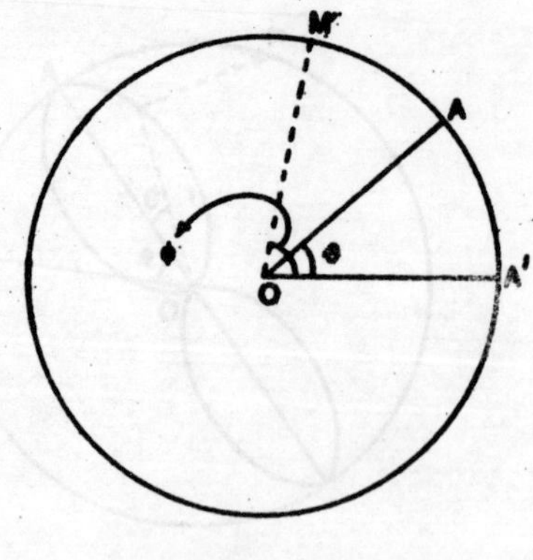

FIG 6.4

Integrating, we get

$$\frac{M}{2}\theta + M'\phi = C.$$

Initially when the man is at A

$$\theta = 0,\ \phi = 0,\ \therefore\ C = 0$$

$$M\theta + 2M'\phi = 0. \tag{6.25}$$

When the man returns to his starting point A,

$$\phi - \theta = 2\phi.$$

Hence from equation (6.25),

$$M\theta + 2M'(\theta + 2\pi) = 0, \quad \text{or,} \quad \theta(M + 2M') = -4\pi M'$$

$$\text{or,} \quad \theta = -\frac{M'}{M + 2M'}4\pi.$$

The negative sign means that the board turns in a direction opposite to the man.

Problem 4. *A circular ring, of mass M and radius a, lies on a smooth horizontal plane, and an insect, of mass m, resting on it starts and walks round it with uniform velocity v relative to the ring. Show that the centre of the ring describes a circle with angular velocity* $\frac{m}{M+2m}\frac{v}{a}$ [MU 81, 84; CU 83H]

Solution: Let initially the insect be resting at the point A of the circular ring of centre O and mass M. If G is the common centre of gravity of the ring and the insect at A, then

$$M \cdot OG = m \cdot AG$$

$$\text{or,} \quad Mx = m(OA - OG)$$

$$= m(a - x)$$

$$\text{or,} \quad x(M + m) = ma$$

$$\therefore \quad x = \frac{am}{M + m}$$

$$\text{and} \quad a - x = \frac{aM}{M + m}.$$

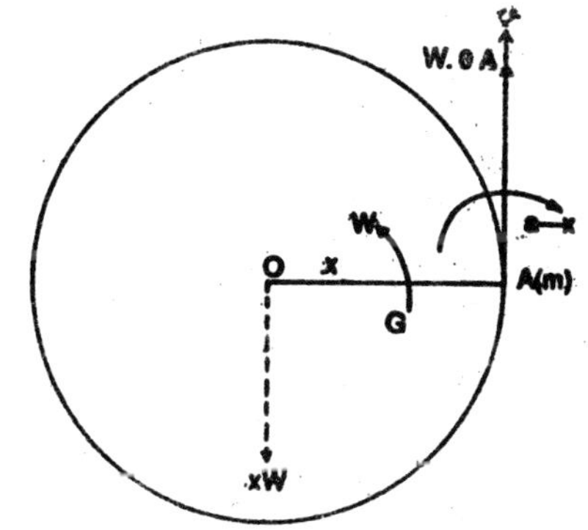

FIG 6.5

Since the external forces, (1) the weight of the ring, (2) the weight of the insect and (3) the reaction of the ring are all vertical, so there is no horizontal component of the force throughout the motion. Hence the total horizontal momentum of the system will remain unaltered throughout the motion. But the initial velocity of G is zero. Therefore the velocity of the common centre of gravity G in the horizontal plane will be zero throughout the motion which means that G is fixed.

Hence the centre O of the ring will describe a circle of radius

$$OG = x = \frac{am}{m+M}$$

about the centre G.

Further since all the external forces acting on the system are vertical, therefore, their moments about the vertical axis through G will be zero, so the moment of momentum about the vertical axis through G will remain constant throughout the motion.

If ω is the angular velocity of the ring about G which was initially at rest, then we have

$$Ma^2\omega + M\omega x^2 + (m\omega GA^2 + mvGA) = 0$$

$$\Rightarrow \quad M\omega(a^2+x^2) + m\omega(a-x)^2 + mv(a-x) = 0$$

$$\Rightarrow \quad \omega[M(a^2+x^2) + m(a-x)^2] = -mv(a-x)$$

$$\Rightarrow \quad \omega[M\{(a-x)^2 + 2ax\} + m(a-x)^2] = -mv(a-x)$$

$$\Rightarrow \quad \omega[(M+m)(a-x)^2 + 2aMx] = -mv(a-x)$$

$$\Rightarrow \quad \omega\left[(M+m)\left(\frac{aM}{m+M}\right)^2 + 2aM\frac{am}{M+m}\right] = -mv\frac{aM}{M+m}$$

$$\Rightarrow \quad \omega\left[\frac{a^2M^2}{(m+M)} + \frac{2a^2Mm}{M+m}\right] = -\frac{avMm}{M+m}$$

$$\Rightarrow \quad \omega a^2M[M+2m] = -avMm$$

$$\Rightarrow \quad \omega = -\frac{m}{M+2m}\frac{v}{a}.$$

The negative sign shows that the ring will turn in a direction opposite to the direction of motion of the insect.

Hence the centre of the ring describes a circle with angular velocity

$$\frac{m}{M+2m}\frac{v}{a}.$$

Type II: Problems based on Section 6.3 and 6.4

Problem 5. *a lamina in the form of an ellipse is rotating in its own plane about one of its foci with angular velocity ω. This focus is set free and the other focus, at the same instant is fixed. Show that the ellipse now rotates about it with angular velocity* $\omega\left(\frac{2-5e^2}{2+3e^2}\right)$. **[MU 77, 81, 82; Raj 81; Agra 77, 82; Garhwal 82]**

Solution: Let C be the centre of an elliptic lamina of mass M rotating with angular velocity ω about the focus S. In this case, the velocity of the centre C will be $\omega CS = \omega ae$, perpendicular to SC in the plane of the lamina.

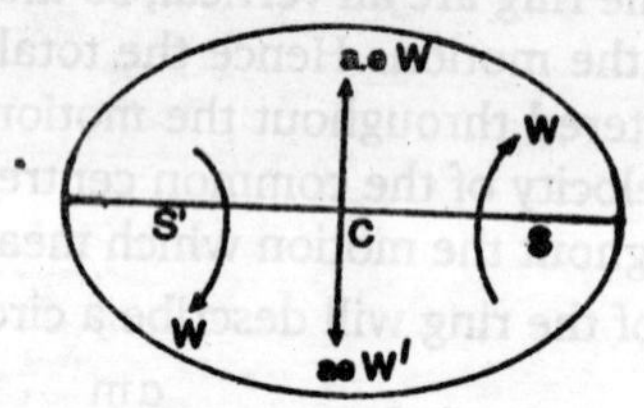

FIG 6.6

The focus S is freed and the other focus S′ is fixed. Let ω_1 be the angular velocity of the lamina about S′, just after its fixture. Since the impulse is at S′, therefore, the moment of this impulse about S′ will be zero.

Hence the angular momentum of the lamina about S′, just after its fixture = angular momentum just before its fixture

$$\Rightarrow \quad MK^2\omega_1 + M\omega_1 S'C \cdot S'C = MK^2\omega - M\omega SC \cdot SC$$

$$\Rightarrow \quad M\left(\frac{a^2+b^2}{4}\right)\omega_1 + M(ae)^2\omega_1 = M\left(\frac{a^2+b^2}{4}\right)\omega - M\omega a^2e^2$$

$$\Rightarrow \quad \omega_1(a^2+b^2+4a^2e^2) = \omega[a^2+b^2-4a^2e^2]$$

$$\Rightarrow \quad \omega_1 = \left[\frac{a^2+b^2-4a^2e^2}{a^2+b^2+4a^2e^2}\right]\omega = \left[\frac{a^2+a^2(1-e^2)-4a^2e^2}{a^2+a^2(1-e^2)+4a^2e^2}\right]\omega$$

$$= \left[\frac{2-e^2-4e^2}{2-e^2+4e^2}\right]\omega = \left[\frac{2-5e^2}{2+3e^2}\right]\omega.$$

Problem 6. *An elliptic lamina is rotating about its centre on a smooth horizontal table. If* $\omega_1, \omega_2, \omega_3$ *be its angular velocities when the extremity of its major axis, its focus and the extremity of its minor axis, respectively become fixed, prove that* $\frac{7}{\omega_1} = \frac{6}{\omega_2} + \frac{5}{\omega_3}$. **[Agra 81; Raj 79, 80]**

Solution: Let an elliptic lamina of mass M be rotating with angular velocity ω about the centre C on a smooth horizontal table.

Firstly, the centre is freed and suddenly an extremity A of the major axis is fixed.

Then ω_1 is the angular velocity of the disc about A.

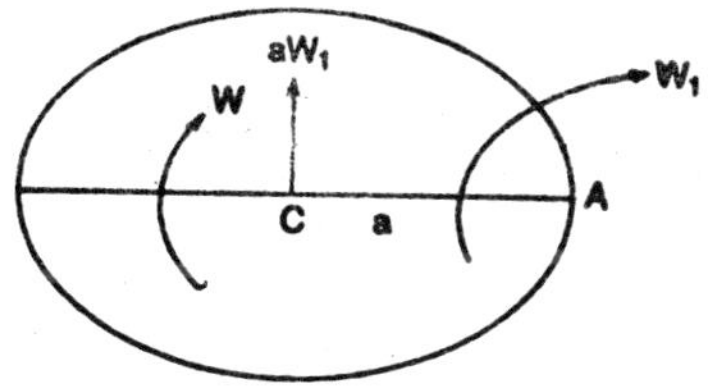

FIG 6.7

Since the impulse is at A, therefore, the moment of the impulse about the vertical axis through A is zero.

Hence, the angular momentum of the disc about the vertical axis through, just after the end A is fixed = angular momentum of the disc about the same axis, just before the end A is fixed.

$$\text{or,}\quad MK^2\omega_1 + Ma\omega_1 AC = MK^2\omega + 0$$

$$\text{or,}\quad M\left(\frac{a^2+b^2}{4}\right)\omega_1 + Ma^2\omega_1 = M\left(\frac{a^2+b^2}{4}\right)\omega$$

$$\text{or,}\quad \frac{\omega_1}{4}(5a^2+b^2) = \frac{\omega}{4}(a^2+b^2).\ \therefore\ \omega_1 = \omega\left(\frac{a^2+b^2}{5a^2+b^2}\right).$$

Secondly, when the centre C is freed and the focus S is fixed, the angular velocity of the disc about S is ω_2.

Since the impulse is at S, therefore, the moment of the impulse about the vertical axis through S is zero.

Hence, the angular momentum of the disc about the vertical axis through S, just after the focus S is fixed = the angular momentum of the disc about the same axis, just before the focus S is fixed.

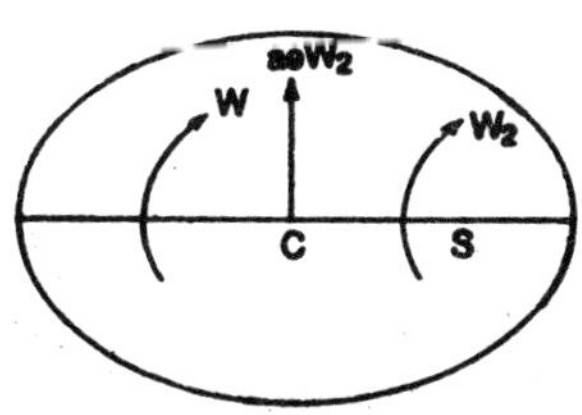

FIG 6.8

$$\Rightarrow\quad MK^2\omega_2 + M\cdot CS\omega_2 CS = MK^2\omega + 0$$

$$\Rightarrow\quad M\left(\frac{a^2+b^2}{4}\right)\omega_2 + Ma^2e^2\omega_2 - M\left(\frac{a^2+b^2}{4}\right)\omega$$

$$\Rightarrow\quad \frac{\omega_2}{4}[a^2+b^2+4a^2e^2] = \frac{\omega}{4}(a^2+b^2)$$

$$\Rightarrow\quad \omega_2[a^2+b^2+4a^2-4b^2] = \omega(a^2+b^2)\ [\because\ b^2 = a^2(1-e^2)]$$

$$\Rightarrow\quad \omega_2[5a^2-3b^2] = \omega(a^2+b^2)$$

$$\therefore\quad \omega_2 = \frac{\omega(a^2+b^2)}{5a^2-3b^2}.$$

Thirdly, when the centre C is freed and the end B of the minor axis is fixed the angular velocity of the disc about B is ω_2.

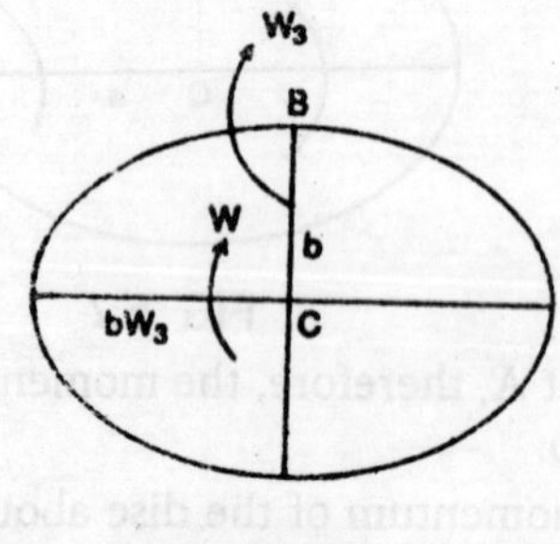

FIG 6.9

Since the impulse is at B, therefore, the moment of the impulse about the vertical axis through C is zero.

Hence, the angular momentum of the disc about the vertical axis through B, just after the end B is fixed = the angular momentum of the disc about the same axis just before the end B is fixed.

$$\Rightarrow \quad MK^2\omega_3 + M\cdot BC\cdot \omega_3 BC = MK^2\omega + 0$$

$$\Rightarrow \quad M\left(\frac{a^2+b^2}{4}\right)\omega_3 + Mb^2\omega_3 = M\left(\frac{a^2+b^2}{4}\right)\omega$$

$$\Rightarrow \quad \frac{\omega_3}{4}(a^2+5b^2) = (a^2+b^2)\frac{\omega}{4}, \text{ or, } \omega_3 = \frac{(a^2+b^2)\omega}{a^2+5b^2}.$$

Now,

$$\text{R.H.S.} = \frac{6}{\omega_2} + \frac{5}{\omega_3} = \frac{6(5a^2-3b^2)}{(a^2+b^2)\omega} + \frac{5(a^2+5b^2)}{(a^2+b^2)\omega}$$

$$= \frac{35a^2+7b^2}{(a^2+b^2)\omega} = \frac{7(5a^2+b^2)}{\omega(a^2+b^2)} = \frac{7}{\omega_1} = \text{L.H.S.}$$

Problem 7. *A uniform circular disc is spinning with angular velocity ω about a diameter when a point P on its rim is suddenly fixed. If the radius vector to P makes an angle α with this diameter, show that the angular velocities after the fixing about the tangent and normal at P are $\frac{1}{5}\omega\sin\alpha$ and $\omega\cos\alpha$.*

Solution: Let the uniform circular disc of mass M and centre O spin with angular velocity ω about the diameter AB. The resolved parts of ω about the tangent PT and normal PO at P are $\omega\sin\alpha$ and $\omega\cos\alpha$, respectively.

The diameter AB is set free and the point P on the circumference is fixed. Let ω_1 and ω_2 be the angular velocities of the circular disc about the tangent PT and normal PO just after the point P is fixed.

Since the impulse is at P, therefore, its moment about the tangent PT and the normal PO at P are zero.

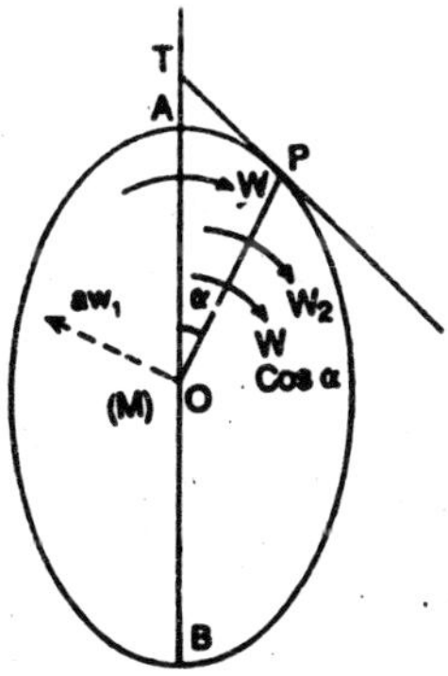

FIG 6.10

Hence, angular momentum of the disc about PT after the point P is fixed = angular momentum of the disc about PT before the point P is fixed.

$$\Rightarrow \quad MK^2\omega_1 + Ma\omega_1 \cdot PO = MK^2\omega \sin\alpha + 0$$

$$\Rightarrow \quad \frac{Ma^2}{4}\omega_1 + Ma^2\omega_1 = \frac{Ma^2}{4}\omega \sin\alpha$$

$$\Rightarrow \quad \frac{5Ma^2}{4}\omega_1 = \frac{Ma^2}{4}\omega \sin\alpha$$

$$\Rightarrow \quad \omega_1 = \frac{\omega}{5}\sin\alpha.$$

Similarly, angular momentum of the disc about PO after the point P is fixed = angular momentum of the disc about PO before the point P is fixed.

$$\text{or,} \quad MK^2\omega_2 + 0 = MK^2\omega\cos\alpha$$

$$\text{or,} \quad \frac{a^2}{4}\omega_2 = \frac{Ma^2}{4}\omega\cos\alpha$$

$$\omega_2 = \omega\cos\alpha.$$

Problem 8. *A right angled triangular lamina is turning about the side AC. Suddenly AC is loosed and BC fixed. If C be the right angle, prove that the angular velocity is $\frac{1}{2}\frac{BC}{CA}$ of its former value.*

Solution: Let M be the mass of the triangular lamina ABC turning with angular velocity ω about AC. The triangle is kinetically equivalent to three particles each of mass $\frac{M}{3}$ placed at the middle points D, E, F of the 3 sides BC, CA and AB, respectively.

Since D and E describe circles of radius $\frac{BC}{2} = \frac{a}{2}$ about AC, so the velocities of D and E about AC is $\frac{1}{2}a\omega$ perpendicular to the lamina.

When AC is released and BC is fixed suddenly let ω' be the angular velocity of the triangle about AC. E and F describe circles of radius $\frac{AC}{2} = \frac{b}{2}$ about BC, so the velocities of E and F about BC are $\frac{1}{2}b\omega$ perpendicular to the lamina. On fixture of BC, the impulse is at BC, therefore, the moment of the impulse about BC is zero.

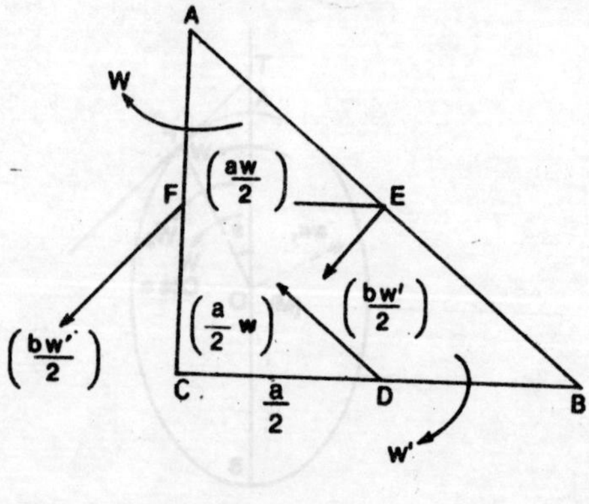

FIG 6.11

$\therefore$ the moment of momentum about BC after its fixture = The moment of momentum about BC before its fixture

$$\Rightarrow \quad \frac{M}{3}\cdot\frac{1}{2}b\omega' CF + \frac{M}{3}\cdot\frac{1}{2}b\omega' DE = \frac{M}{3}\cdot\frac{1}{2}a\omega DE$$

$$\Rightarrow \quad b\omega'\frac{b}{2} + b\omega'\frac{b}{2} = a\omega\cdot\frac{b}{2}$$

$$\Rightarrow \quad b^2\omega' = \frac{ab\omega}{2}$$

$$\Rightarrow \quad \omega' = \frac{a\omega}{2b} = \frac{1}{2}\frac{BC}{CA}\omega.$$

Problem 9. *A uniform square lamina of mass M and side 2a is moving freely about a diagonal with uniform angular velocity ω when one of the corners not in that diagonal becomes fixed. Show that the new angular velocity is $\frac{1}{7}\omega$ and that the impulse of the force on the fixed point is $\frac{\sqrt{2}}{7}Ma\omega$.* [MU]

Solution: Let ABCD be the square lamina of mass M and side 2a, moving freely about the diagonal BD with uniform angular velocities. When the diagonal BD is set free and the corner A not in diagonal BD is fixed, let ω' be the angular velocity about AL, a line through A and parallel to BD.

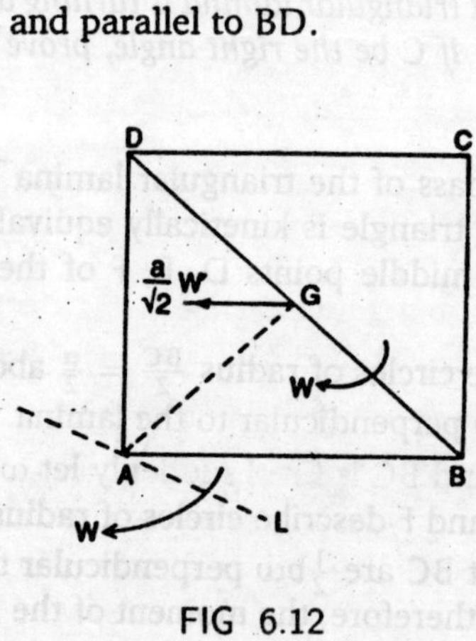

FIG 6.12

Initially, the velocity of G is zero as BD is fixed. When A is fixed, G describes a circle of radius $AG = a\sqrt{2}$, with A as centre, so its velocity, then is $a\sqrt{2}\omega'$, perpendicular to the lamina. Since A is fixed, therefore the impulse is at A and its moment about AL is zero.

$\therefore$ the angular momentum about AL after the fixture of A = the angular momentum about AL before the fixture of A.

$$\Rightarrow M\frac{a^2}{3}\omega' + Ma\sqrt{2}\omega' \cdot AG = MK^2 + 0. \tag{6.26}$$

$\therefore$ K^2 about BD $= \frac{1}{3}a^2\cos^2 45° + \frac{1}{3}a^2\sin^2 45° = 2\frac{a^2}{3}\frac{1}{2} = \frac{a^2}{3}$.

Thus, from equation (6.26), we get

$$\frac{a^2}{3}\omega' + a\sqrt{2}\omega' \cdot a\sqrt{2} = \frac{a^2}{3}\omega$$

$$\Rightarrow \frac{\omega'}{3} + 2\omega' = \frac{\omega}{3}$$

$$\Rightarrow \frac{7}{3}\omega' = \frac{\omega}{3}$$

$$\omega' = \frac{\omega}{7}.$$

Now, the impulse at A = change in momentum

$$= M(a\sqrt{2}\omega' - 0) - Ma\sqrt{2}\frac{1}{7}\omega - \frac{\sqrt{2}}{7}Ma\omega.$$

Problem 10. *A cube is turning with angular velocity ω about a diagonal when suddenly the diagonal is let go and one of the edges which does not meet this diagonal is fixed. Show that the resulting angular velocity about this edge is $\frac{\omega\sqrt{3}}{12}$.* **[MU 82]**

Solution: Let M be the mass and $2a$ the side of the cube turning with angular velocity ω about the diagonal OP. OA, OB, OC are taken as axes of x, y, z, respectively. Then the direction cosines of the diagonal OP are $\frac{1}{\sqrt{3}}, \frac{1}{\sqrt{3}}, \frac{1}{\sqrt{3}}$.

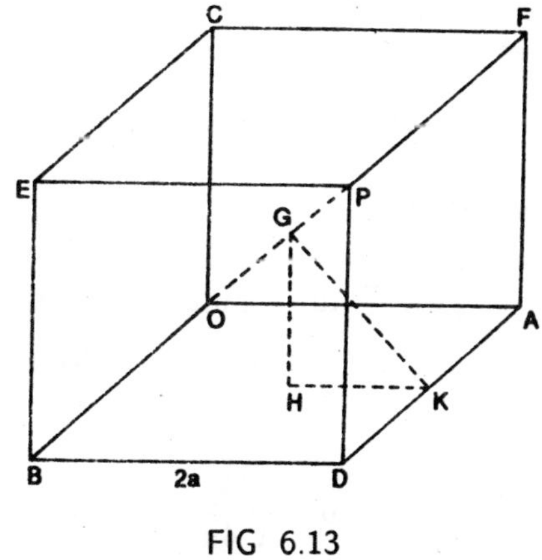

FIG 6.13

Resolved parts of ω about OA, OB, OC are $\frac{\omega}{\sqrt{3}}, \frac{\omega}{\sqrt{3}}$, respectively.

Length of the diagonal OP of the cube

$$= \sqrt{OD^2 + DP^2} = \sqrt{OA^2 + AD^2 + DP^2} = 2\sqrt{3}a.$$

Also

$$\begin{aligned} GH &= HK = a \\ GK &= \sqrt{GH^2 + HK^2} = \sqrt{2}a. \end{aligned}$$

Now the diagonal OP is freed and the edge AD which does not meet diagonal OP is fixed. Let ω' be the angular velocity of the cube about AD just after its fixture.

On fixture of AD, G will describe a circle of radius GK with K as centre. Hence velocity of G about AD is $\sqrt{2}a\omega'$ about AD, perpendicular to GK.

On fixture of AD, the impulse is on AD and so its moment about AD is zero.

Since the moment of momentum remains the same before and after the fixture, therefore, we have

$$M\frac{2}{3}a^2\omega' + M \cdot \sqrt{2}a\omega' GK = M\frac{2}{3}a^2\frac{\omega}{\sqrt{3}} + 0$$

$$\Rightarrow \quad \frac{2}{3}a^2\omega' + \sqrt{2}a\omega'\sqrt{2}a = \frac{2a^2}{3\sqrt{3}}\omega$$

$$\Rightarrow \quad \frac{2}{3}a^2\omega' + 2a^2\omega' = \frac{2a^2}{3\sqrt{3}}\omega$$

$$\frac{8}{3}a^2\omega' = \frac{2a^2}{3\sqrt{3}}\omega$$

$$\omega' = \frac{\omega}{4\sqrt{3}} = \frac{\omega\sqrt{3}}{12}.$$

Problem 11. *A uniform rod AB of mass 3m has a small hook at B and swings through an angle 60° on each side of the vertical about a fixed horizontal axis through the end A. When B is passing through its lowest point the hook picks up a particle of mass m. Show that the system will afterwards swing through an angle θ on each side of the vertical given by* $\cos\theta = \frac{17}{20}$.

Solution: Let ω be the angular velocity of the rod when B is in the lowest position, i.e., when AB is vertical.

From energy equation we have

$$\frac{1}{2}3m\frac{4a^2}{3}\omega^2 = 3mg(a - a\cos 60°)$$

$$\Rightarrow \quad 2a^2\omega^2 = 3ga\left(1 - \frac{1}{2}\right) = \frac{3ga}{2}.$$

$$\omega = \sqrt{\frac{3g}{4a}}. \tag{6.27}$$

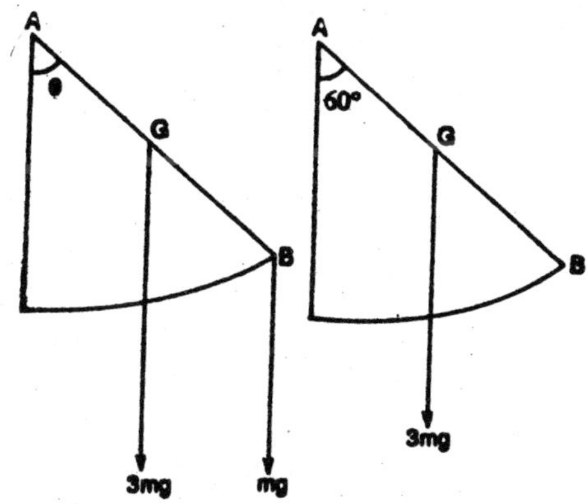

FIG. 6.14

In the lowest position a mass m is picked up by the hook at B, so that there is impulsive action at A.

As the moment of the impulsive action at A is zero about A, therefore, moment of momentum about A remains unchanged before and after impulse. If ω' be the velocity after impulse, then

$$3m\left(\frac{a^2}{3}\omega' + a^2\omega'\right) + m2a\omega' 2a = 3m\left(\frac{a^2}{3}\omega + a^2\omega\right)$$

$$\Rightarrow \quad 3m\frac{4}{3}a^2\omega' + 4a^2 m\omega' = 3m\frac{4}{3}a^2\omega$$

$$\Rightarrow \quad 8a^2 m\omega' = 4ma^2\omega$$

$$\Rightarrow \quad \omega' = \frac{\omega}{2} = \frac{1}{2}\sqrt{\frac{3g}{4a}}. \qquad \text{[From equation (6.27)]}$$

Let θ be the angle through which the rod turns after picking up m, then energy equation gives

$$\frac{1}{2}3m\frac{4a^2}{3}\omega'^2 + \frac{1}{2}m4a^2\omega'^2 = 3mg(a - a\cos\theta)$$
$$+mg(2a - 2a\cos\theta)$$

$$\Rightarrow \quad 2a^2 m\omega'^2 + 2a^2 m\omega'^2 = 5mga(1 - \cos\theta)$$

$$\Rightarrow \quad 4a^2 m\omega'^2 = 5mga(1 - \cos\theta)$$

$$\Rightarrow \quad 4a\frac{1}{4}\frac{3g}{4a} = 5g(1 - \cos\theta)$$

$$\Rightarrow \quad \frac{3g}{4} = 5g(1 - \cos\theta)$$

$$\Rightarrow \quad \frac{3}{20} = 1 - \cos\theta$$

$$\Rightarrow \quad \cos\theta = 1 - \frac{3}{20} = \frac{17}{20}.$$

Type III: Problems based on Sections 6.4 and 6.5

Problem 12. *A uniform straight rod of length 2a, has two small rings at its ends which can respectively slide on thin smooth horizontal and vertical wires OX and OY. The rod starts at an angle α to the horizon with angular velocity* $\sqrt{\left\{\frac{3g}{2a}(1-\sin\alpha)\right\}}$ *and moves downwards.*

Show that it will strike the horizontal wire at the end of the time

$$2\sqrt{\frac{a}{3g}}\log\left\{\cot\left(\frac{\pi}{8}-\frac{\alpha}{4}\right)\tan\frac{\pi}{8}\right\}.$$

[MU 83; Agra 81; Raj 81]

Solution: Let AB be the rod of mass M and length 2a. At time t, let the rod be inclined at an angle θ to the horizontal. Referred to OX and OY as axes of coordinates of C.G. at G of AB are $(a\cos\theta, a\sin\theta)$.

$$\therefore \text{ (velocity)}^2 \text{ of G} = (-a\sin\theta\dot{\theta})^2 + (a\cos\theta\dot{\theta})^2 = a^2\dot{\theta}^2.$$

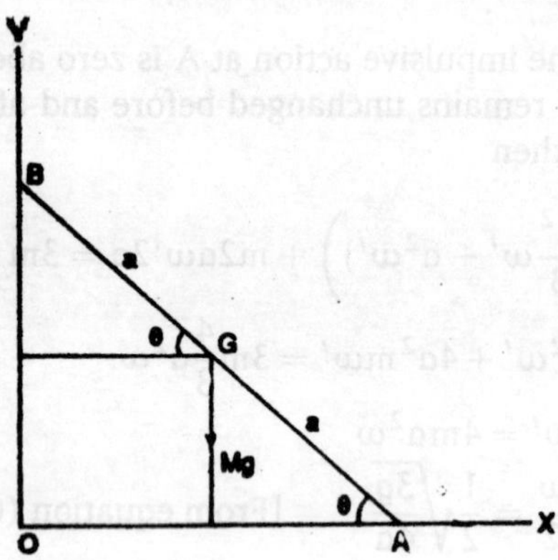

FIG 6.15

∴ kinetic energy of the rod at time t

$$= \frac{1}{2}M\frac{1}{3}a^2\dot{\theta}^2 + \frac{1}{2}Ma^2\dot{\theta}^2 = \frac{Ma^2\dot{\theta}^2}{2}\left(\frac{1}{3}+1\right)$$

$$= \frac{1}{2}Ma^2\dot{\theta}^2\frac{4}{3} = \frac{2}{3}Ma^2\dot{\theta}^2.$$

But, initially

$$\dot{\theta} = \sqrt{\frac{3g}{2a}(1-\sin\alpha)}.$$

∴ initial kinetic energy of the rod

$$= \frac{2}{3}Ma^2\frac{3g}{2a}(1-\sin\alpha) = Mga(1-\sin\alpha).$$

Hence energy equation gives

$$\text{change in K.E.} = \text{work done by the gravity}$$

$$\Rightarrow \quad \frac{2}{3}Ma^2\dot{\theta}^2 - Mga(1 - \sin\alpha) = Mg(a \sin\alpha - a\sin\theta)$$

$$\Rightarrow \quad \frac{2}{3}a\dot{\theta}^2 = g[1 - \sin\alpha + \sin\alpha - \sin\theta]$$

$$\Rightarrow \quad \frac{2}{3}a\dot{\theta}^2 = g(1 - \sin\theta)$$

$$\Rightarrow \quad \dot{\theta}^2 = \frac{3g}{2a}(1 - \sin\theta)$$

$$\therefore \quad \dot{\theta} = \frac{d\theta}{dt} = -\sqrt{\frac{3g}{2a}(1 - \sin\theta)}.$$

Here, for θ negative sign is taken because motion is towards θ decreasing.

$$\Rightarrow dt = -\sqrt{\frac{2a}{3g}}\frac{d\theta}{\sqrt{1 - \sin\theta}}$$

$$\Rightarrow \int_0^t dt = -\sqrt{\frac{2a}{3g}}\int_\alpha^0 \frac{d\theta}{\sqrt{1 - \sin\theta}}$$

$$\Rightarrow t = -\sqrt{\frac{2a}{3g}}\int_\alpha^0 \frac{d\theta}{\sqrt{\cos^2\frac{\theta}{2} + \sin^2\frac{\theta}{2} - 2\sin\frac{\theta}{2}\cos\frac{\theta}{2}}}$$

$$= -\sqrt{\frac{2a}{3g}}\int_\alpha^0 \frac{d\theta}{\sqrt{\left(\cos\frac{\theta}{2} - \sin\frac{\theta}{2}\right)^2}} = \sqrt{\frac{2a}{3g}}\int_0^\alpha \frac{d\theta}{\cos\frac{\theta}{2} - \sin\frac{\theta}{2}}$$

$$= \sqrt{\frac{2a}{3g}}\frac{1}{\sqrt{2}}\int_0^\alpha \frac{d\theta}{\sin\left(\frac{\pi}{4} - \frac{\theta}{2}\right)} = \sqrt{\frac{a}{3g}}\int_0^\alpha \operatorname{cosec}\left(\frac{\pi}{4} - \frac{\theta}{2}\right)d\theta$$

$$= 2\sqrt{\frac{a}{3g}}\left[\log\tan\left(\frac{\pi}{8} - \frac{\theta}{4}\right)\right]_0^\alpha$$

$$= -2\sqrt{\frac{a}{3g}}\left[\log\tan\left(\frac{\pi}{8} - \frac{\alpha}{4}\right) - \log\tan\frac{\pi}{8}\right]$$

$$= 2\sqrt{\frac{a}{3g}}\log\left[\tan\frac{\pi}{8}\cot\left(\frac{\pi}{8} - \frac{\alpha}{4}\right)\right].$$

Problem 13. *A uniform rod of length 2a, is placed with one end in contact with a smooth horizontal table and is then allowed to fall. If α be its initial inclination to the vertical, show that its angular velocity when it is inclined at an angle θ is* $\left\{\frac{6g}{a}\cdot\frac{\cos\alpha - \cos\theta}{1 + 3\sin^2\theta}\right\}^{1/2}$.

Find also the reaction of the table.

[IAS 59, 89; Agra 76, 82; Raj 80; Garhwal 84; MU 74, 76, CU 95H, 00H]

Solution: Let AB be the rod of mass M and length $2a$ placed with end A, in contact with a smooth table, inclined at an angle α to the vertical. Then the rod is allowed to fall. Since there is no horizontal force on the rod, therefore, its C.G. at G will move down in a vertical line through G. Let the vertical line through G cut the horizontal plane at O, then O is the fixed point.

If at time t the rod AB is inclined at an angle θ to the vertical and referred to O as origin, horizontal and vertical lines OX, OY through O as axes, the coordinates of G are $(0, a\cos\theta)$.

The $(\text{velocity})^2$ of G

$$= 0^2 + (-a\sin\theta\dot{\theta})^2 = a^2\sin^2\theta\dot{\theta}^2.$$

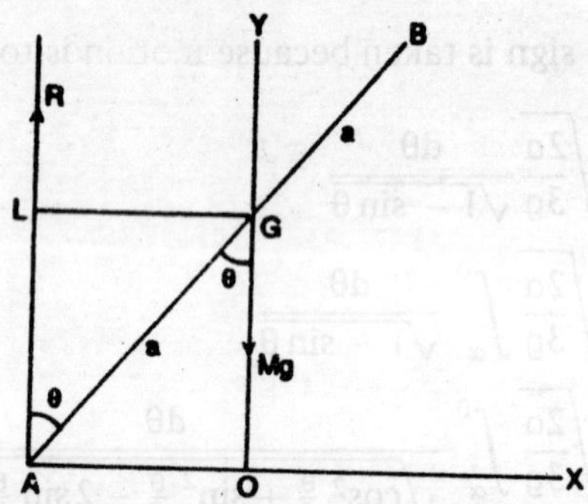

FIG 6.16

Hence, the energy equation gives

$$\frac{1}{2}M\frac{1}{3}a^2\dot{\theta}^2 + \frac{1}{2}M(a^2\sin^2\theta\dot{\theta}^2) = Mg(a\cos\alpha - a\cos\theta)$$

$$\Rightarrow \quad \frac{a^2\dot{\theta}^2}{6}[1 + 3\sin^2\theta] = ga(\cos\alpha - \cos\theta)$$

$$\Rightarrow \quad \dot{\theta}^2 = \frac{6g}{a}\frac{\cos\alpha - \cos\theta}{1 + 3\sin^2\theta},$$

$$\text{i.e.,} \quad \dot{\theta} = \left[\frac{6g}{a}\frac{\cos\alpha - \cos\theta}{1 + 3\sin^2\theta}\right]^{1/2}. \tag{6.28}$$

To find the reaction at A

If R is the reaction at A, then taking moment about G, we get

$$M\frac{1}{3}a^2\ddot{\theta} = R.GL = R.a\sin\theta, \text{ or, } R = \frac{Ma}{3\sin\theta}\ddot{\theta}. \tag{6.29}$$

Differentiating equation (6.28) with respect to t, we get

$$2\dot{\theta}\ddot{\theta} = \frac{6g}{a}\frac{[\sin\theta(1+3\sin^2\theta) - 6\sin\theta\cos\theta(\cos\alpha - \cos\theta)]\dot{\theta}}{(1+3\sin^2\theta)^2}$$

$$\text{or, } \ddot{\theta} = \frac{3g}{a}\frac{\sin\theta[1+3\sin^2\theta - 6\cos\theta(\cos\alpha - \cos\theta)]}{(1+3\sin^2\theta)^2}$$

$$= \frac{3g}{a}\frac{\sin\theta[1+3\sin^2\theta + 6\cos^2\theta - 6\cos\alpha\cos\theta]}{(1+3\sin^2\theta)^2}$$

$$= \frac{3g}{a}\frac{\sin\theta[1+3+3(\cos^2\theta - 2\cos\theta\cos\alpha)]}{(1+3\sin^2\theta)^2}$$

$$= \frac{3g}{a}\frac{\sin\theta[1+3+3\{(\cos\alpha - \cos\theta)^2 - \cos^2\alpha\}]}{(1+3\sin^2\theta)^2}$$

$$= \frac{3g}{a}\frac{\sin\theta[1+3+3(\cos\alpha - \cos\theta)^2 - 3\cos^2\alpha]}{(1+3\sin^2\theta)^2}$$

$$\ddot{\theta} = \frac{3g}{a}\frac{\sin\theta[1+3\sin^2\alpha + 3(\cos\alpha - \cos\theta)^2]}{(1+3\sin^2\theta)^2}.$$

Substituting the value of $\ddot{\theta}$ in equation (6.29), we get

$$R = \frac{Ma}{3\sin\theta}\frac{3g}{a}\frac{\sin\theta[1+3\sin^2\alpha + 3(\cos\alpha - \cos\theta)^2]}{(1+3\sin^2\theta)^2}$$

$$= Mg\frac{[1+3\sin^2\alpha + 3(\cos\alpha - \cos\theta)^2]}{(1+3\sin^2\theta)^2}.$$

Problem 14. *A hemisphere of mass M and radius a, is placed with its plane base on a smooth table and a heavy rod, of mass m, is constrained to move in a vertical line with one end P on the curved surface of the hemisphere. If at any time t, the radius to P makes an angle θ with the vertical, show that* $a\dot{\theta}^2[M\cos^2\theta + m\sin^2\theta] = 2mg(\cos\alpha - \cos\theta)$.

Solution: Let C be the centre, G′ the C.G. of the hemisphere of mass M and radius a. Let PQ be the rod of length 2a, mass m, be constructed to move in a vertical line with one end P on the curved surface of the hemisphere. Let the vertical line along which the rod moves cut the horizontal plane in the point O. Then O is the fixed point. At time t, $\angle CPO = \theta$.

Taking O as origin the horizontal and vertical lines through O as axes the coordinates of G, the C.G. of the rod, are $(0, a\cos\theta + a)$ and coordinates of G′, the C.G. of hemisphere are $\left(a\sin\theta, \frac{3a}{8}\right)$.

Hence energy equation gives

$$\frac{1}{2}M(a\cos\theta\dot{\theta})^2 + \frac{1}{2}m(-a\sin\theta\dot{\theta})^2$$

$$= mg(a\cos\alpha - a\cos\theta) + Mg\cdot 0$$

$$\Rightarrow a\dot{\theta}^2(M\cos^2\theta + m^2\sin^2\theta) = 2mg(\cos\alpha - \cos\theta).$$

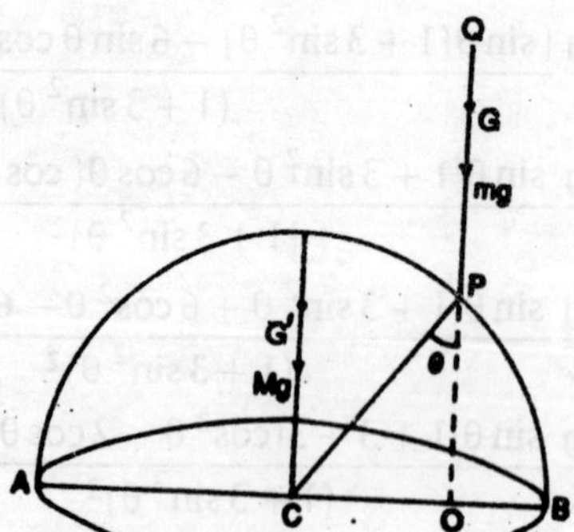

FIG 6.17

Problem 15. *A sphere of radius* b, *rolls without slipping down the cycloid* $x = a(\theta + \sin\theta)$, $y = a(1 - \cos\theta)$. *It starts from rest with its centre on the horizontal line* $y = 2a$. *Show that the velocity* V *of its centre when at its lowest point is given by* $V^2 = \frac{10}{7}g(2a - b)$.

Solution: Let O be the vertex andAB the base (i.e., horizontal line $y = 2a$) of the given cycloid. Let C be the centre and M the mass of the sphere of radius b rolling without slipping down the cycloid, starting from rest when its centre C was on the line AB (i.e., line $y = 2a$).

Initially, A′ coincides with A. At time t, let CA′ make an angle ϕ with the vertical and let P be the point of contact of the sphere with the cycloid. Let the tangent at P make an angle ψ with the horizontal and let arc OP $= s$ and arc OA $= c$.

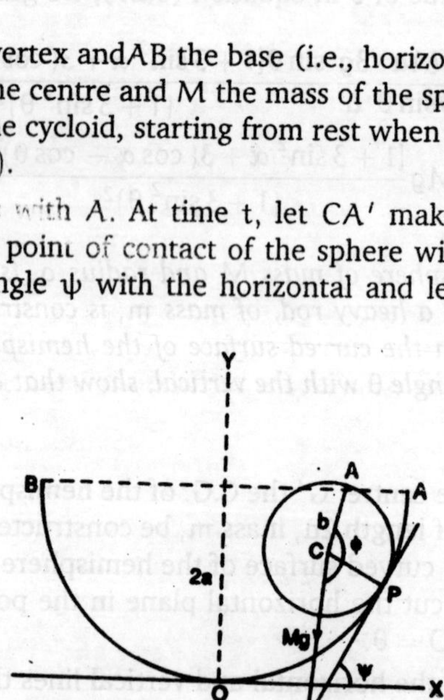

FIG 6.18

Since the sphere rolls without slipping, arc AP = arc A′P,

$$\text{i.e., } c - s = b(\phi - \psi)$$

$$s = c - b(\phi - \psi)$$

$$\dot{s} = -b(\dot{\phi} - \dot{\psi}). \qquad (6.30)$$

If V is the velocity of the centre C of sphere, then

$$V = \text{velocity of P+ velocity of C relative to P}$$
$$V = \dot{s} - b\dot{\psi} = -b(\dot{\phi} - \dot{\psi}) - b\dot{\psi} = -b\dot{\phi} \text{ [From equation (6.30)]} \quad (6.31)$$

Now energy equation gives

$$\text{i.e., } \frac{1}{2}M\frac{2}{5}b^2\dot{\phi}^2 + \frac{1}{2}MV^2 = Mg(2a - h),$$

where $h = CL =$ height of centre C of sphere at time t.

$$\frac{1}{5}b^2\dot{\phi}^2 + \frac{1}{2}b^2\dot{\phi}^2 = g(2a - h).$$

From equation (6.31),

$$\frac{V^2}{5} + \frac{V^2}{2} = g(2a - h)$$
$$\text{or, } \frac{7}{10}V^2 = g(2a - h)$$
$$\therefore V^2 = \frac{10}{7}g(2a - h).$$

When the sphere is at the lowest point O, then $V = V$ and $h = b$.

$$\therefore V^2 = \frac{10}{7}g(2a - b).$$

Problem 16. *A uniform rod of length 2a, hangs in a horizontal position being supported by vertical strings, each of length l, attached to its ends, the other extremities being attached to fixed points. The rod is given an angular velocity ω about a vertical axis through its centre. Find its angular velocity when it has turned through an angle and show that it will rise through a distance* $\frac{a^2\omega^2}{6g}$.

Prove also that the time of a small oscillation about the position of equilibrium is $2\pi\sqrt{\frac{l}{3g}}$.

Solution: Let A′B′ be the initial horizontal position of the rod of length 2a and mass M, supported by two vertical strings CA′ and DB′ each of length l attached to its ends. At time t, let the rod turn through an angle θ and its new position be AB. Thus at time t the strings are in the position CA and DB. If O′ is the middle point of the rod and MON′ a horizontal line, then M, A, O′, N, B are all in the same horizontal plane.

$$\therefore \angle ANC = \angle BND = 90°.$$

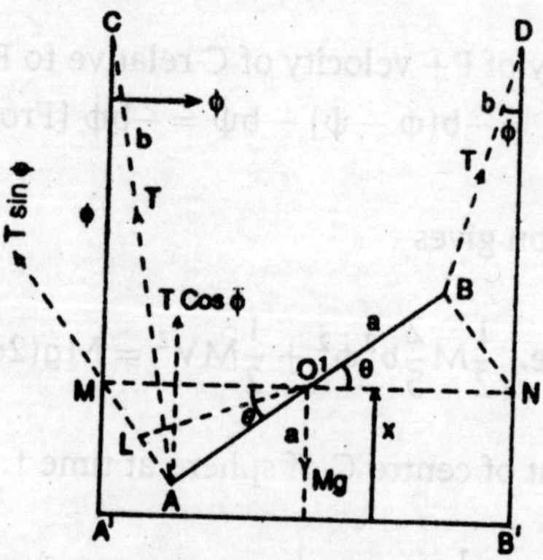

FIG 6.19

If the rod rises through a height x in time t, then

$$x = OO' = MA' = CA' - CM = l - l\cos\phi. \qquad (6.32)$$

In right angled ΔAMC,

$$\sin\phi = \frac{AM}{AC} = \frac{AM}{l}$$

$$AM = l\sin\phi.$$

Also $O'M = O'A$.

Hence from right angled $\Delta O'AL$,

$$\sin\frac{\theta}{2} = \frac{AL}{O'A} = \frac{AL}{a}$$

$$\text{or,} \quad AL = a\sin\frac{\theta}{2}$$

$$\therefore AM = 2AL = 2a\sin\frac{\theta}{2}$$

$$\therefore l\sin\phi = 2a\sin\frac{\theta}{2}. \qquad (6.33)$$

Differentiating, we get

$$l\cos\phi\dot{\phi} = a\cos\frac{\theta}{2}\dot{\theta}$$

$$\dot{\phi} = \frac{a\cos\frac{\theta}{2}}{l\cos\phi}\dot{\theta}.$$

From equation (6.32),

$$\dot{x} = l\sin\phi\dot{\phi} = \frac{l\sin\phi\, a\cos\frac{\theta}{2}}{l\cos\phi}\dot{\theta} = \frac{2a\sin\frac{\theta}{2}a\cos\frac{\theta}{2}\dot{\theta}}{l\cos\phi} \quad \text{[From equation (6.33)]}$$

$$= \frac{a^2\sin\theta\dot{\theta}}{l\sqrt{1-\sin^2\phi}}$$

$$\dot{x} = \frac{a^2\sin\theta\dot{\theta}}{\sqrt{l^2-4a^2\sin^2\frac{\theta}{2}}}. \tag{6.34}$$

Equation of energy gives

$$\frac{1}{2}m\left(\dot{x}^2+\frac{a^2}{3}\dot{\theta}^2\right)-\frac{1}{2}m\frac{a^2}{3}\omega^2 = -mgx$$

$$\Rightarrow \frac{1}{2}m\left[\frac{a^4\sin^2\theta\dot{\theta}^2}{\left(l^2-4a^2\sin^2\frac{\theta}{2}\right)}+\frac{a^2}{3}\dot{\theta}^2\right]-\frac{1}{2}m\frac{a^2}{3}\omega^2 = -mgx$$

$$\Rightarrow a^2\dot{\theta}^2\left[\frac{a^2\sin^2\theta}{l^2-4a^2\sin^2\frac{\theta}{2}}+\frac{1}{3}\right]-\frac{1}{3}a^2\omega^2 = -2gx. \tag{6.35}$$

This gives the angular velocity of the rod in any position. The rod will come to instantaneous rest when $\dot{\theta} = 0$.

Thus putting $\dot{\theta} = 0$ in equation (6.35), we get

$$2gx = \frac{a^2\omega^2}{3}; \quad \therefore\ x = \frac{a^2\omega^2}{6g}.$$

Now we proceed to find the time of small oscillation.

We consider the motion of AB in the vertical direction.

$$2T\cos\phi - mg = m\ddot{x}. \tag{6.36}$$

But

$$x = l - l\cos\phi = l(1-\cos\phi)$$

$$= l2\sin^2\frac{\phi}{2} = 2l\frac{\phi^2}{4} = \frac{l\phi^2}{2}.$$

Since $\sin\frac{\phi}{2}$ is small, so $\sin\frac{\phi}{2} = \frac{\phi}{2}$.

Also from equation (6.33),

$$l\sin\phi = 2a\sin\frac{\theta}{2}$$

$$\Rightarrow l\phi = 2a\frac{\theta}{2}$$

$$\Rightarrow \phi = \frac{a\theta}{l}$$

$$\therefore\ x = \frac{l}{2}\frac{a^2\theta^2}{l^2} = \frac{a^2\theta^2}{2l}.$$

$$\therefore \dot{x} = \frac{a^2}{2l}2\theta\dot{\theta} = \frac{a^2}{l}\theta\dot{\theta}$$
$$\text{and } \ddot{x} = \frac{a^2}{l}\left[\dot{\theta}^2 + \theta\ddot{\theta}\right].$$

Hence from equation (6.36), we get

$$2T\cos\phi - mg = \frac{ma^2}{l}\left(\dot{\theta}^2 + \theta\ddot{\theta}\right).$$

Since $\dot{\theta}$ and $\ddot{\theta}$ are small, so they can be neglected, ϕ being small gives $\cos\phi = 1$.

$$2T - mg = 0.$$

Again taking moments about the vertical axis through O, we have

$$m\frac{a^2}{3}\ddot{\theta} = -2T\sin\phi \times \text{ perpendicular from O on BN}$$
$$\Rightarrow m\frac{a^2}{3}\ddot{\theta} = -2T\sin\phi\, a\cos\frac{\theta}{2}$$
$$\Rightarrow m\frac{a^2}{3}\ddot{\theta} = -mg\frac{2a\sin\frac{\theta}{2}}{l}a\cos\frac{\theta}{2} \text{ [From equation (6.33)]}$$
$$[\because 2T = mg]$$
$$\Rightarrow \frac{\ddot{\theta}}{3} = -\frac{g\sin\theta}{l}$$
$$\therefore \ddot{\theta} = -\frac{3g}{l}\sin\theta = -\frac{3g}{l}\theta.$$

Since θ is small which is a S.H.M, so time of one complete oscillation $= 2\pi\sqrt{\frac{l}{3g}}$.

Problem 17. *A uniform rod, of length 2a, is freely jointed at one end to a small ring, whose mass is equal to that of the rod. The ring is free to slide on a smooth horizontal wire and initially is at rest and the rod is vertical and below the ring and rotating with angular velocity $\sqrt{\frac{3g}{a}}$ in a vertical plane passing through the wire. When the rod is inclined at an angle θ to the vertical, show that its angular velocity is $\sqrt{\frac{3g}{a}\cdot\frac{1+4\cos\theta}{8-3\cos^2\theta}}$ and find the velocity of the ring then.*

Solution: Let M be the mass of the rod AB of length 2a. Initially, let the ring of equal mass be at O and the rod hanging at rest below it and rotating with angular velocity $\sqrt{\frac{3g}{a}}$. At time t, let the ring move to A such that OA = x and let θ be the angle through which the rod has turned during this time.

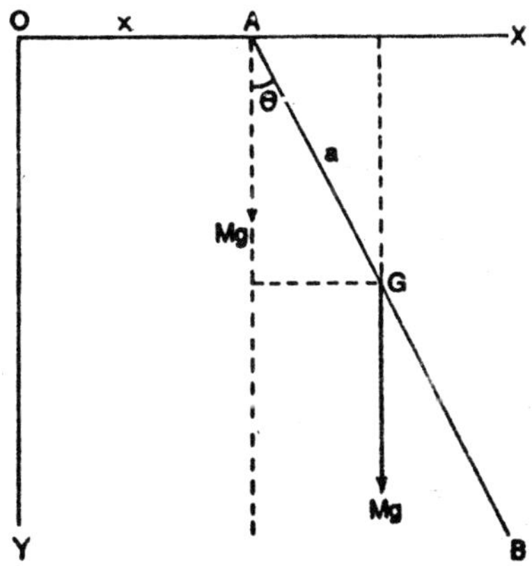

FIG 6.20

Referred to O as origin, horizontal and vertical lines OX, OY through O as axes, the coordinates of A are $(x, 0)$ and the coordinates of C.G. at G of AB are given by $(x + a \sin \theta, a \cos \theta)$.

So $(\text{velocity})^2$ of G

$$= (\dot{x} + a \cos \theta \dot{\theta})^2 + (-a \sin \theta \dot{\theta})^2 = \dot{x}^2 + 2\dot{x}\dot{\theta} a \cos \theta + a^2 \dot{\theta}^2.$$

At time t, the total horizontal momentum

$$= M\dot{x} + M(\dot{x} + a \cos \theta \dot{\theta}) = M(2\dot{x} + a \cos \theta \dot{\theta}). \tag{6.37}$$

Initially,

$$\dot{\theta} = \sqrt{\frac{3g}{a}}; \; \dot{x} = 0 \text{ and } \theta = 0.$$

Thus, from equation (6.37), we get the total initial horizontal momentum

$$= Ma\sqrt{\frac{3g}{a}} = M\sqrt{3ag}.$$

Since there is no horizontal force on the system, therefore, the momentum in horizontal direction remains the same throughout.

$$\therefore \quad M\dot{x} + M(\dot{x} + a \cos \theta \dot{\theta}) = M\sqrt{3ag}$$
$$2\dot{x} = \sqrt{3ag} - a \cos \theta \dot{\theta}. \tag{6.38}$$

Now the total kinetic energy at time t

$$= \text{K.E. of the ring} + \text{K.E. of the rod}$$
$$= \frac{1}{2}M\dot{x}^2 + \frac{1}{2}M\frac{1}{3}a^2\dot{\theta}^2 + \frac{1}{2}M(\dot{x}^2 + a^2\dot{\theta}^2 + 2\dot{x}\dot{\theta}a \cos \theta)$$
$$= \frac{1}{2}M\left[2\dot{x}^2 + \frac{4}{3}a^2\dot{\theta}^2 + 2a\dot{x}\dot{\theta} \cos \theta\right]. \tag{6.39}$$

But initially,

$$\dot{\theta} = \sqrt{\frac{3g}{a}},\ \dot{x} = 0 \text{ and } \theta = 0.$$

Substituting in equation (6.39), the total initial kinetic energy

$$= \frac{1}{2}M\frac{4}{3}a^2\frac{3g}{a} = 2Mga.$$

Hence the energy equation gives

$$\frac{1}{2}M\left[\frac{4}{3}a^2\dot{\theta}^2 + 2\dot{x}^2 + 2a\dot{x}\dot{\theta}\cos\theta\right] - 2Mga = -Mg(a - a\cos\theta)$$

$$\Rightarrow \frac{1}{2}\left[\frac{4}{3}a^2\dot{\theta}^2 + \frac{2}{4}\left\{\sqrt{3ag} - a\cos\theta\dot{\theta}\right\}^2 + a\dot{\theta}\left\{\sqrt{3ag} - a\cos\theta\dot{\theta}\right\}\cos\theta\right]$$

$$= ga(1 + \cos\theta)$$

$$\Rightarrow \frac{4}{3}a^2\dot{\theta}^2 + \frac{1}{2}\left[3ag + a^2\cos^2\theta\dot{\theta}^2 - 2\sqrt{3ag}a\cos\theta\dot{\theta}\right]$$

$$+a\dot{\theta}\cos\theta\left\{\sqrt{3ag} - a\cos\theta\dot{\theta}\right\} = 2ag(a + \cos\theta)$$

$$\Rightarrow \frac{4}{3}a^2\dot{\theta}^2 + 3ag - \frac{a^2}{2}\cos^2\theta\dot{\theta}^2 = 2ag(1 + \cos\theta)$$

$$\Rightarrow a^2\dot{\theta}^2\left[\frac{4}{3} - \frac{1}{2}\cos^2\theta\right] = -\frac{3ag}{2} + 2ag\cos\theta + 2ag$$

$$\Rightarrow \frac{a^2}{6}\dot{\theta}^2[8 - 3\cos^2\theta] = \frac{ag}{2} + 2ag\cos\theta$$

$$\Rightarrow \dot{\theta}^2(8 - 3\cos^2\theta) = \frac{ag6}{2a^2}(1 + 4\cos\theta) = \frac{3g}{a}(1 + 4\cos\theta)$$

$$\therefore \quad \dot{\theta} = \sqrt{\frac{3g}{a}\cdot\frac{1 + 4\cos\theta}{8 - 3\cos^2\theta}}.$$

Problem 18. *A hoop of radius a rolling on a horizontal rod with velocity V comes into collision with a rough inelastic kerb of height h, which is perpendicular to the plane of the hoop. Show that, if the hoop is to clear the kerb without jumping,* $\frac{2a}{2a-h}\sqrt{g(a-h)} > v > \frac{2a}{2a-h}\sqrt{gh}$.

Solution: Let AO be the kerb of height h. When the hoop strikes the kerb, let Ω be its angular velocity and velocity of the centre is given V,

$$\therefore \quad V = a\Omega. \tag{6.40}$$

After striking the kerb let ω be its angular velocity; so the velocity of the centre C will be $a\omega$ at right angles to the radius AC.

Since the impulse acts at A, so its moment at A is zero, hence moment of momentum about A remains unchanged.

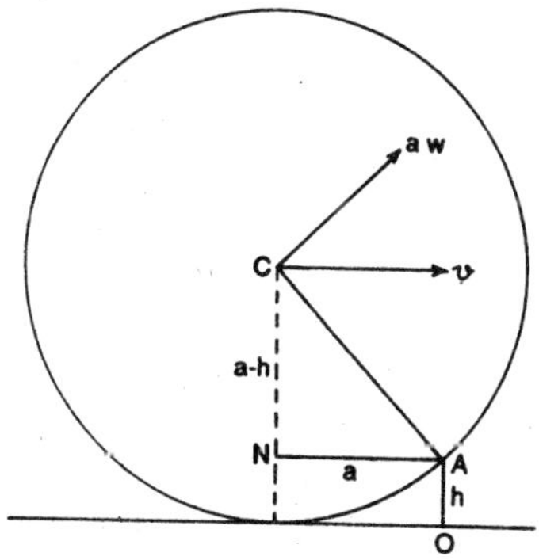

FIG 6.21

Hence moment of momentum about A before and after the impulsive action are $ma^2\Omega + mV(a-h)$ and $ma^2\omega + ma\omega a$.

So by principle of moment of momentum, we have

$$ma^2\Omega + mV(a-h) = ma^2\omega + ma^2\omega,$$

$$\text{i.e.,}\quad maV + mV(a-h) = 2a^2m\omega \text{ [From equation (6.40)]}$$

$$\text{or,}\quad 2a^2\omega = V(2a-h).$$

After striking, when the hoop is turning about A and rising, let θ be the angle which the radius CA makes with the horizontal. Then energy equation gives

$$\frac{1}{2}m(a^2\dot{\theta}^2 + a^2\dot{\theta}^2) - \frac{1}{2}m(a^2\omega^2 + a^2\omega^2) = -mg(h + a\sin\theta - a)$$

$$\text{or,}\quad a^2\dot{\theta}^2 - a^2\omega^2 = -g(h + a\sin\theta - a)$$

$$\dot{\theta}^2 = \omega^2 - \frac{g}{a^2}(h + a\sin\theta - a). \tag{6.41}$$

Let R be the normal reaction at this instant. Then considering motion of the hoop, we have

$$ma\dot{\theta}^2 = mg\sin\theta - R. \tag{6.42}$$

So that

$$R = mg\sin\theta - ma\dot{\theta}^2$$

$$= mg\sin\theta - ma\left[\omega^2 - \frac{g}{a^2}(h + a\sin\theta - a)\right] \text{ [From equation (6.41)]}$$

$$R = \frac{mg}{a}(h - a + 2a\sin\theta) - ma\omega^2. \tag{6.43}$$

In order that the hoop may clear (surmount) the kerb, $\dot{\theta}$ must not vanish before the hoop gets to the highest point, i.e., $\dot{\theta}^2$ must be positive when $\theta = \frac{\pi}{2}$.

Hence from equation (6.41), we get

$$\omega^2 > \frac{g}{a^2}(h+a-a)$$

$$\Rightarrow \omega^2 > \frac{gh}{a^2}$$

$$\Rightarrow \frac{V^2(2a-h)^2}{4a^4} > \frac{gh}{a^2}$$

$$\Rightarrow V^2 > \frac{4a^2}{(2a-h)^2}gh$$

$$\therefore V > \frac{2a}{2a-h}\sqrt{gh}.$$

And in order that the hoop may remain in contact with the kerb, R must be positive when it is least,

i.e., R must be positive when $\sin\theta = \frac{a-h}{a}$.

Hence from equation (6.43),

$$R = \frac{mg}{a}\left[h - a + \frac{2a(a-h)}{a}\right] - ma\omega^2 > 0$$

$$\Rightarrow \frac{g}{a}[h-a+2a-2h] - a\omega^2 > 0$$

$$\Rightarrow \frac{g}{a}(a-h) < a\omega^2$$

$$\Rightarrow a\omega^2 < \frac{g}{a}(a-h)$$

$$\Rightarrow w < \frac{1}{a}\sqrt{g(a-h)}$$

$$\Rightarrow \frac{V(2a-h)}{2a^2} < \frac{1}{a}\sqrt{g(a-h)}$$

$$\Rightarrow V < \frac{2a}{2a-h}\sqrt{g(a-h)}.$$

Hence in order that the hoop may clear the kerb without jumping,

$$\frac{2a}{2a-h}\sqrt{g(a-h)} > v > \frac{2a}{2ah}\sqrt{gh}.$$

Problem 19. *A uniform hollow cylinder of radius* a *is at rest inside a cylinder of internal radius* b*, the axes of both cylinders being horizontal. The system is started impulsively from rest so that the cylinder of radius* b *moves in a horizontal direction at right angles to its axis with velocity* V*, and is then forced to continue moving horizontally with the same velocity and without rotation. The cylinder of radius* a *rolls without slipping on the inner surface of the cylinder of radius* b*. Prove that it makes complete revolution right round that surface if* $V^2 > 12g(b-a)$.

Solution: Let O be the centre of the hollow cylinder of internal radius b and C the centre of hollow cylinder of radius a inside the first cylinder. At time t, let the inner cylinder be at the point P of the outer cylinder such that CB makes an angle ϕ with the vertical while B coincided with A initially.

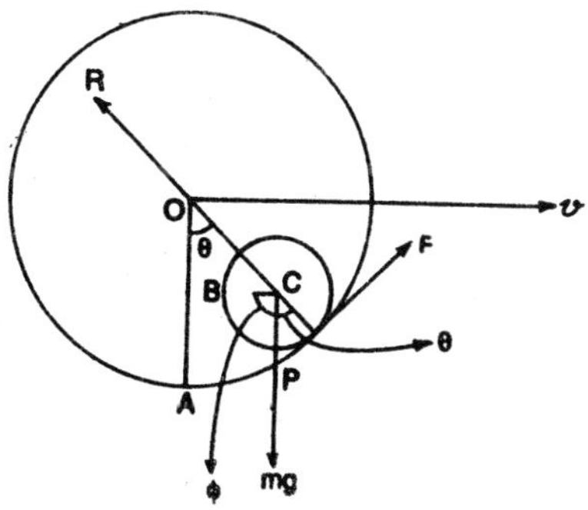

FIG 6.22

Let $\angle AOP = \theta$, since there is no slipping arc AP = arc BP

$$\begin{aligned} \therefore \quad & b\theta = a(\phi + \theta) \\ \text{or,} \quad & (b-a)\theta = a\phi \\ \text{or,} \quad & (b-a)\dot{\theta} = a\dot{\phi} \\ \text{and} \quad & (b-a)\ddot{\theta} = a\ddot{\phi}. \end{aligned} \tag{6.44}$$

Let (x, y) be the coordinates of O at any time t, then $\dot{x} = v$, $\dot{y} = 0$ and $\ddot{x} = 0$, $\ddot{y} = 0$, i.e., acceleration of O is zero.

Since C describes a circle of radius $OC = OP - OC = b - a = c$ (say) with centre at O, therefore, its accelerations along and perpendicular to CO are $C\dot{\theta}^2$ and $C\ddot{\theta}$, respectively.

Hence equations of motion of the inner cylinder are

$$Mc\dot{\theta}^2 = R - Mg\cos\theta (\text{along CO}) \tag{6.45}$$

$$\text{and} \quad Ma\ddot{\theta} = F - Mg\sin\theta (\text{perpendicular to CO}) \tag{6.46}$$

$$\text{and} \quad Ma^2\ddot{\phi} = -Fa (\text{moment about C}). \tag{6.47}$$

Since the inner cylinder was initially at rest, therefore the impulsive action on it will be at the point of contact A. So the moment of the impulsive action about A is zero.

$\therefore$ moment of momentum about A after impulse = moment of momentum about A before impulse,

$$\begin{aligned} \text{i.e.,} \quad & Ma^2\dot{\phi} + Ma(b-a)\dot{\theta} + MaV = 0 \\ \text{or,} \quad & (b-a)\dot{\theta} + (b-a)\dot{\theta} + V = 0 \text{ [From equation (6.44)]} \\ \text{or,} \quad & 2(b-a)\dot{\theta} + V = 0 \\ \therefore \quad & (b-a)\dot{\theta} = c\dot{\theta} = -\frac{V}{2} \quad (\because b - a = c) \text{ (Initially)}. \end{aligned} \tag{6.48}$$

From equation (6.46),

$$F = Ma\dot{\theta} - Mg\sin\theta.$$

Thus from equation (6.47), we get

$$Ma^2\ddot{\phi} = -[MC\ddot{\theta} + Mg\sin\theta]a$$
$$\Rightarrow \quad a\ddot{\phi} = -(C\ddot{\theta} + g\sin\theta)$$
$$\Rightarrow \quad (b-a)\ddot{\theta} = -c\ddot{\theta} - g\sin\theta \text{ [From equation (6.44)]}$$
$$\Rightarrow \quad 2c\ddot{\theta} = -g\sin\theta \quad [\because b-a=c]$$
$$\Rightarrow \quad 2c\dot{\theta}\ddot{\theta} = -g\sin\theta\dot{\theta}$$
$$\Rightarrow \quad cd(\dot{\theta}^2) = -g\sin\theta\dot{\theta}.$$

Integrating, we get

$$c\dot{\theta}^2 = g\cos\theta + C_1.$$

Initially, when $\theta = 0$, $\dot{\theta} = -\frac{V}{2c}$. From equation (6.48)

$$\frac{cV^2}{4c^2} = C_1 + g$$
$$\therefore C_1 = \frac{V^2}{4c} - g$$
$$\therefore c\dot{\theta}^2 = g\cos\theta + \frac{V^2}{4c} - g. \tag{6.49}$$

Hence from equation (6.45),

$$R = Mc\dot{\theta}^2 + Mg\cos\theta = M\left(g\cos\theta + \frac{V^2}{4c} - g\right) + Mg\cos\theta$$
$$R = 2Mg\cos\theta + M\left(\frac{V^2}{4c} - g\right).$$

The inner cylinder will make complete revolution if $R > 0$, when $\theta = \pi$, i.e., if

$$2Mg\cos\pi + M\left(\frac{V^2}{4c} - g\right) > 0$$
$$\Rightarrow \quad -2Mg + M\left(\frac{V^2}{4c} - g\right) > 0$$
$$\Rightarrow -2g + \frac{V^2}{4c} - g > 0 \Rightarrow \frac{V^2}{4c} > 3g$$
$$\Rightarrow \quad V^2 > 12cg, \Rightarrow V^2 > 12g(b-a).$$

6.12 Exercise 6

Type I

1. A small insect moves along a uniform bar of mass equal to itself and of length $2a$, the ends of which are constrained to remain on the circumference of a fixed circle, whose radius is $\frac{2a}{\sqrt{3}}$. If the insect starts from the middle point of the bar and moves along the bar with relative velocity v, show that the bar in time t will turn through an angle $\frac{1}{\sqrt{3}}\tan^{-1}\frac{Vt}{a}$.

[MU 76, 80, 82; Raj 78, 81; Agra 76, 79; CU 90H]

2. A rod OA can turn freely in a horizontal plane about the end O and lies at rest. An insect, whose mass is one-third that of the rod, alights on the end A and commences crawling along the rod with uniform velocity V. At the same instant the rod is set in rotation about O in such a way that the initial velocity of A is V. When the insect reaches O prove that the rod has rotated through a right angle, and that the angular velocity of the rod is then twice the initial angular velocity.

3. A heavy circular disc is revolving in a horizontal plane about its centre which is fixed. An insect, of mass $\frac{1}{n}$th that of the disc, walks from the centre along a radius and then flies away. Show that the final angular velocity is $\frac{n}{n+2}$ times the original angular velocity of the disc. **[MU 76, 77, 80, 82; CU 82H]**

4. A circular ring, of mass M and radius a, lies on a smooth horizontal plane, and an insect, of mass m, resting on it starts and walks round it with uniform velocity v relative to the ring. Show that the centre of the ring describes a circle with angular velocity $\frac{m}{M+2m}\frac{V}{a}$. **[CU 83H]**

Type II

5. An elliptic area of eccentricity e, is rotating with angular velocity ω about one latus rectum; suddenly this latus rectum is loosed and the other fixed. Show that the new angular velocity is $\frac{\omega(1-4e^2)}{1+4e^2}$.

[Raj 80; CU 89H; LNMU 93, CU 96H, 02H]

6. A circular plate rotates about an axis through its centre perpendicular to its plane with angular velocity ω. The axis is set free and a point in the circumference of the plate fixed. Show that the resulting angular velocity is $\frac{\omega}{3}$. **[MU 73, 74, 76, 80–82, CU 91H]**

7. A circular plate is turning in its own plane about a point A on its circumference. Suddenly A is forced and a point B also on the circumference is fixed. Show that the plate will be reduced to rest if the arc AB is one-third of the circumference.

8. A rod of length $2a$, is moving about one end with uniform angular velocity upon a smooth horizontal plane. Suddenly this end is set free and a point, distant b from this end, is fixed; find the motion considering the cases when $b <=> \frac{4a}{3}$.

[*Ans.* $\omega' = \frac{3a\left(\frac{4}{3}a-b\right)}{a^2+3(a-b)^2}\omega$, determines the motion when $b < \frac{4}{3}a$, ω' is positive,

when $b = \frac{4}{3}a$, ω' is zero,

when $b > \frac{4}{3}a$, ω' is negative.]

9. An equilateral triangular lamina is rotating in its plane with uniform angular velocity about an axis through one vertex. If this vertex is released and one of the other vertices fixed, show that the new angular velocity is $\frac{1}{5}$ of its former value. **[IAS 59; MU 79, 83]**

10. **A uniform cube is spinning freely with angular velocity ω about a diagonal through a corner O, when suddenly the diagonal through O of one of the faces through O becomes fixed, show that the new angular velocity is $\frac{2\sqrt{2}}{5\sqrt{3}}\omega$.**

11. **If an octant of an ellipsoid bounded by the three principal planes be rotating about the axis of a with angular velocity ω and if this axis suddenly becomes freed while the axis b becomes suddenly fixed at the same instant, show that the new angular velocity is $\frac{2ab\omega}{\pi(a^2+c^2)}$.**

12. **Three equal uniform rods placed in a straight line are freely jointed and move with a velocity v perpendicular to their lengths. If the middle point of the middle rod be suddenly fixed, show that the ends of the other two rods will meet in time $\frac{4\pi a}{9v}$, where a is the length of each rod. [MU 79; Agra 76; VU 89H]**

Type III

13. **A straight uniform rod of mass m is placed at right angles to a smooth plane of inclination α with one end in contact with it, the rod is then released. Show that, when its inclination to the plane is ϕ, the reaction of the plane will be $mg\frac{3(1-\sin\phi)^2+1}{(3\cos^2\phi+1)^2}\cos\alpha$. [MU 77, 80–82]**

14. **Two like rods AB and BC, each of length $2a$, are freely jointed at B. AB can turn round the end A, and C can move freely on a vertical straight line through A, and they are then released. Initially the rods are held in a horizontal line, C begins in coincidence with A, and they are then released. Show that when the rods are inclined at an angle θ to the horizontal, the angular velocity of either is $\sqrt{\frac{3g}{a}\cdot\frac{\sin\theta}{1+3\cos^2\theta}}$. [MU 78]**

15. **A uniform plank of thickness $2h$ rests across the top of a fixed, circular cylinder of radius a whose axis is horizontal. Prove that if it be set in motion, the equation of energy is**

$$\frac{1}{2}[K^2+h^2+a^2\theta^2]\dot{\theta}^2+g[a\theta\sin\theta-(a+h)(1-\cos\theta)]=\text{constant}.$$

on the assumption that the motion is of pure rolling. Hence show that if $a > h$, the horizontal position is stable and the period of a small oscillation is the same as of a simple pendulum of length $\frac{K^2+h^2}{a-h}$.

16. **A bead of mass M can slide on a smooth straight horizontal wire and a particle of mass m is attached to the body by a light string of length l. The particle is held in contact with the wire with the string taut and is then let fall. Prove that when the string is inclined to the wire at an angle θ, the bead will have slipped a distance $\frac{ml(1-\cos\theta)}{M+m}$ along the wire and that the angular velocity ω of the string will be given by the equation**

$$(M+m\cos^2\theta)l\omega^2=2(M+m)g\sin\theta.$$

17. **A sphere of radius a, and of radius of gyration K about any axis through its centre rolls with linear velocity v on a horizontal plane, the direction of**

motion being perpendicular to a vertical face of a fixed rectangular block of height h, where $h < a$. The sphere strikes the block. The sphere and the block are perfectly rough and perfectly elastic. Show that the sphere will surmount the block if

$$(a^2 - ah + K^2)^2 V^2 > 2gha^2(a^2 + K^2).$$ **[MU 78]**

18. A thin uniform circular wire of radius a, can revolve freely about a vertical diameter, and a bead slides freely along the wire. If initially the wire was rotating with angular velocity Ω, and the bead was at rest relatively to the wire very close to the highest point, show that when the bead is at its greatest distance from the axis of rotation, the angular velocity of the wire is $\Omega\frac{n}{n+2}$ and that the angular velocity of the bead relative to the wire is $\sqrt{\left(\frac{n\Omega^2}{n+2} + \frac{2g}{a}\right)}$, the mass of the wire being n times that of the bead.

19. If the earth, supposed to be a uniform sphere, had in a certain period contracted slightly, so that its radius was less by $\frac{1}{n}$th than before. Show that the length of the day would have shortened by $\frac{48}{n}$ hours.

[MU 84, 85; MU 94H]

20. If the earth, supposed to be a uniform solid sphere, has in a certain period contracted slightly so that its radius is less by one percent than what it is now. Show that the length of a day before contraction is taken to be 24 hours.

[CU 94H]

Chapter 7

Motion of a Rigid Body

7.1 Motion of a Rigid Body about a Fixed Axis

Let a given rigid body rotate about a fixed axis and let ω be the angular velocity at any time t. Then the kinetic energy of the rigid body will be given as

$$T = \tfrac{1}{2}\omega^2 \Sigma m_i r_i^2$$

where m_i is an elementary mass with radius vector r_i. If I be the moment of inertia of the rigid body about fixed axis, then

$$T = \tfrac{1}{2} I\omega^2 \tag{7.1}$$

The angular momentum of an elementary particle with mass m_i and radius vector $\vec{r}$, will be given as $m_i \vec{r}_i \omega^2$ and so the angular momentum of the rigid body will be

$$H = \omega \Sigma m_i r_i^2 = I\omega \tag{7.2}$$

Hence by the principle of angular momentum and the principle of energy, we have

$$I\dot{\omega} = N \tag{7.3}$$

$$\text{and} \quad \tfrac{1}{2} I\omega^2 + V = E. \tag{7.4}$$

7.2 Motion of a Rigid Body about a Fixed Point

[TMBU 93H, 97H; MU 95H, 97H, 05H; BBABU 95H]

Consider a rigid body constrained to rotate about a fixed point O. Let t_1, t_2 be two instants. In the time interval $t_2 - t_1$, the body receives a displacement which is equivalent to a rotation about O. If we keep t_1, fixed and let t_2 approach t_1, the direction of $\vec{n}$ will approach some limiting directions which we denote by $\vec{\omega}$. The ratio of the angle of rotation n to n to the time interval $t_2 - t_1$, will approach a limiting value $\vec{\omega}$. The vector $\vec{\omega} = \omega\vec{\omega}$ is called the *angular velocity* of the body at the instant t_1. The line through O in the direction of $\vec{\omega}$ about which the rigid body is rotating at this instant is called the *instantaneous axis of rotation.*

The rate of turning is ω radians per unit time and is a rotation in the positive sense about the instantaneous axis.

7.3 Velocity of an arbitrary point P when a rigid body is rotating about a fixed point O

[TMBU 95H, MU 95H, 97H, 01H]

Or Relation between angular velocity and linear velocity of a point of the body.

[TMBU 93H, 98H, BBABU 95H, MU 95H, 97H, 01H]

Let us consider a rigid body rotating with the angular velocity $\vec{\omega}$ about the instantaneous axis passing through the fixed point O. Let P be an arbitrary point of this body and $\vec{r}$ be the position vector of P, i.e., $\overrightarrow{OP} = \vec{r}$. Clearly the linear velocity of the point P is $r\sin\theta\vec{\omega}$ which is always perpendicular to the plane OMP. Let C be the plane of the circle described by P due to the rotation and so $OA \perp C$ and consequently the plane $OMP \perp C$. Since MP is the line of intersection of the planes C and OMP and it is perpendicular to $\vec{v}$, and so $\vec{v}$ is perpendicular to the plane OMP. Hence we can write $\vec{v} = \vec{\omega} \times \vec{r}$.

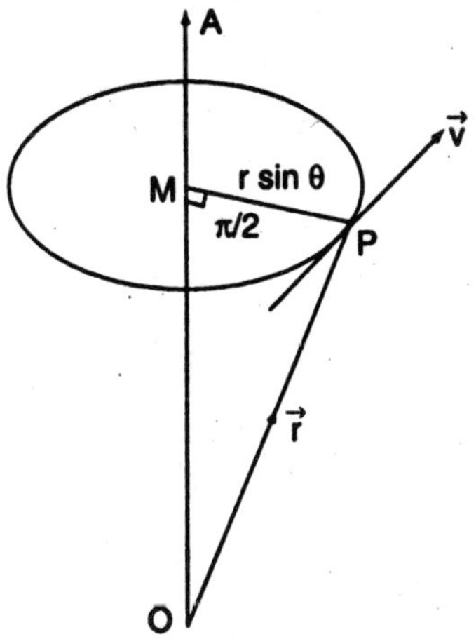

FIG 7.1

If $\omega_x, \omega_y, \omega_z$ be the components of $\vec{\omega}$ along x, y, z-axes respectively, then

$$\vec{v} = \vec{\omega} \times \vec{r} = \begin{vmatrix} \vec{\imath} & \vec{\jmath} & \vec{k} \\ \omega_x & \omega_y & \omega_z \\ x & y & z \end{vmatrix}$$

$$= (\omega_y z - \omega_z y)\vec{\imath} + (\omega_z x - \omega_x z)\vec{\jmath} + (\omega_x y - \omega_y x)\vec{k}$$

where $\vec{\imath}, \vec{\jmath}, \vec{k}$ are unit vectors along x, y, z-axes, respectively and $\vec{r} = x\vec{\imath} + y\vec{\jmath} + z\vec{k}$. Thus the resolved parts of $\vec{v}$ are

$$v_x = \omega_y z - \omega_z y$$
$$v_y = \omega_z x - \omega_x z$$
$$v_z = x_x y - \omega_y x.$$

7.4 Moment of Momentum or Angular Momentum of a Rigid Body about a Fixed Point

[TMBU 96H, 98H, BBABU 96H, 02H, MU 99H, 04H]

The angular momentum of a particle of mass m, moving with velocity $\vec{v}$ about a fixed point O is the moment of linear momentum $m\vec{v}$ about O, i.e., $\vec{H} = \vec{r} \times m\vec{v}$, where $\vec{r}$ is the position vector of the particle relative to O.

In general if m be the mass of the particle, $\vec{r}$ be the position vector and $\vec{v}$ be the velocity, then angular moment of a girid body about a fixed point O is given by $\vec{H} = \Sigma \vec{r} \times m\vec{v}$, where Σ extends to all particles of the system.

$$
\begin{aligned}
\therefore\ \vec{H} &= \Sigma \vec{r} \times m(\vec{\omega} \times \vec{r}) \quad (\because\ \vec{v} = \vec{\omega} \times \vec{r})\\
&= \Sigma m\vec{r} \times (\vec{\omega} \times \vec{r})\\
&= \Sigma m[(\vec{r}\cdot\vec{r})\,\vec{\omega} - (\vec{r}\cdot\vec{\omega})\,\vec{r}]\\
&= \Sigma m\,[r^2\vec{\omega} - (\vec{r}\cdot\vec{\omega})\vec{r}]\\
&= (\Sigma m r^2)\,\vec{\omega} - \{\Sigma m(\vec{r}\cdot\vec{\omega})\}\,\vec{r}\\
&= [\Sigma m(x^2+y^2+z^2)]\left(\omega_x\vec{\imath} + \omega_y\vec{\jmath} + \omega_z\vec{k}\right)\\
&\quad - \left\{\Sigma m(x\omega_x + y\omega_y + z\omega_z)(x\vec{\imath} + y\vec{\jmath} + z\vec{k}\right\}\\
&= \Sigma m\left[x^2\omega_x\vec{\imath} + x^2\omega_y\vec{\jmath} + x^2\omega_z\vec{k} + y^2\omega_x\vec{\imath} + y^2\omega_y\vec{\jmath} + y^2\omega_z\vec{k}\right.\\
&\quad + z^2\omega_x\vec{\imath} + z^2\omega_y\vec{\jmath} + z^2\omega_z\vec{k} - x^2\omega_x\vec{\imath} - xy\omega_y\vec{\imath} - xz\omega_z\vec{\imath}\\
&\quad \left. - xy\omega_x\vec{\jmath} - y^2\omega_y\vec{\jmath} - yz\omega_z\vec{\jmath} - zx\omega_x\vec{k} - zy\omega_y\vec{k} - z^2\omega_z\vec{k}\right]\\
&= \Sigma m\,[(y^2\omega_x + z^2\omega_x - xy\omega_y - xz\omega_z)\vec{\imath}\\
&\quad + (x^2\omega_y + z^2\omega_y - xy\omega_x - yz\omega_z)\vec{\jmath}\\
&\quad + (x^2\omega_z + y^2\omega_z - xz\omega_x - yz\omega_y)\vec{k}]\\
&= \Sigma m\,[\{y^2 + z^2)\omega_x - xy\omega_y - xz\omega_z\}\,\vec{\imath}\\
&\quad + \{(z^2 + x^2)\omega_y - xy\omega_x - yz\omega_z\}\,\vec{\jmath}\\
&\quad + \{(x^2 + y^2)\omega_z - xz\omega_x - yz\omega_y\}\,\vec{k}]\\
\Rightarrow\ \vec{H} &= (A\omega_x - F\omega_y - E\omega_z)\vec{\imath} + (B\omega_y - F\omega_x - D\omega_z)\vec{\jmath}\\
&\quad + (C\omega_z - E\omega_x - D\omega_y)\vec{k},
\end{aligned}
$$

where A, B, C, F, G, H are the moments and product of inertia with respect to the triad $\vec{\imath}, \vec{\jmath}, \vec{k}$.

In case $\vec{\imath}, \vec{\jmath}, \vec{k}$ are taken as principal axes of inertia at O, then $F = G = H = 0$, and (1) becomes $\vec{H} = A\omega_x\vec{\imath} + B\omega_y\vec{\jmath} + C\omega_z\vec{k}$.

7.5 Kinetic Energy of a Rigid Body Rotating about a Fixed Point

[TMBU 92, 94H, BBABU 97H, 02H, MU 96H, 98H, 2000H, 02H]

We consider a rigid body rotating with angular velocity $\vec{\omega}$ about the fixed point O. Then velocity $\vec{v}$ of the particle $P = \vec{\omega} \times \vec{r}$ where $\overrightarrow{OP} = \vec{r}$.

If m be the mass of the particle; the kinetic energy T of the system is given by $T = \frac{1}{2}\Sigma m\vec{v}^2$.

Now

$$\vec{r} = \overrightarrow{OP} = x\vec{i} + y\vec{j} + z\vec{k}$$
$$\vec{\omega} = \omega_x\vec{i} + \omega_y\vec{j} + \omega_z\vec{k}$$

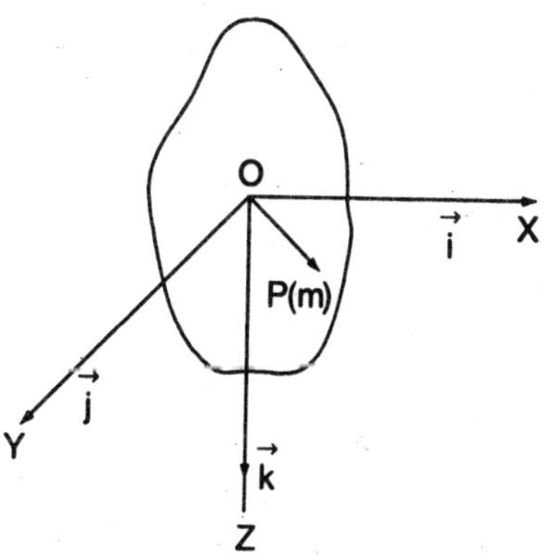

FIG 7.2

where $\omega_x, \omega_y, \omega_z$ are the components of the angular velocity $\vec{\omega}$ in the directions of OX, OY and OZ respectively

$$\therefore \quad \vec{v} = \vec{\omega} \times \vec{r} = \begin{vmatrix} \vec{i} & \vec{j} & \vec{k} \\ \omega_x & \omega_y & \omega_z \\ x & y & z \end{vmatrix}$$

$$= (\omega_y z - \omega_z y)\vec{i} + (\omega_z x - \omega_x z)\vec{j} + (\omega_x y - \omega_y x)\vec{k}$$

$$\therefore \quad 2T = \Sigma m\left[(\omega_y z - \omega_z y)^2 + (\omega_z x - \omega_x z)^2 + (\omega_x y - \omega_y x)^2\right]$$

$$= \Sigma m[(y^2+z^2)\omega_x^2 + (z^2+x^2)\omega_y^2 + (x^2+y^2)\omega_z^2 - 2\omega_y\omega_z yz - 2\omega_z\omega_x zx - 2\omega_x\omega_y xy]$$

$$= \omega_x^2\Sigma m(y^2+z^2) + \omega_y^2\Sigma m(z^2+x^2) + \omega_z^2\Sigma m(x^2+y^2) - 2\omega_y\omega_z\Sigma myz - 2\omega_z\omega_x\Sigma mzx - 2\omega_x\omega_y\Sigma mxy$$

$$\Rightarrow \quad 2T = A\omega_x^2 + B\omega_y^2 + C\omega_z^2 - 2F\omega_y\omega_z - 2G\omega_z\omega_x - 2H\omega_x\omega_y,$$

where A, B, C, D, E, F, G, H are the moments and products of inertia for x, y, z axes.

In the case when the axes of x, y, z are taken as principal axis of inertia, than $F = G = H = 0$. So (1) becomes $2T = A\omega_x^2 + B\omega_y^2 + C\omega_z^2$.

7.6 Relation Between Angular Momentum and Kinetic Energy of a Rigid Body Rotating about a Fixed Point [MU 97H]

We have

$$\vec{\omega} = \omega_x\vec{\imath} + \omega_y\vec{\jmath} + \omega_z\vec{k}$$

and
$$\vec{H} = A\omega_x\vec{\imath} + B\omega_y\vec{\jmath} + C\omega_z\vec{k}$$

$$\therefore \quad \vec{\omega}\cdot\vec{H} = A\omega_x^2 + B\omega_y^2 + C\omega_z^2 = 2T$$

$$\Rightarrow \quad T = \frac{1}{2}\left(\vec{\omega}\cdot\vec{H}\right)$$

Objective Type Questions

I. Choose (a), (b), (c) or (d), whichever is correct.

1. Moment of inertia of a rod of mass M and length $2a$ about a line through the mid-point perpendicular to its length is
   (a) $\frac{1}{2}Ma^2$ (b) $\frac{1}{3}Ma^2$ (c) $\frac{4}{3}Ma^2$ (d) none of these.

2. Moment of inertia of a rod of mass M and length $2a$ about a line through one of its extremities perpendicular to its length is
   (a) $\frac{1}{2}Ma^2$ (b) $\frac{1}{3}Ma^2$ (c) $\frac{4}{3}Ma^2$ (d) none of these.

3. Moment of inertia of a rectangular lamina of mass M, whose edges are of lengths $2a$ and $2b$ about a line through its centre parallel to one of its edges of length $2b$ is
   (a) $\frac{1}{3}Mb^2$ (b) $\frac{1}{3}Ma^2$ (c) $\frac{1}{3}M(a^2+b^2)$ (d) none of these.

4. Moment of inertia of a rectangular lamina of mass M, whose edges are of lengths $2a$ and $2b$, about a line through its centre perpendicular to its plane is
   (a) $\frac{1}{3}Ma^2$ (b) $\frac{1}{3}Mb^2$ (c) $\frac{1}{3}M(a^2+b^2)$ (d) none of these.

5. Moment of inertia of a circular ring of mass M and radius a about a diameter is
   (a) Ma^2 (b) $\frac{1}{2}Ma^2$ (c) $\frac{1}{3}Ma^2$ (d) none of these.

6. Moment of inertia of a circular ring of mass M and radius a about a line through its centre perpendicular to its plane is
   (a) $\frac{1}{3}Ma^2$ (b) $\frac{1}{2}Ma^2$ (c) Ma^2 (d) none of these.

7. Moment of inertia of a thin uniform circular disc of mass M and radius a about its diameter is
   (a) Ma^2 (b) $\frac{1}{2}Ma^2$ (c) $\frac{1}{3}Ma^2$ (d) $\frac{1}{4}Ma^2$.

8. Moment of inertia of a thin uniform circular disc of mass M and radius a about a line through its centre perpendicular to its plane is
   (a) Ma^2 (b) $\frac{1}{2}Ma^2$ (c) $\frac{1}{3}Ma^2$ (d) $\frac{1}{4}Ma^2$.

9. Moment of inertia of an elliptic disc, whose major and minor axes are of lengths $2a$ and $2b$, about its minor axis is
   (a) $\frac{1}{4}Ma^2$ (b) $\frac{1}{4}Mb^2$ (c) $\frac{1}{3}Ma^2$ (d) $\frac{1}{3}Mb^2$.

10. Moment of inertia of a hollow sphere of mass M and radius a about any diameter is

(a) Ma^2 (b) $\frac{2}{3}Ma^2$ (c) $\frac{1}{3}Ma^2$ (d) none of these.

11. Moment of inertia of a solid sphere of mass M and radius a about any diameter is

(a) $\frac{1}{5}Ma^2$ (b) $\frac{2}{5}Ma^2$ (c) $\frac{3}{5}Ma^2$ (d) none of these.

12. Moment of inertia of a rectangular parallelepiped of mass M, whose edges are of lengths $2a$, $2b$ and $2c$ parallel to the rectangular axes OX, OY and OZ, where O is the centre of the parallelepiped about a line through O parallel to the edge $2b$ is

(a) $\frac{M}{3}(a^2+b^2)$ (b) $\frac{M}{3}(b^2+c^2)$ (c) $\frac{M}{3}(c^2+a^2)$ (d) none of these.

13. Moment of inertia of an ellipsoid of mass M and volume enclosed by the surface $\frac{x^2}{a^2}+\frac{y^2}{b^2}+\frac{z^2}{c^2}=1$ about

(a) $\frac{1}{5}M(a^2+b^2)$ (b) $\frac{1}{5}M(b^2+c^2)$ (c) $\frac{1}{5}M(c^2+a^2)$ (d) none of these.

14. Moment of inertia of a cube of S idea and mass M about any of its edges is

(a) $\frac{1}{3}Ma^2$ (b) $\frac{2}{3}Ma^2$ (c) $\frac{4}{3}Ma^2$ (d) $\frac{8}{3}Ma^2$.

15. Moment of inertia of a right circular cylinder of height h, radius of its base a and mass M about its axis is

(a) $\frac{1}{2}Mh^2$ (b) $\frac{1}{2}Ma^2$ (c) $\frac{1}{2}M(a^2+h^2)$ (d) none of these.

16. Product of inertia of a right-angled triangle about sides a, b, containing the right-angle is

(a) $\frac{1}{3}Mab$ (b) $\frac{1}{4}Mab$ (c) $\frac{1}{12}Mab$ (d) none of these.

17. Moment of inertia of a thin uniform circular plate of mass M and radius a about an axis through a point on the circumference and perpendicular to its plane is

(a) $\frac{1}{2}Ma^2$ (b) Ma^2 (c) $\frac{3}{2}Ma^2$ (d) none of these.

18. Any triangle of mass M is kinematically equivalent to three particles placed at the mid-points of the sides and each of mass equal to

(a) $\frac{1}{4}M$ (b) $\frac{1}{3}M$ (c) $\frac{1}{2}M$ (d) none of these.

19. Moment of inertia of an equilateral triangle of side $2a$ and mass M about one of its sides is

(a) $\frac{1}{4}Ma^2$ (b) $\frac{1}{3}Ma^2$ (c) $\frac{1}{2}Ma^2$ (d) none of these.

20. If a rigid body be rotating about a fixed axis and I be its moment of inertia about the axis of rotation, then the moment of the effective forces of the rigid body about the axis of rotation is

(a) $I(\dot{\theta})^2$ (b) $I\ddot{\theta}$ (c) $\frac{1}{2}I(\dot{\theta})^2$ (d) none of these.

21. The time of a complete small oscillation of a compound pendulum in terms of h and k, the distance of the C.G. of the body from the fixed axis and the radius of gyration of the body about the fixed axis respectively is

(a) $2\pi\sqrt{\frac{gh}{k^2}}$ (b) $2\pi\sqrt{\frac{gk}{h^2}}$ (c) $2\pi\sqrt{\frac{k^2}{gh}}$ (d) none of these.

II. True and False Type Questions

Write T *or* F *according as the following statement is true or false:*

1. If I denotes the moment of inertia of a body of mass M about a line and $I = Mk^2$, then k is called the radius of gyration of the body about that line.

2. Moment of inertia of a rod of mass M and length $2a$ about a line through one of its extremities perpendicular to its length is $\frac{1}{3}Ma^2$.

3. Moment of inertia of a rectangular lamina of mass M, whose sides are of lengths $2a$ and $2b$, about a line thorough its centre and perpendicular to its plane is $\frac{M}{3}(a^2+b^2)$.

4. Moment of inertia if an elliptic disc of major and minor axes $2a$ and $2b$ respectively and mass M about its major axis is $\frac{1}{4}Ma^2$.

5. Moment of inertia of an ellipsoid of mass M, whose volume is enclosed by the surface $\frac{x^2}{a^2}+\frac{y^2}{b^2}+\frac{z^2}{c^2}=1$ about x-axis is $\frac{1}{5}(b^2+c^2)M$.

6. Routh's rule is *Moment of inertia* about an axis of symmetry

$$= \frac{M\times \text{(sum of squares of perpendicular semi-axes)}}{D},$$

where $D = 3, 4$ or 5 according as the body is rectangular, circular or spherical.

7. The moment of inertia of a hollow sphere of mass M about a diameter, its external and internal radii being a and b, is $\frac{2}{5}\frac{M(a^5+b^5)}{a^3+b^3}$.

8. The moment of inertia of a right circular cylinder of height h, mass M and radius of whose base is a is $\frac{1}{2}Ma^2$.

9. The product of inertia of a right-angled triangle of mass M about sides a, b containing the right angle is $\frac{1}{24}Mab$.

10. Moment of inertia about three rectangular axes are such that the sum of any two of them is greater when the third.

11. For every body there exists at every point O, a set of three mutually perpendicular axes such that the products of inertia of a body about these axes taken two at a time vanish.

12. The momental ellipsoid at the centre of the ellipsoid

$$\frac{x^2}{a^2}+\frac{y^2}{b^2}+\frac{z^2}{c^2}=1$$

is $(b^2+c^2)x^2+(c^2+a^2)y^2+(a^2+b^2)z^2 = \text{constant}$.

13. Any triangle of mass M is kinetically equivalent to three particles each of mass $\frac{M}{3}$ placed at the middle point of the sides.

14. The moment of inertia of a triangular lamina about one side is $\frac{1}{6}Mh^2$, where h is the altitude of the triangular lamina and M its mass.

15. The principal axes at a corner A of a square of side $2a$ is the diagonal through A and the other, a line through A making an angle of 45° with the above diagonal and lying in the plane of the square.

16. Kinetic energy of a rigid body rotating about a fixed axis is $\frac{1}{2}I(\dot{\theta})^2$, where the symbols have their usual meanings.

17. Equation of motion of a rigid body about the axis of rotation is $I\ddot{\theta} = L$, where I is the moment of inertia of the rigid body about the axis of rotation and L is the sum of the moments of the external forces acting on the body about the axis of rotation.

18. Motion of a body about its centre of inertia is the same as it would be if the centre of inertia were fixed and the same forces acted on the body.

19. The time of a complete small oscillation of a compound pendulum in terms of h and k, the distance of the C.G. of the body from the fixed axis and the radius of gyration of the body about the fixed axis respectively, is $2\pi\sqrt{\frac{gh}{k^2}}$.

20. Centres of suspension and oscillation are reversible.

III. Fill in the Blanks Type Questions

Fill in the blanks in the following:

1. If I denotes the moment of inertia of a body of mass M about a line and $I = Mk^2$, then k is called the —— of the body about that line.

2. Moment of inertia of a rod of mass M and length $2a$ about a line through the mid-point perpendicular to its lengths is ——.

3. Moment of inertia of a rectangular lamina of mass M whose sides are of lengths $2a$ and $2b$ about a line through its centre parallel to one edge of length $2b$ is ——.

4. Moment of inertia of a thin uniform circular disc of radius a and mass M about any of its diameter is ——.

5. Moment of inertia of a spherical shell of radius a and mass M about any of its diameter is ——.

6. Moment of inertia of a rectangular parallelepiped of mass M, whose edges are of lengths $2a, 2b, 2c$ parallel to the axes OX, OY, OZ respectively, where O is the centre of the parallelepiped about a line through O, parallel to the edge $2a$, is ——.

7. Moment of inertia of a hollow sphere of mass M about a diameter, its external and internal radii being a and b is

$$\frac{2}{5}\frac{M(\text{———})}{a^3 - b^3}.$$

8. Moment of inertia of a right circular cylinder of mass M about a line through its centre of gravity perpendicular to its axis is $\frac{1}{2}M(\text{——})$.

9. The product of inertia of a right-angled triangle of mass M about sides a, b containing the right-angle is $(\text{——})abM$.

10. Moment of inertia of a lamina about any line OL through a point O in its plane, when the moments and product of inertia of the lamina about rectangular axes OX and OY through O are given as A, B and C and the line OL is inclined to OX at an angle θ is $A\cos^2\theta + B\sin^2\theta - \text{——}$.

11. Moment of inertia of the lamina given in question 10 above about the line OL and a line ON perpendicular to OL is $\frac{1}{2}(\quad)\sin 2\theta + F\cos 2\theta$.

12. The sum of moments of inertia about any three rectangular axes meeting at a given point is always constant and equal to —— the moment of inertia about that point.

13. The momental ellipsoid at the centre of the ellipsoid $\frac{x^2}{a^2} + \frac{y^2}{b^2} + \frac{z^2}{c^2} = 1$ is

$$(b^2 + c^2)x^2 + (c^2 + a^2)y^2 + (a^2 + b^2)z^2 = \text{——}$$

14. If α, β, γ be the distances of the vertices of a triangle of mass M from any straight line in its plane, then the moment of inertia of the triangle about this line is $\frac{1}{6}M(\text{——})$.

15. The length AB and AD of the sides of a rectangle ABCD are $2a$ and $2b$, then the inclination θ of one of the principal axes to AB is given by $\tan 2\theta = \frac{\text{——}}{2(a^2 - b^2)}$.

16. Kinetic energy of a rigid body rotating about a fixed axis is $\frac{1}{2}I(\text{——})^2$, where the symbols have their usual meanings.

17. D' Alembert's principle is *the impressed forces acting on the system together with the* —— *forces at each point of the system are in equilibrium.*

18. The centre of inertia of a body —— as if all the mass of the body were concentrated; at if and as if all the external forces acting on the body were acting at it in the directions parallel to those in which they act.

19. The time of complete oscillation of a compound pendulum in terms of h and k, the distance of the C.G. of the body from the fixed axis and the radius of gyration of the body about the fixed axis respectively is $2\pi\sqrt{\text{——}}$.

20. Centre of suspension and oscillation are ——.

Answers to Objective Questions

I. Multiple Choice Type

1. (b). **2.** (c). **3.** (b). **4.** (c). **5.** (b). **6.** (c). **7.** (d). **8.** (b). **9.** (a). **10.** (b). **11.** (b). **12.** (c). **13.** (b). **14.** (d). **15.** (b). **16.** (c). **17.** (c). **18.** (b). **19.** (c). **20.** (c). **21.** (c).

II. True and False Type

1. T. **2.** F. **3.** T. **4.** F. **5.** T. **6.** T. **7.** F. **8.** T. **9.** F. **10.** T. **11.** T. **12.** T. **13.** F. **14.** T. **15.** F. **16.** T. **17.** T. **18.** T. **19.** F. **20.** T.

III. Fill in the Blanks Type

1. radius of gyration. **2.** $\frac{1}{3}Ma^2$. **3.** $\frac{1}{3}Ma^2$. **4.** $\frac{1}{4}Ma^2$. **5.** $\frac{2}{3}Ma^2$. **6.** $\frac{M}{3}(b^2+c^2)$. **7.** a^5-b^5. **8.** $3a^2+h^2$. **9.** $\frac{1}{12}$. **10.** $-2F\sin\theta\cos\theta$. **11.** $A-B$. **12.** twice. **13.** constant. **14.** $\alpha^2+\beta^2+\gamma^2+\alpha\beta+\beta\gamma+\gamma\alpha$. **15.** $3ab$. **16.** $\dot{\theta}$ **17.** reversed effective. **18.** moves. **19.** $\frac{k^2}{gh}$. **20.** reversible.

Some exciting gifts are up for grabs with this book!!

Rigid Dynamics

Dr M. M. Rahman

FREE Gifts Rs. 1000/-

DEAR READERS

By now you must have gone through this book in the entirety. So we invite your assessment on this title with absolute impartiality and candid observations. After all, we publishers blindly depend on your benchmarking. Friends, please fill up the open spaces below with CAPITAL LETTERS. You may add some extra sheets of paper if the space provided is not sufficient for you.

i. What are the important topics of your syllabus which are not adequately covered in this book?

ii. In which chapters and on what topics the treatment of the subject-matter by the author is not satisfactory and why?

iii. Have you come across the typos, factual inaccuracies, inconsistencies, and other forms of mistakes in the book? If yes, then please specify the mistakes and page numbers.

iv. Name top three books on the same subject, which, in your consideration, are better than this book. Could you provide some reasons?

Your 1st Preference: ....

Your 2nd Preference: ....

Your 3rd Preference: ....

v. Any suggestion to improve the quality of illustrations, diagrams etc.?

vi. How did you like the cover design? Any suggestions on colours?

vii. Suggestions for further improvement of production standard:

viii. Who recommended you this book:

☐ Teacher ☐ Friend ☐ Parents ☐ Bookseller ☐ Librarian

ix. Name of the person who has recommended:

Designation: ....

Address: ....

Phone No.: ....

e-mail: ....

x. Name and address of the bookseller from whom you have purchased this book:

....

....

....

xi. Name and address of the library where you would like to find the book:

....

....

xii. Name and address of your institution (Please mention the University or Board):

....

....

xiii. Your name, e-mail, and complete postal address:

....

....

....

xiv. A few other titles published by us that you would like to read:

....

....

The best assessment will be rewarded half-yearly. The award will be in the form of our publications, as decided by the Editorial and Marketing Operations Division, amounting to Rs. 1000/- (Rupees one thousand) only.

....

Please mail the filled up coupon at your earliest to :

Editorial & Marketing Operations Division

NEW CENTRAL BOOK AGENCY (P) Ltd

8/1 Chintamoni Das Lane, Kolkata 700 009

West Bengal, INDIA

Ph. No.: 033 + 22418411, 22418573, 22419094

Fax: 033 + 22419573, 22413852

e-mail: ncbapvtltd@eth.net, ncba@cal2.vsnl.net.in

Website: www.centralgroup.com